Universitext

Universitext is a series of textbooks that presents material from a wide variety of mathematical disciplines at master's level and beyond. The books, often well class-tested by their author, may have an informal, personal, or even experimental approach to their subject matter. Some of the most successful and established books in the series have evolved through several editions, always following the evolution of teaching curricula, into very polished texts.

Thus as research topics trickle down into graduate-level teaching, first textbooks written for new, cutting-edge courses may find their way into *Universitext*.

Kai Köhler

Differential Geometry and Homogeneous Spaces

 Springer

Kai Köhler
Mathematisches Institut
Heinrich-Heine-University Düsseldorf
Düsseldorf, Germany

ISSN 0172-5939   ISSN 2191-6675 (electronic)
Universitext
ISBN 978-3-662-69720-7   ISBN 978-3-662-69721-4 (eBook)
https://doi.org/10.1007/978-3-662-69721-4

Mathematics Subject Classification (2020): 53-01, 53C30, 53C35, 53C20, 83C05, 53C05

This Springer imprint is published by the registered company Springer-Verlag GmbH, DE, part of Springer Nature.
The registered company address is: Heidelberger Platz 3, 14197 Berlin, Germany

If disposing of this product, please recycle the paper.

Preface

Riemannian geometry studies a relatively general class of spaces, the smooth manifolds, in combination with Riemannian metrics. This means that angles and distances can be measured on the manifolds. Examples are surfaces in Euclidean $\mathbb{R}^3$ without vertices and edges, or the spheres $S^n \subset \mathbb{R}^{n+1}$ as sets of all points with a fixed distance to the origin, or the tori $\mathbb{R}^n/\mathbb{Z}^n$ as quotients of Euclidean space.

A typical question in Riemannian geometry is, for example: Is there a canonical distinguished metric on a given manifold? More generally, one can ask: What are the relations between topology and (metric) geometry? One can also associate Riemannian manifolds to algebraic objects such as Lie algebras and get further possibilities to study the algebraic objects. In Riemannian geometry, this often involves infinitesimal data arising from the Riemannian metric. For example, various curvature terms are used that arise from second derivatives of the metric. These infinitesimal data can then be related to the topological structure of the manifold. A typical classical result in which the geometry allows conclusions about the topology is the Hadamard–Cartan Theorem 6.5.9: if the sectional curvature on a Riemannian manifold M is non-positive everywhere, then M is diffeomorphic to a quotient of $\mathbb{R}^n$ by a discrete group. Another example is the Poincaré–Hopf Theorem 4.2.8, which shows, among other things, that the integral over a certain curvature invariant must be an integer and that it has an intuitive geometric interpretation.

Riemannian geometry has numerous interactions with other areas within and outside of mathematics, such as

- algebraic topology and differential topology, where in particular manifolds without metrics are studied,
- algebraic geometry, as projective varieties over $\mathbb{C}$ are common objects of study,
- analysis, for instance via C^∞-solutions of differential equations,
- group theory, for example via the study of Lie groups and their discrete subgroups,
- physics, where Riemannian geometry is used in mechanics, general relativity, and quantum field theory.

The aim of this book is to provide, within the scope of a two-semester lecture course, the most important foundations of Riemannian geometry with all necessary intermediate results, and to present in detail the central example class of homogeneous spaces. Homogeneous spaces are Riemannian manifolds with a transitively acting isometry group. Alternatively, they can be described as quotients of Lie groups by subgroups. Homogeneous spaces play an important role in many areas of mathematics, for example as moduli spaces whose points parametrize solutions of a mathematical problem. Symmetric spaces, i.e. spaces which allow point reflection at every point, are treated as a special case in a separate chapter. In the last chapter, as an important application of Riemannian geometry, some foundations of general relativity are deduced axiomatically.

The more special differential geometry of curves and surfaces in two- and three-dimensional Euclidean space can partly be described in a more elementary way, and there are quite a few results in this direction which are not or only marginally discussed in this book (cf. [doC], [Kl1], [Kü], [Bär]).

My primary goal is that the reader can use this book to teach or learn the contents with complete proofs within the limited time of two semesters. The content of the book is based on lecture courses I gave more or less in this form in the academic years 2006/07, 2009/10, 2013/14, 2018/19 and 2023/24 at the Heinrich Heine University Düsseldorf to students of the 5th and 6th semester.

Visual ideas are characteristic for the mathematical field of geometry, and accordingly the graphics contained in the book are intended to encourage the reader to make up similar pictorial outlines as often as possible while reading the book. Exercises are given at the end of each section. For the exercises marked with "*", the appendix contains solution sketches.

Results from the usual first three semesters of undergraduate study are assumed, and much is recapitulated; in some cases in detail, such as submanifolds of $\mathbb{R}^n$, tensor products, and exterior algebra. In this book, the development of the theory is as rigorous as possible. However, from Section 6.5 onwards, in some passages the existence of a universal covering is used, which is not proven. The proof can be found in almost any introductory lecture course or textbook on algebraic topology, such as [Ha]. To provide a further outlook, we also depart from this principle towards the end of the Section 7.6.

Section 2.4 on cohomology and Chapter 4 on the Poincaré–Hopf Theorem are not essential for understanding the rest of the book. Similarly, Chapters 6 and 7 on homogeneous and symmetric spaces are largely irrelevant to the final Chapter 8 on relativity. The following figure shows in more detail how the sections build on each other (except for occasional definitions, examples, and exercises).

There are many other textbooks on Riemannian geometry, some of which focus on different results or would require much more extensive lecture course cycles. A major influence on this book were the textbooks of O'Neill [ON2], Berline, Getzler, and Vergne [BGV], Gallot, Hulin, and Lafontaine [GHL], Cheeger and Ebin [ChEb], Helgason [Hel], Kobayashi and Nomizu [KoN], Lee [L], Besse [Besse] and Klingenberg [Kl2].

This book is essentially a direct translation of the second German-language edition. However, hundreds of minor improvements have been included and the figures have been re-created in color. I would like to thank the staff at Springer for their cooperation, especially Ms. Ruhmann for the idea of including keyword lists at the beginning of each section and the copy editor. I would also like to express my sincere gratitude for the suggestions and ideas of Matthias Dellweg, Daniel Grieser, Wolfgang Kühnel, Jens Piontkowski, Wilhelm Singhof, the referees, many participants in my lecture courses, and other attentive readers.

Düsseldorf, May 2024   *Kai Köhler*

Interdependence of the Sections

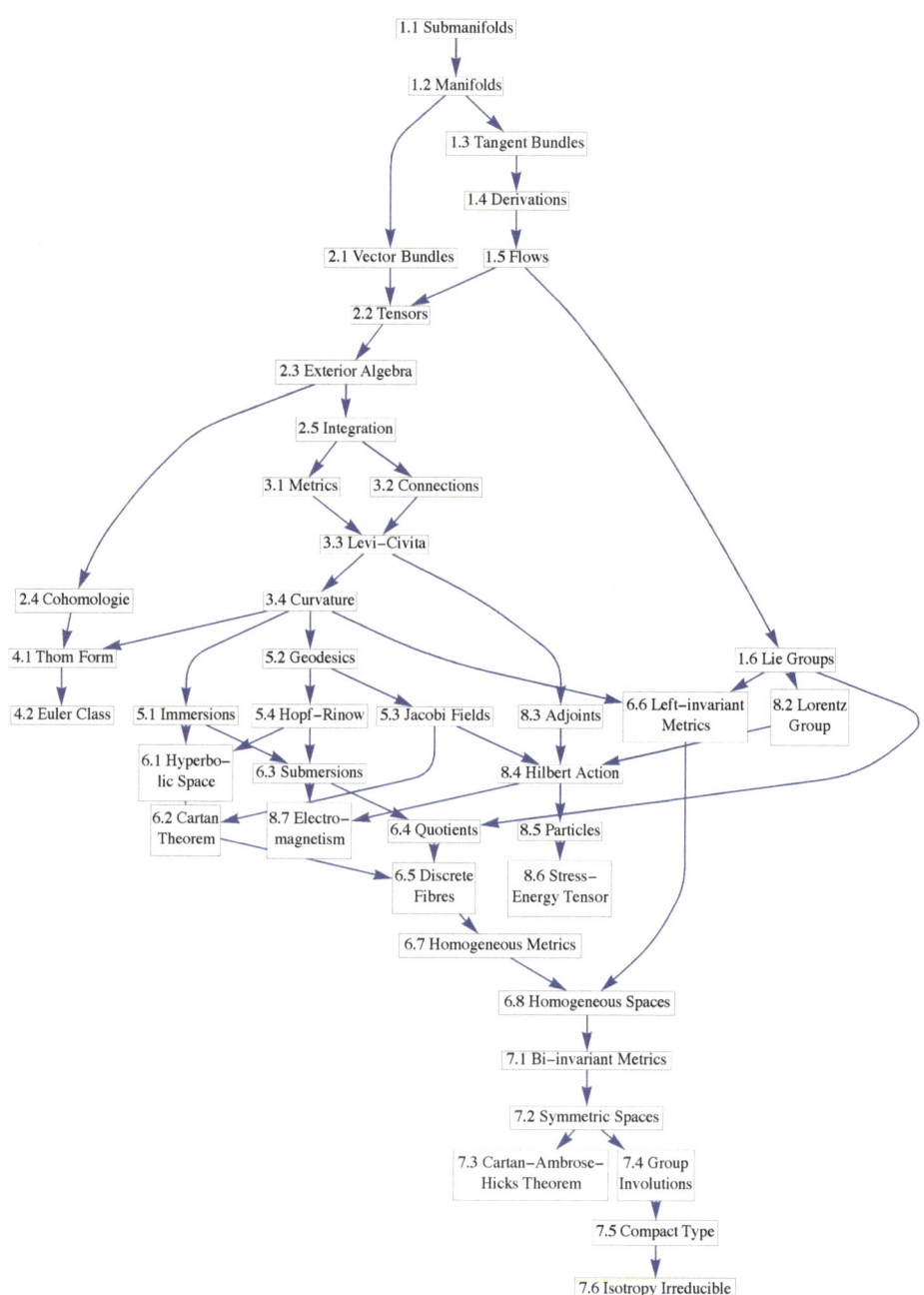

Contents

Chapter 1
Manifolds

The objects on which geometric structures such as metrics, curvatures and isometries are studied in this book are differentiable manifolds. In essence, these are topological spaces that are locally diffeomorphic to an $\mathbb{R}^n$. They are defined in this chapter together with the associated morphisms. As a basis for differential geometry in later chapters, some facts about the differential topology of manifolds are also presented in this and the next chapter. In contrast to purely topologically defined manifolds, differentiable manifolds carry a tangent space which plays such a central and important role that it will be described in this chapter successively in three very different ways: First, locally by comparison with vectors on the $\mathbb{R}^n$, second, by a purely algebraic notion of derivations acting on real-valued functions, and third, by families of diffeomorphisms. Finally, as a guiding example, manifolds are considered which additionally carry a group structure. As a motivation and important as well as intuitive example, we start with submanifolds of an $\mathbb{R}^n$, which are often treated in lecture courses on basic analysis.

1.1 Submanifolds of Euclidean Space

submanifold of $\mathbb{R}^{n+k}$	Clifford torus
parametrization	stereographic projections
sphere	special orthogonal group
special linear group	

Before defining manifolds in general, we consider submanifolds of an $\mathbb{R}^{n+k}$, which are easier to grasp and introduce. These are subsets which locally look like an n-dimensional subspace in $\mathbb{R}^{n+k}$, in the sense that they together with a neighborhood are said to be diffeomorphic to such a subspace with a neighborhood. Differentiable in this book will always mean C^∞ for simplicity.

© The Editor(s) (if applicable) and The Author(s), under exclusive license to Springer-Verlag GmbH, DE, part of Springer Nature 2024
K. Köhler, *Differential Geometry and Homogeneous Spaces*, Universitext,
https://doi.org/10.1007/978-3-662-69721-4_1

Definition 1.1.1 *A subset $M \subset \mathbb{R}^{n+k}$ is called an n-**dimensional** (C^∞) **submani-fold**[1] of $\mathbb{R}^{n+k}$ if for every point $p \in M$ there exist a neighborhood $U \overset{\text{open}}{\subset} \mathbb{R}^{n+k}$ of p, a neighborhood $W \overset{\text{open}}{\subset} \mathbb{R}^{n+k}$ of 0 and a C^∞ diffeomorphism $h : U \to W$ such that $h(U \cap M) = W \cap (\mathbb{R}^n \times \{0_{\mathbb{R}^k}\}).$*

Example 1.1.2 i) Every open subset of $\mathbb{R}^n$ is a submanifold of $\mathbb{R}^n$; choose h as the identity map.

ii) For $\widetilde{U} \overset{\text{open}}{\subset} \mathbb{R}^n$ and a C^∞ map $g : \widetilde{U} \to \mathbb{R}^k$ the graph

$$\{(x, g(x)) \mid x \in \widetilde{U}\}$$

is a submanifold. For $h : \widetilde{U} \times \mathbb{R}^k \to \widetilde{U} \times \mathbb{R}^k, (x, y) \mapsto (x, y-g(x))$ is a diffeomorphism with inverse $h^{-1} : (x, y) \mapsto (x, y + g(x))$.

Submanifolds can also be regarded as inverse images of differentiable functions at regular points, or as a generalization of the last example, as graphs. In the latter description, one has to be a bit more careful that the topology of the submanifold corresponds to the subset topology induced by the topology of $\mathbb{R}^{n+k}$.

Lemma 1.1.3 *The following are equivalent (Fig. 1.1):*

1) *$M \subset \mathbb{R}^{n+k}$ is an n-dimensional submanifold.*

2) *For all points $p \in M$, there exist an open neighborhood $U \overset{\text{open}}{\subset} \mathbb{R}^{n+k}$ of p as well as a C^∞ map $f : U \to \mathbb{R}^k$ with $f^{-1}(\{0\}) = U \cap M$ and surjective derivative f' at every point of $f^{-1}(\{0\})$.*

3) *For all points $p \in M$, there exist an open neighborhood $U \overset{\text{open}}{\subset} \mathbb{R}^{n+k}$ of p, an open subset $V \subset \mathbb{R}^n$ and a C^∞ map $\gamma : V \to U$ with $\gamma(V) = U \cap M$ and rang $\gamma' \equiv n$, such that γ is a homeomorphism of V and $U \cap M$. Such a γ is called a **(local) parametrization** of M at p.*

The homeomorphism condition of course enforces the injectivity of γ, but more subtly it prevents the problematic possibility from Fig. 1.2: In a neighborhood of $0 \in M$ there is no suitable h.

Proof. (1)$\Rightarrow$(2) Let $\pi : \mathbb{R}^{n+k} \to \mathbb{R}^k$ be the projection to the last k coordinates and choose $f := \pi \circ h$.

(2)$\Rightarrow$(3): Let $f : U \to \mathbb{R}^k$ be defined around $p = \binom{x_0}{y_0}$ with f' being surjective on $f^{-1}(\{0\}) = U \cap M$. Without loss of generality let $f'_{|\binom{x_0}{y_0}}$ be surjective on $\{0_{\mathbb{R}^n}\} \times \mathbb{R}^k$, i.e. the Jacobian matrix $(\frac{\partial f_j}{\partial y_\ell})_{\substack{j=1,\dots,k \\ \ell=1,\dots,k}}$ is invertible. By the Implicit Function Theorem there exist $V \subset \mathbb{R}^n, g : V \to U' \subset \mathbb{R}^k \ C^\infty$ with $f(\binom{x}{g(x)}) = 0$ for all $x \in V$, and

$$\gamma : V \to V \times U'$$

$$x \mapsto \binom{x}{g(x)}$$

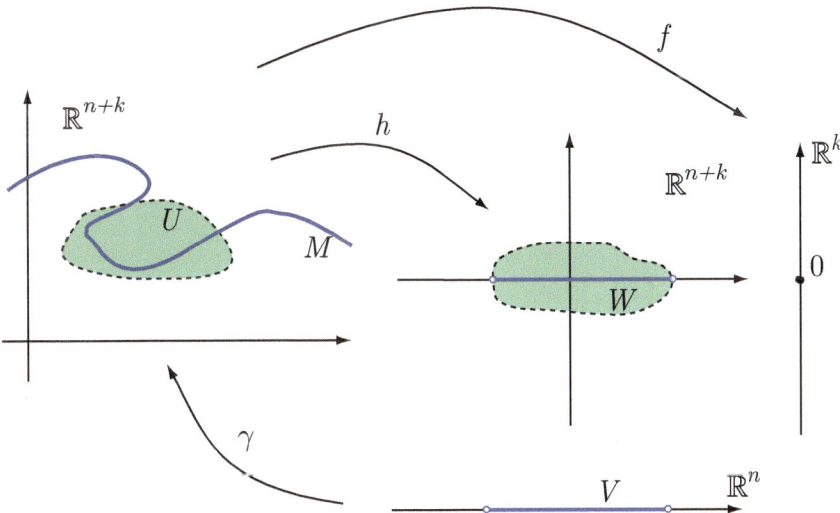

Fig. 1.1 Characterization of submanifolds.

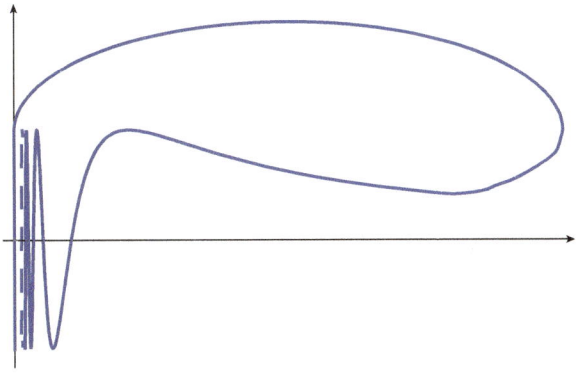

Fig. 1.2 This is not a submanifold of $\mathbb{R}^2$.

has the desired properties. Then

$$\gamma^{-1} : (V \times U') \cap M \to V$$
$$\begin{pmatrix} x \\ y \end{pmatrix} \mapsto x$$

is continuous.

$(3) \Rightarrow (1)$ Set $\gamma(x_0) = p$ and let $A : \mathbb{R}^k \to [\gamma'_{|x_0}(\mathbb{R}^n)]^\perp$ be a vector space isomorphism. Then setting

$$\widetilde{h} : (V - x_0) \times \mathbb{R}^k \to \mathbb{R}^{n+k}$$
$$\begin{pmatrix} x \\ y \end{pmatrix} \mapsto \gamma(x + x_0) + Ay,$$

$\widetilde{h}'_{|\binom{0}{0}} = (\gamma'_{|x_0}, A)$ is invertible. Thus, according to the Inverse Function Theorem, neighborhoods $V' \subset (V - x_0)$, $V'' \subset \mathbb{R}^k$ of 0 exist such that $\widetilde{h}_{|V' \times V''} : V' \times V'' \to \widetilde{h}(V' \times V'')$ is invertible. As $\gamma^{-1} : U \cap M \to V$ is continuous, there is a neighborhood $\widehat{U} \subset \widetilde{h}(V' \times V'')$ of p with $\gamma(V' + x_0) = M \cap \widehat{U}$. Then $h := (\widetilde{h}^{-1})_{|\widehat{U}}$ has the desired property.   □

Thus, for a graph as in Example 1.1.2(ii) $f : \widetilde{U} \times \mathbb{R}^k \to \mathbb{R}^k$, $\begin{pmatrix} x \\ y \end{pmatrix} \mapsto y - g(x)$, $\gamma : \widetilde{U} \to \widetilde{U} \times \mathbb{R}^k, x \mapsto \begin{pmatrix} x \\ g(x) \end{pmatrix}$ are archetypal examples of maps as in (2),(3).

Example 1.1.4 i) The **sphere** $S^n \subset \mathbb{R}^{n+1}$ is a submanifold, as can be seen by choosing $f(\mathbf{x}) = \|\mathbf{x}\|^2 - 1$. Since $f'_{|(x_0,\dots,x_n)^t} = (2x_j)_j \neq 0$ except at $0 \notin S^n$, it follows that rang $f' = 1$ on S^n.

ii) The **special linear group** $\mathrm{SL}(n, \mathbb{R}) := \{A \in \mathbb{R}^{n \times n} \mid \det A = 1\} \subset \mathbb{R}^{n \times n}$ is a submanifold. Since for $f(A) = \det A - 1$ we have $f'_{|A}(X) = \det A \cdot \mathrm{Tr}\,(A^{-1}X)$ for invertible A. In particular $f'_{|A}(A \cdot \mathbb{R}) = \det A \cdot n \cdot \mathbb{R} = \mathbb{R}$, so f' is surjective on $\mathrm{SL}(n, \mathbb{R})$.

iii) The **Clifford torus** $T^n \subset \mathbb{R}^{2n}$ is defined using $f((x_1, y_1, \dots, x_n, y_n)^t) = (x_1^2 + y_1^2 - 1, \dots, x_n^2 + y_n^2 - 1)^t$ on $(\mathbb{R}^2 \setminus \{0\})^n$. Then

$$f'_{|(x_1,\dots,y_n)^t} = \begin{pmatrix} 2x_1\ 2y_1 & & & \\ & 2x_2\ 2y_2 & & 0 \\ & & \ddots & \\ & 0 & & 2x_n\ 2y_n \end{pmatrix}$$

is surjective on $(\mathbb{R}^2 \setminus \{0\})^n$.

Remark For any local parametrizations $\gamma_1 : V_1 \to U_1, \gamma_2 : V_2 \to U_2$ and h_1, h_2 as in Definition 1.1.1 the composition $\gamma_2^{-1} \circ \gamma_1 = h_2 \circ h_1^{-1} : \gamma_1^{-1}(U_1 \cap U_2 \cap M) \to \gamma_2^{-1}(U_1 \cap U_2 \cap M)$ is a diffeomorphism (Fig. 1.3). This property is the basic trick to define C^∞ structures on more general topological spaces.

Of course, verifying that a map is a parametrization of M becomes easier if it is already known that M is a submanifold.

Lemma 1.1.5 *Let* $M \subset \mathbb{R}^{n+k}$ *be an* n-*dimensional submanifold and consider* $V \overset{\text{open}}{\subset} \mathbb{R}^n$. *Let* $\gamma \in C^\infty(V, \mathbb{R}^{n+k})$ *be injective with* $\gamma(V) \subset M$ *and* rang $\gamma' \equiv n$. *Then* γ *is a parametrization.*

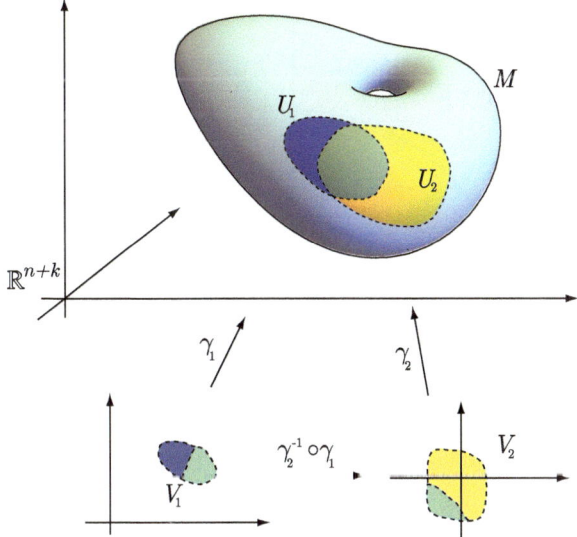

Fig. 1.3 Composition of local parametrizations.

Proof. At $p \in V$ choose a diffeomorphism $h : U \to W$ locally around $\gamma(p)$ as in Definition 1.1.1. Then with the projection $\pi : \mathbb{R}^{n+k} \to \mathbb{R}^n$, $f := \pi \circ h \circ \gamma|_{\gamma^{-1}(U) \cap V}$ is injective, smooth and rank $f' \equiv n$. So f is a diffeomorphism onto its image U', and the latter is open. Thus $\gamma = h^{-1} \circ \begin{pmatrix} f \\ 0_{\mathbb{R}^k} \end{pmatrix}$ is a homeomorphism on $h^{-1}(\pi^{-1}(U')) \cap M$. As an injective local homeomorphism, γ is a homeomorphism and hence a parametrization. □

Example For the Clifford torus T^n and an interval $I \overset{\text{open}}{\subset} \mathbb{R}$ of length $< 2\pi$,

$$\gamma : I^n \to T^n, (\vartheta_1, \ldots, \vartheta_n)^t \mapsto (\cos \vartheta_1, \sin \vartheta_1, \ldots, \cos \vartheta_n, \sin \vartheta_n)^t$$

provides a parametrization.

Exercises

Exercise 1.1.6 Let M be the sphere $S^n \subset \mathbb{R}^{n+1}$, $U_+ := S^n \setminus \{(1, 0, \ldots, 0)\}$, , $U_- := S^n \setminus \{(-1, 0, \ldots, 0)\}$. Show that the **stereographic projections** (Fig. 1.4)

$$\varphi_+ : U_+ \to \mathbb{R}^n, (x_0, \ldots, x_n) \mapsto \frac{(x_1, \ldots, x_n)}{1 - x_0},$$

$$\varphi_- : U_- \to \mathbb{R}^n, (x_0, \ldots, x_n) \mapsto \frac{(x_1, \ldots, x_n)}{1 + x_0}$$

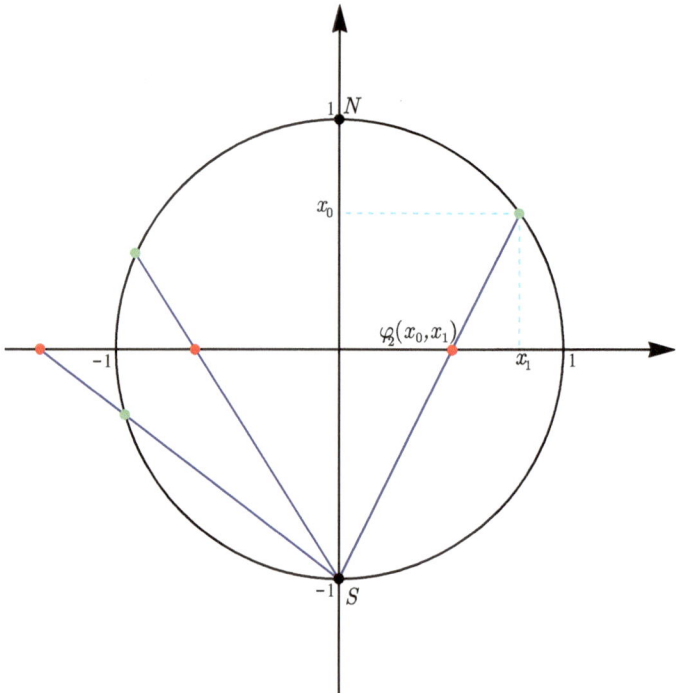

Fig. 1.4 Stereographic projection.

are inverse maps of local parametrizations γ_+, γ_-, and compute the latter. Determine $\varphi_- \circ \varphi_+^{-1}$.

Exercise 1.1.7 Prove that the **special orthogonal group**

$$SO(n) := \{A \in \mathbb{R}^{n \times n} \mid \det A = 1, AA^t = \text{id}\}$$

$(n \in \mathbb{N})$ is a submanifold of $\mathbb{R}^{n^2}$.

Exercise* 1.1.8 Show that for any $R > r$, the set M defined by

$$M = \left\{ \begin{pmatrix} x \\ y \\ z \end{pmatrix} \in \mathbb{R}^3 \,\middle|\, (\sqrt{x^2 + y^2} - R)^2 + z^2 = r^2 \right\}.$$

is a submanifold of $\mathbb{R}^3$.

Exercise 1.1.9 Let $M \subset \mathbb{R}^2$ be the image of the map $g :\,] - \frac{\pi}{2}, \frac{\pi}{2} [\to \mathbb{R}^2, \vartheta \mapsto \cos \vartheta \cdot \sin \vartheta \cdot (\cos \vartheta, \sin \vartheta)$. Check whether M is a submanifold. *Hint: How many connected components does $M \setminus \{(0, 0)\}$ have in each sufficiently small neighborhood of $(0, 0)$?*

Exercise 1.1.10 1) Show that the product $M \times N = \{(x, y) \mid x \in M, y \in N\}$ of two submanifolds M, N is again a submanifold.

2) Show that $S^n \times \mathbb{R}$ is a submanifold of $\mathbb{R}^{n+1}$.
3) Prove that every n-dimensional manifold M which is a product of (several) spheres can be represented as a submanifold of $\mathbb{R}^{n+1}$.

1.2 Smooth Manifolds

topological manifold	connected
topological space	dimension
Hausdorff space	differentiable structure
base of a topology	differentiable manifold
second-countable	torus
Lindelöf space	projective space
paracompact	submanifold
locally finite open refinement	coordinate
line with two origins	submersion
long line	immersion
C^∞ atlas	local diffeomorphism
transition map	diffeomorphism
change of coordinates	embedding
chart	Klein's bottle
parametrization	

To define a more general manifold structure on a set M without a surrounding $\mathbb{R}^{n+k}$, one must first determine which subsets of M should be open, i.e., one needs a topology. The following further properties of the topology have been found to be useful requirements:

Definition 1.2.1 *A (topological) manifold[2] is a second-countable Hausdorff space[3] M, which is locally homeomorphic to $\mathbb{R}^n$.*

More precisely: Every point $p \in M$ has a neighborhood U for which there is an n such that U is homeomorphic to $\mathbb{R}^n$.

This means: M is a **topological space**. That is, M is a pair $(\widehat{M}, O)$ where $\widehat{M}$ is a set and O is a set of subsets of $\widehat{M}$ (i.e. $O \subset \mathcal{P}(\widehat{M})$), such that

(O1) For any set J and sets $U_j \in O$ for $j \in J$ we have $\bigcup_{j \in J} U_j \in O$. In particular, $\emptyset \in O$.

(O2) For a finite set J and sets $U_j \in O$ for $j \in J$ we have $\bigcap_{j \in J} U_j \in O$. In particular, we find $\widehat{M} \in O$.

M is **Hausdorff**, i.e.

(T2) for any $x, y \in \widehat{M}, x \neq y$ there are neighborhoods U_x, U_y such that $U_x \cap U_y = \emptyset$.

---

[2] 1854, Riemann.
[3] 1914, Felix Hausdorff, 1868–1942.

A system of open sets is a **base** for O if every $U \in O$ is a union of sets from the base. A topological space is called **second-countable** if it has a countable base for the topology. Since topologies are often constructed using bases, this notion is transmitted particularly well in such cases.

Example 1) If (M, d) is a metric space, then $\{B_r(p) \mid r > 0, p \in M\}$ is a base for the topology.

2) $\mathbb{R}^n$ equipped with the standard topology is a topological manifold. A countable base is given for example by the balls $\{B_r(p) \mid r \in \mathbb{Q}^+, p \in \mathbb{Q}^n\}$.

Lemma 1.2.2 *The following conditions imply that a topological space M is second-countable:*

1) $M = \bigcup_{m \in \mathbb{N}} U_m$ *where* $U_m \overset{\text{open}}{\subset} M$ *is homeomorphic to an open subset* V_m *of an* $\mathbb{R}^n$,
2) M *is a subset of a second-countable space,*
3) M *is a finite product of second-countable spaces.*

Proof. 1) Given homeomorphisms $\varphi_m : U_m \to V_m$, $V_m \subset \mathbb{R}^n$,

$$\{\varphi_m^{-1}(V_m \cap B_r(p)) \mid m \in \mathbb{N}, r \in \mathbb{Q}^+, p \in \mathbb{Q}^n\}$$

is a countable base for the topology.

2),3) The restriction or product of the bases provides a base again.   □

(3) also holds for countable products. The purpose of the second-countability condition is to keep manifolds "small" enough to allow, for example, the concept of an integral or the existence of a metric. Second countability is often used in the equivalent form of paracompactness,

Definition 1.2.3 *A topological space M is called a **Lindelöf space**[4] if every open cover $(U_j)_{j \in J}$ of M can be reduced to a countable one. The space M is called **paracompact** if every open cover $(U_j)_{j \in J}$ has a **locally finite open refinement** $(\widetilde{U}_k)_{k \in K}$. More precisely, there should exist an open cover $(\widetilde{U}_k)_{k \in K}$ such that*

1) *for any $k \in K$ there exists a $j \in J$ such that $\widetilde{U}_k \subset U_j$,*
2) *for all $p \in M$ there is a neighborhood V of p such that $\{k \in K \mid \widetilde{U}_k \cap V \neq \emptyset\}$ is finite.*

In particular, every compact space is Lindelöf and paracompact. Not every compact topological space is second-countable, but for spaces locally homeomorphic to $\mathbb{R}^n$ this also follows directly from Lemma 1.2.2(1). Locally finite covers behave in some respects like finite ones:

Proposition 1.2.4 *For any locally finite open cover $(\widetilde{U}_k)_{k \in K}$ of a topological space M and $L \subset K$, $\overline{\bigcup_{k \in L} \widetilde{U}_k} = \bigcup_{k \in L} \overline{\widetilde{U}_k}$.*

---

[4] Ernst Leonard Lindelöf, 1870–1946.

Proof. Let $p \in M$. Then p has a neighborhood V that intersects only finitely many $\widetilde{U}_k$, $k \in L$. Therefore

$$V \cap \bigcup_{k \in K} \overline{\widetilde{U}_k} = V \cap \bigcup_{k \in L} \overline{\widetilde{U}_k} = V \cap \overline{\bigcup_{k \in L} \widetilde{U}_k} = V \cap \overline{\bigcup_{k \in K} \widetilde{U}_k}. \qquad \square$$

Definition 1.2.5 *A topological space M is called* **connected** *if there are no two open subsets* $U, V \subset M$, $U \neq \emptyset \neq V$ *such that* $U \cap V = \emptyset$, $U \cup V = M$.

Theorem 1.2.6 *For a Hausdorff space M which is locally homeomorphic to* $\mathbb{R}^n$ *for some n, the following statements are equivalent:*

1) M is second-countable,
2) M is a Lindelöf space,
3) M is paracompact and it has countably many connected components.

(1)$\Rightarrow$(2) holds for any topological space.

Proof. (1)$\Rightarrow$(2): Let $(U_j)_{j \in J}$ be an open cover and let $(V_m)_{m \in \mathbb{N}}$ be a countable base of M. For every $p \in M$ choose a $V_{m(p)} \subset M$ with $p \in V_{m(p)}$ such that there exists a $j_{m(p)}$ with $V_{m(p)} \subset U_{j_{m(p)}}$. Set $N := \operatorname{im} m \subset \mathbb{N}$. Then $M = \bigcup_{m \in N} V_m \subset \bigcup_{m \in N} U_{j_m}$, hence $(U_{j_m})_{m \in N}$ is a countable subcover.

(2)$\Rightarrow$(3): The Lindelöf property immediately implies that M has countably many connected components.

Consider an open cover $(U^j)_{j \in J}$. For every $p \in M$ choose a neighborhood U_p from the cover and an open neighborhood V_p with $\overline{V_p} \subset U_p$ (via local homeomorphisms to $\mathbb{R}^n$). Reduce the cover $(V_p)_{p \in M}$ with the Lindelöf property to a countable one $(V_{p_j})_{j \in \mathbb{N}}$ and for $k \in \mathbb{N}$ set $\widetilde{U}_k := U_{p_k} \setminus \bigcup_{\ell < k} \overline{V_{p_\ell}}$. For each $p \in M$ and $m \in \mathbb{N}$ minimal such that $p \in \overline{V_{p_m}}$, we have $p \in \widetilde{U}_m$. For $n \in \mathbb{N}$ we get $\widetilde{U}_k \cap V_{p_n} = \emptyset$ for all $k > n$.

(3)$\Rightarrow$(1): Without loss of generality let M be connected. For all $p \in M$, choose a second-countable neighborhood V_p with compact $\overline{V_p}$ (via local homeomorphisms to $\mathbb{R}^n$) and reduce this cover to a locally finite one $(\widetilde{U}_k)_{k \in K}$. Each $\overline{\widetilde{U}_k}$ can be covered by finitely many $\widetilde{U}_\ell$. For all $m \in \mathbb{N}_0$, choose a finite $K_m \subset K$ with $\bigcup_{k \in K_m} \overline{\widetilde{U}_k} \subset \bigcup_{k \in K_{m+1}} \widetilde{U}_k$. Then $N := \bigcup_{\substack{m \in \mathbb{N}_0 \\ k \in K_m}} \widetilde{U}_k$ is a countable union of second-countable spaces, hence second-countable. We have

$$\overline{N} \overset{\text{Prop. 1.2.4}}{=} \bigcup_{\substack{m \in \mathbb{N}_0 \\ k \in K_m}} \overline{\widetilde{U}_k} \subset N,$$

thus $N = M$. $\qquad \square$

An example of a space that satisfies all conditions for manifolds except the Hausdorff condition is the **line with two origins** (Fig. 1.5): $\mathbb{R}$ with an additional point $0'$ at 0, where open sets around these two points contain none, one, or both points.

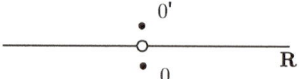

Fig. 1.5 Line with two origins.

Fig. 1.6 The long line.

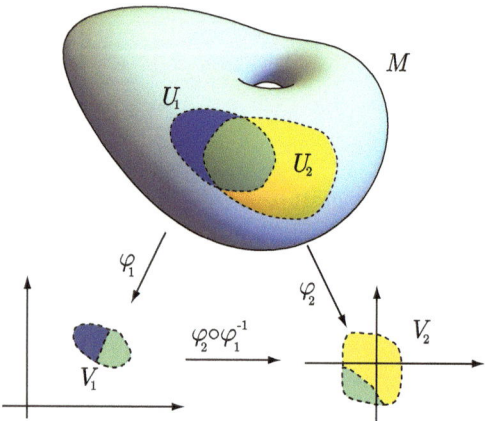

Fig. 1.7 Transition maps on manifolds.

A disjoint union of uncountably many copies of $\mathbb{R}^n$ is not second-countable. The simplest example of a connected space that satisfies all conditions except second countability is the **long line**: For the smallest uncountable ordinal ω_1, ω_1-many intervals are joined one after another, starting with the interval $]0,1[$, both to the right and to the left. This is only sketched here in Fig. 1.6.

In the remaining part of this book we will use the same symbol for $\widehat{M}$ and $M = (\widehat{M}, O)$ for the sake of simplicity. Using the compatibility of two parametrizations of submanifolds mentioned at the end of the last section, we can now define what a differentiable structure on a topological manifold should be. For one of the local homeomorphisms of M to an $\mathbb{R}^n$ it is not (yet) possible to define what differentiability should be; but for two homeomorphisms a C^∞ compatibility can be formulated.

Definition 1.2.7 *A C^∞ **atlas**[5] for a topological manifold M is a set of homeomorphisms $\mathfrak{A} = \{\varphi_j : U_j \to V_j \mid U_j \subset M \text{ open}, V_j \subset \mathbb{R}^n \text{ open}, n \in \mathbb{N}, j \in J\}$ with $M = \bigcup_{j \in J} U_j$ (i.e. the U_j form an open cover of M), such that for all $j, k \in J$ the* **transition maps** *(or* **changes of coordinates**) $\varphi_k \circ \varphi_j^{-1} : \varphi_j(U_j \cap U_k) \to \varphi_k(U_j \cap U_k)$ are C^∞ diffeomorphisms (Fig. 1.7). The φ_j are called* **charts**, *the φ_j^{-1}* **parametrizations**.

---

[5] 1891, Felix Klein, for Riemann surfaces.

Lemma 1.2.8 *For a connected manifold with C^∞ atlas all charts have the same dimension. This number is then called the **dimension** of M.*

Proof. Suppose M had charts $\varphi : U \to V \subset \mathbb{R}^n$, $\psi : U' \to V' \subset \mathbb{R}^m$ with $m \neq n$. If there existed an $x \in U \cap U'$, then $\psi \circ \varphi^{-1} : \varphi(U \cap U') \to \psi(U \cap U')$ would be a diffeomorphism and $(\psi \circ \varphi^{-1})'_{|\varphi(x)} : \mathbb{R}^n \to \mathbb{R}^m$ would be a vector space isomorphism $\notlightning$.

Thus $W_1 :=$ union of the domains of all charts of dimension n, $W_2 :=$ the same for the other charts, are disjoint. But $W_1 \neq \emptyset \neq W_2$ and $W_1 \cup W_2 = M$ in contradiction with the connectedness of M. $\qquad\square$

Of course, one does not want to take M to be a different differentiable manifold with each new choice of a C^∞ atlas. Therefore we divide by the following equivalence relation.

Definition 1.2.9 *Two C^∞ atlases $\mathfrak{A} - \{\varphi_j \mid j \in J\}$, $\mathfrak{A}' - \{\psi_k \mid k \in K\}$ are equivalent if $\mathfrak{A} \cup \mathfrak{A}'$ is again an atlas, i.e. if for any $j \in J, k \in K$ the map $\varphi_j \circ \psi_k^{-1}$ is a C^∞ diffeomorphism. A **differentiable (or C^∞) structure** on the topological manifold M is an equivalence class of atlases[6]. A **differentiable (or C^∞) manifold** is a topological manifold together with a differentiable structure.*

Example 1.2.10 i) The space $\mathbb{R}^n$ together with the atlas consisting of the single chart $\varphi = \mathrm{id}_{\mathbb{R}^n}$ is a C^∞ manifold.

ii) The quotient $\mathbb{R}^n/\mathbb{Z}^n$ is a manifold, when equipped with the quotient topology ($U \subset \mathbb{R}^n/\mathbb{Z}^n$ is open $:\Leftrightarrow \pi^{-1}(U)$ is open using the canonical projection π) and the charts

$$\varphi_{I,x} : (x + I^n)/\mathbb{Z}^n \xrightarrow{\mathrm{id}} (x + I^n)$$

for $I :=] - \frac{1}{4}, \frac{1}{4}[$ and each $x \in \mathbb{R}^n$, namely a **torus**.

For $x = (x_1, \ldots, x_n)^t$, $y = (y_1, \ldots, y_n)^t$ and $1 \leq j \leq n$, $(x_j + I) - (y_j + I) = x_j - y_j +] - \frac{1}{2}, \frac{1}{2}[$ contains at most one integer $k_j \in \mathbb{Z}$. So there is at most one $k \in \mathbb{Z}^n$ such that $(x + I^n) \cap (y + I^n + k) \neq \emptyset$. Then the transition map is given by $\varphi_y \circ \varphi_x^{-1} = \mathrm{id}_{\mathbb{R}^n} - k$ and is therefore smooth.

As an image of the compactum $[0, 1]^n$ under the continuous map π, $\mathbb{R}^n/\mathbb{Z}^n$ is compact.

iii) Given $K = \mathbb{R}, \mathbb{C}, \mathbb{H}$, consider $x, y \in K^{n+1} \setminus \{0\}$ as equivalent, $x \sim y :\Leftrightarrow$ there exists a $\lambda \in K$ such that $x = \lambda y$. So, the equivalence class of a point consists of the straight line through that point and through 0. The set of equivalence classes (i.e., K-straight lines in K^{n+1}) is the **projective space** $\mathbb{P}^n K := K^{n+1} \setminus \{0\}/\sim$ (Fig. 1.8). The topology is again given by the quotient topology. As an atlas we choose $U_k := \{[(x_0, \ldots, x_n)] \in \mathbb{P}^n K \mid x_k \neq 0\}$, $V_k := K^n$,

$$\varphi_k : U_k \to V_k$$
$$[(x_0, \ldots, x_n)] \mapsto (x_k^{-1}x_0, \ldots, x_k^{-1}x_{k-1}, x_k^{-1}x_{k+1}, \ldots, x_k^{-1}x_n).$$

---

[6] 1895, Poincaré.

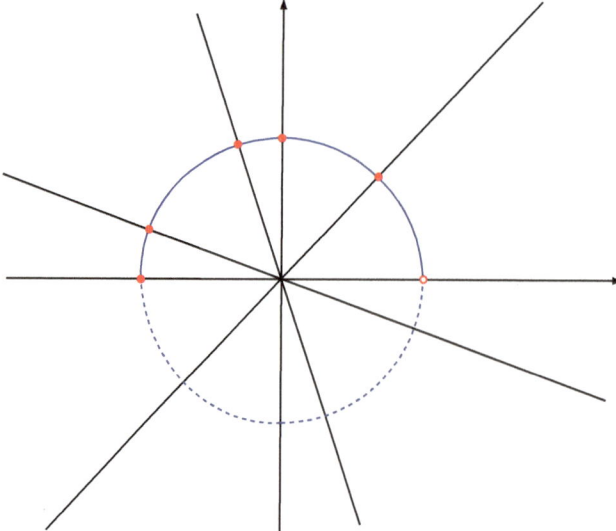

Fig. 1.8 $\mathbb{P}^1\mathbb{R}$ as the set of straight lines in $\mathbb{R}^2$ parametrized by a semicircle.

Then

$$\varphi_j \circ \varphi_k^{-1}(x_0, \ldots, x_{k-1}, x_{k+1}, \ldots, x_n)$$
$$= (x_j^{-1}x_0, \ldots, x_j^{-1}x_{j-1}, x_j^{-1}x_{j+1}, \ldots, x_j^{-1}, \ldots, x_j^{-1}x_n).$$
$$\nwarrow k\text{th element}$$

Points on $\mathbb{P}^n K$ are written as $[(x_0, \ldots, x_n)] =: (x_0 : \cdots : x_n)$. One obtains $\dim \mathbb{P}^n K = n \cdot \dim_{\mathbb{R}} K$. Analogous to (ii), applying the continuous map $\pi : K^{n+1} \setminus \{0\} \to \mathbb{P}^n K$, the space $\mathbb{P}^n K = \pi(S^{(n+1)\cdot\dim_{\mathbb{R}} K - 1})$ is seen to be compact.

In this example, it is no longer obvious at all how one could describe this space as a submanifold. While it is possible to do so, for many purposes it would be more unwieldy and unnatural than the above description.

Henceforth "manifold" shall mean connected C^∞ manifold. M is often written as M^n, where n denotes not a power but the dimension.

Definition 1.2.11 *A subset $N \subset M^n$ of a manifold M is called a **submanifold** of M if at each $p \in N$ there exists a chart $\varphi : U \to V \subset \mathbb{R}^n$ such that $\varphi(U \cap N)$ is a submanifold of $\mathbb{R}^n$.*

Lemma 1.2.12 *Submanifolds and products of C^∞ manifolds are C^∞ manifolds.*

Proof. Equip $N \subset M$ with the subspace topology induced by M ($O := N \cap$ open subsets of M) and the product $M_1 \times M_2$ with the product topology generated by products of open sets. Construct the atlases in the same way from atlases of M, M_1, M_2. $\square$

A C^∞ structure corresponds to being able to define differentiability for maps in the following way:

Definition 1.2.13 *A map $f : M^m \to N^n$ between C^∞ manifolds is C^k if around all $x \in M$, $f(x) \in N$ charts $\varphi : U \to V \subset \mathbb{R}^m$, $\psi : U' \to V' \subset \mathbb{R}^n$ exist such that $\psi \circ f \circ \varphi^{-1} : V \to V'$ is a C^k map.*

This definition is independent of the choice of charts, since the transition maps are C^∞ (the following diagram commutes):

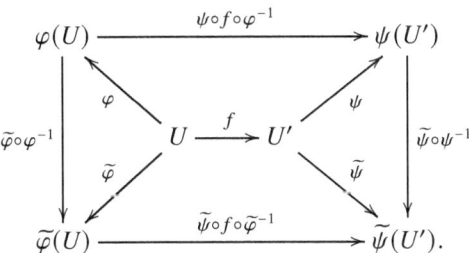

Therefore weaker differentiability than smoothness can be defined, but not stronger differentiability. To define analytic maps, one would need an atlas with analytic transition maps. For a C^ℓ manifold, correspondingly C^k differentiability of maps could be defined only for $\ell \geq k$.

Remark Thus we have a definition of differentiability, but not yet a definition of derivatives!

Example Each chart $\varphi : U \to V$ is C^∞, and so is each **coordinate** $\varphi^j : U \to \mathbb{R}$ and the parametrization $\varphi^{-1} : V \to U$.

Applying the notion of a differentiable map, it is now possible to define when two C^∞ manifolds are said to be isomorphic.

Definition 1.2.14 *A C^∞ map $F : U \to \mathbb{R}^n$, $U \subset \mathbb{R}^m$, is called a **submersion/ immersion/ local diffeomorphism** at $x \in U$ if its derivative $F'_{|x} : \mathbb{R}^m \to \mathbb{R}^n$ is surjective/ injective/ bijective there. A C^∞ map $f : M^m \to N^n$ is called a **submersion/immersion/local diffeomorphism** if for all $x \in M$ there exist charts φ, ψ around x, $f(x)$ such that $\psi \circ f \circ \varphi^{-1}$ at $\varphi(x)$. The map f is called a **diffeomorphism** if it is bijective with smooth inverse. It is called an **embedding** if it is an immersion and $f : M \to f(M)$ is a homeomorphism.*

For example, closed or open injective immersions are embeddings.

Example The Clifford torus in $\mathbb{R}^{2n}$ is diffeomorphic to $\mathbb{R}^n / \mathbb{Z}^n$ (Exercise 1.2.21).

Example Consider $M = \mathbb{R}$ with the atlas $\{\varphi : M \to \mathbb{R}, x \mapsto x\}$ and $N = \mathbb{R}$ with the atlas $\{\psi : N \to \mathbb{R}, x \mapsto x^3\}$. The atlases are not equivalent, as $(\psi \circ \varphi^{-1})'_{|0} = 0$. Still $f : M \to N, x \mapsto \sqrt[3]{x}$ is a diffeomorphism, since f is bijective and $\psi \circ f \circ \varphi^{-1} = x$. Thus it is C^∞ with smooth inverse map.

Lemma 1.2.15 *For an embedding $\iota : M \to N$, ι is a diffeomorphism onto its image.*

Proof. Let $p \in M$, $\varphi : U \to V$ be a chart around p and $\psi : U' \to V'$ be a chart around $\iota(p)$. Then $\psi \circ \iota \circ \varphi^{-1}$ is a parametrization as in Lemma 1.1.3, so the inverse map onto the image is locally a chart and hence C^∞.   □

Lemma 1.2.16 *Let M, N be C^∞-manifolds, M be compact, and $f : M \to N$ be an injective immersion. Then f is an embedding.*

Proof. We need to show that f^{-1} is continuous. Let $(y_n)_n$ be a sequence in $f(M)$ with $y_n \to y$ and set $x_n := f^{-1}(y_n)$, $x := f^{-1}(y)$. Suppose that x_n does not converge to x, i.e. x has a neighborhood U not met by almost all x_n. Since $M \setminus U$ is compact, there exists a subsequence $(x_{n_k})_k$ convergent to an x_0. But this implies $f(x_{n_k}) \to f(x_0) \neq y$.   □

Be careful! A topological manifold can admit several non-diffeomorphic differentiable structures, or possibly none. For example, according to Donaldson theory, there are infinitely many different C^∞ structures on $\mathbb{R}^4$ with the standard topology (but only one on $\mathbb{R}^n$ with $n \neq 4$, [Go1], [Go2], [DoK]). And there are 28 different ones on the 7-dimensional sphere ([M]).

Remark The strong Whitney Embedding Theorem states that for $n > 0$ every n-dimensional smooth manifold can be embedded into an $\mathbb{R}^{2n}$ ([Whit, Th. 5]). The term "manifold" therefore does not provide more objects than the term "submanifold of a Euclidean space", but allows these to be considered independently of embeddings, which often offers advantages. For example, in the case of manifolds constructed as quotients, such as projective spaces, the construction of a suitable embedding is not necessary.

Exercises

Exercise* 1.2.17 Show that there are open neighborhoods $U_r \subset \mathbb{R}$ around all points $r \in \mathbb{Q} \subset \mathbb{R}$ with $\bigcup_{r \in \mathbb{Q}} U_r \neq \mathbb{R}$.

Exercise* 1.2.18 Verify the Lindelöf condition for submanifolds without using second countability, instead applying the definition in a more direct way.

Exercise 1.2.19 Prove the following facts about projective spaces:

1) The spaces $\mathbb{P}^1 \mathbb{R}$ and $\mathbb{P}^1 \mathbb{C}$ are diffeomorphic to the spheres S^1 and S^2, respectively. (Analogously $\mathbb{P}^1 \mathbb{H}$ is diffeomorphic to S^4.)
2) The map $f : S^n \to \mathbb{P}^n \mathbb{R}, x \mapsto [x]$ is surjective and a local diffeomorphism. Determine the preimage of each point.

Exercise* 1.2.20 Show that for all $R > r$ the submanifold M of $\mathbb{R}^3$ determined by

$$\left\{ \begin{pmatrix} x \\ y \\ z \end{pmatrix} \in \mathbb{R}^3 \,\middle|\, (\sqrt{x^2 + y^2} - R)^2 + z^2 = r^2 \right\}$$

(Exercise 1.1.8) is diffeomorphic to the torus $\mathbb{R}^2/\mathbb{Z}^2$.

Exercise 1.2.21 Prove that the n-dimensional Clifford torus is diffeomorphic to the torus $\mathbb{R}^n/\mathbb{Z}^n$.

Exercise 1.2.22 Show that in $\{A \in \mathbb{R}^{3\times 3} \,|\, A^t = A, \operatorname{Tr} A = 1\} \cong \mathbb{R}^5$ the matrices with $A^2 = A$ form a submanifold M which is diffeomorphic to $\mathbb{P}^2\mathbb{R}$.

Exercise 1.2.23 Let M be the square $M := \{ \begin{pmatrix} x \\ y \end{pmatrix} \,|\, x, y \in [-1, 1], |x| = 1 \text{ or } |y| = 1\}$ with the subspace topology induced by $\mathbb{R}^2$. Find a C^∞ atlas for M.

Exercise 1.2.24 Let **Klein's bottle** M be (as a topological space) the quotient of $\mathbb{R}^2$ by the equivalence relation $(s, t) \sim (s + 2\pi n, (-1)^n t + 2\pi m)$ for all $n, m \in \mathbb{Z}$.

1) Show that M carries the structure of a C^∞ manifold.
2) For $r > 1$, prove that

$$\begin{pmatrix} \varphi \\ \psi \end{pmatrix} \mapsto \begin{pmatrix} \cos \varphi \cdot (r + \cos \psi) \\ \sin \varphi \cdot (r + \cos \psi) \\ \cos \frac{\varphi}{2} \cdot \sin \psi \\ \sin \frac{\varphi}{2} \cdot \sin \psi \end{pmatrix}$$

is an embedding of Klein's bottle into $\mathbb{R}^4$.

(Remark: No embedding into $\mathbb{R}^3$ exists.)

1.3 First Description of the Tangent Bundle: Via Charts

representative of a tangent vector	derivative
tangent vector	tangent map
base point	vector field
tangent space	orientable
tangent bundle	

Although the definition of differentiability of a function f was given in the last section, it is still not clear what the derivative of f should be. After all, the derivative $(\psi \circ f \circ \varphi^{-1})'$ depends on the choice of the charts φ, ψ. If we use the notion of directional derivative to define derivatives as infinitesimal variations of f in a given direction, we notice that we must first clarify what exactly a direction or a tangent vector is supposed to be. Since this notion is so fundamental to differentiation and thus to all of calculus on manifolds, we describe it in three different ways to cover as many aspects as possible.

In the first approach, a chart $\varphi : U \to V$ is used to identify a tangent vector on M with a tangent vector on $V \subset \mathbb{R}^n$. The set of all directions at each point of V corresponds to an $\mathbb{R}^n$. For a different chart this description changes, but the disparity can be divided out by an equivalence relation.

Definition 1.3.1 *A **representative of a tangent vector** an $x \in M^n$ is a pair $(\varphi, u) \in \mathfrak{A} \times \mathbb{R}^n$ where φ is a chart around x and $u \in \mathbb{R}^n$. Two representatives $(\varphi, u), (\psi, v)$ shall be equivalent if $(\psi \circ \varphi^{-1})'_{|\varphi(x)}(u) = v$. The equivalence classes $[(\varphi, u)]$ are called **tangent vectors at the base point** x. The tangent space $T_x M$ at x is the set consisting of all these tangent vectors.*

Remark For an open subset M of $\mathbb{R}^m$ and maps $\varphi, \psi : M \to \mathbb{R}^m$ this corresponds to the behavior of directional derivatives: For a tangent vector $X \in \mathbb{R}^m$ of M at the point x and $u := \varphi'_{|x}(X), v := \psi'_{|x}(X)$, we obtain $(\psi \circ \varphi^{-1})'_{|\varphi(x)}(u) = v$.

Lemma 1.3.2 *With the arithmetic operations $[(\varphi, u)] + [(\varphi, v)] := [(\varphi, u + v)]$, $\lambda \cdot [(\varphi, u)] := [(\varphi, \lambda u)]$ for $\lambda \in \mathbb{R}, u, v \in \mathbb{R}^n$, $T_x M$ becomes an n-dimensional $\mathbb{R}$-vector space.*

Proof. Well-definedness: for any given chart ψ, $(\psi \circ \varphi^{-1})'_{|\varphi(x)} \in \mathbb{R}^{n \times n}$ is a linear map, so it is compatible with addition and scalar multiplication. The choice of the chart is arbitrary in the equivalence relation, so $(\varphi, \mathbb{R}^n)$ is a complete representative system for any fixed chart φ and $\dim T_x M = n$. $\square$

Definition 1.3.3 *If M^n is a manifold, let $TM := \{(x, X) \mid x \in M, X \in T_x M\}$ be the disjoint union of the tangent spaces. Let the **tangent bundle** $\pi : TM \twoheadrightarrow M$, $(x, X) \mapsto x$ be the projection to the base point.*

Example Given $V \overset{\text{open}}{\subset} \mathbb{R}^n$, $TV \cong V \times \mathbb{R}^n$ via the canonical chart id_V.

Lemma 1.3.4 *For M n-dimensional, TM is a 2n-dimensional manifold in a canonical way and π is a C^∞ submersion.*

Proof. Given an atlas $\mathfrak{A} = \{\varphi_j : U_j \to V_j \mid j \in J\}$ of M, set

$$\widetilde{\mathfrak{A}} := \left\{ T\varphi_j : \begin{array}{c} TU_j \\ (x, [(\varphi_j, u)]) \end{array} \to TV_j \overset{\text{via kanon. chart}}{=} \begin{array}{c} V_j \times \mathbb{R}^n \\ (\varphi_j(x), u) \end{array} \;\middle|\; j \in J \right\}.$$

Then for each $\varphi, \psi \in \mathfrak{A}$

$$
\begin{aligned}
(T\psi \circ (T\varphi)^{-1})((y, u)) &= T\psi(\varphi^{-1}(y), [(\varphi, u)]) \\
&= T\psi(\varphi^{-1}(y), [(\psi, (\psi \circ \varphi^{-1})'_{|y}(u))]) \\
&= ((\psi \circ \varphi^{-1})(y), (\psi \circ \varphi^{-1})'_{|y}(u)),
\end{aligned}
$$

so this is a C^∞ diffeomorphism. As the topology on TM choose the one induced by $T\varphi$ (i.e. preimages of open sets under $T\varphi$ are open).

TM is Lindelöf: For an open cover $(W_k)_{k \in K}$ of TM and domains U_j of charts, $(\pi(W_k) \cap U_j)_{j,k}$ is a cover of M. Reduce this to a countable one $(Z_\ell)_\ell$ and cover $\pi^{-1}(Z_\ell) \cong Z_\ell \times \mathbb{R}^n$ with countably many of the W_k. For all ℓ together these provide a countable cover of TM.

TM is Hausdorff: For two points p, q with $\pi(p) \neq \pi(q)$ and disjoint neighborhoods $U, V \subset M$ of $\pi(p), \pi(q)$, the preimages $\pi^{-1}(U), \pi^{-1}(V)$ yield disjoint neighborhoods. Otherwise, for a chart W around $\pi(p) = \pi(q)$ the points p, q lie in the manifold $\pi^{-1}(W) \cong W \times \mathbb{R}^n$.

Furthermore

$$(\varphi \circ \pi \circ (T\varphi)^{-1})(y, u) = (\varphi \circ \pi)(\varphi^{-1}(y), [(\varphi, u)]) = y,$$

and thus π is submersion. $\qquad\square$

Example Using the charts $\varphi_I : e^{i\vartheta} \mapsto \vartheta$ of S^1, we find $\varphi_I \circ \varphi_{I'}^{-1} = \mathrm{id}_{I \cap I'}$. Thus $TS^1 \cong T(\mathbb{R}/\mathbb{Z}) \to (\mathbb{R}/\mathbb{Z}) \times \mathbb{R} \cong S^1 \times \mathbb{R}, (x, [(\varphi_I, u)]) \mapsto (x, u)$ is a well-defined diffeomorphism. Hence TS^1 is diffeomorphic to a cylinder.

Now it is also possible to define what the derivative of a differentiable map should be.

Lemma and Definition 1.3.5 *Given charts* φ, ψ *at* $x, f(x)$, *the* **derivative** (*or* **tangent map**) $Tf : TM \to TN$ *of a* C^∞ *map* $f : M \to N$ *at* $x \in M$ *is the linear map*

$$T_x f : T_x M \to T_{f(x)} N$$
$$[(\varphi, u)] \mapsto [(\psi, (\psi \circ f \circ \varphi^{-1})'_{|\varphi(x)}(u))] \qquad (u \in \mathbb{R}^n).$$

As a function of x, $Tf : TM \to TN$ *is a* C^∞ *map.*

Proof. Well-definedness: for two charts $\varphi, \widetilde{\varphi}$ around x one finds

$$\begin{aligned} T_x f([(\varphi, u)]) &= T_x f([(\widetilde{\varphi}, (\widetilde{\varphi} \circ \varphi^{-1})'_{|\varphi(x)}(u))]) \\ &= [(\psi, (\psi \circ f \circ \widetilde{\varphi}^{-1})'_{|\widetilde{\varphi}(x)}(\widetilde{\varphi} \circ \varphi^{-1})'_{|\varphi(x)}(u))] \\ &= [(\psi, (\psi \circ f \circ \varphi^{-1})'_{|\varphi(x)}(u))]; \end{aligned}$$

analogously for two charts $\psi, \widetilde{\psi}$ around $f(x)$.

Smoothness: For a suitable open subset $U \subset \mathbb{R}^n$ we get

$$T\psi \circ Tf \circ (T\varphi)^{-1} : TU \to T\mathbb{R}^n$$
$$(y, u) \mapsto ((\psi \circ f \circ \varphi^{-1})(y), (\psi \circ f \circ \varphi^{-1})'_{|y}(u));$$

in particular Tf is smooth. $\qquad\square$

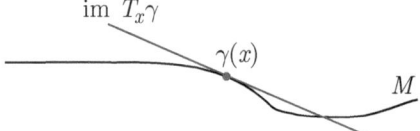

Fig. 1.9 Tangent space of a parametrized submanifold of $\mathbb{R}^{n+k}$.

Example Let $M \subset \mathbb{R}^{n+k}$ be a submanifold, let $\iota : M \hookrightarrow \mathbb{R}^{n+k}$ be the canonical embedding and let $\gamma : V \to \mathbb{R}^{n+k}$ be a local parametrization. Then with respect to the chart γ^{-1} of M and the canonical chart id of $\mathbb{R}^{n+k}$

$$T\iota : TM \to T\mathbb{R}^{n+k}$$
$$(\gamma(x), [(\gamma^{-1}, u)]) \mapsto (\iota(\gamma(x)), [(\mathrm{id}, (\mathrm{id} \circ \iota \circ \gamma)'_{|\gamma^{-1}(\gamma(x))}(u))])$$
$$= (\gamma(x), [(\mathrm{id}, \gamma'_{|x}(u))]).$$

Thus $T\iota$ embeds TM as im $\gamma' \subset T\mathbb{R}^{n+k}$ (Fig. 1.9). This corresponds to the intuitive notion that $T_x M$ should be tangent to M at the point x in the sense of Euclidean geometry.

Lemma 1.3.6 (Chain rule) *If $M \xrightarrow{f} N \xrightarrow{g} P$ are C^∞ maps and $x \in M$, then $T_x(g \circ f) = T_{f(x)}g \circ T_x f$ holds.*

Proof. For φ, ψ, ω charts around $x, f(x), g(f(x))$, one finds for any $u \in \mathbb{R}^m$

$$T_x(g \circ f)([(\varphi, u)]) = [(\omega, (\omega \circ g \circ f \circ \varphi^{-1})'(u))]$$
$$= [(\omega, (\omega \circ g \circ \psi^{-1})' \cdot (\psi \circ f \circ \varphi^{-1})'(u))]$$
$$= T_{f(x)}g([(\psi, (\psi \circ f \circ \varphi^{-1})'(u))]) = T_{f(x)}g \circ T_x f. \qquad \square$$

In other words,

$$T : \begin{Bmatrix} \text{manifolds} \\ C^\infty \text{ maps} \end{Bmatrix} \to \begin{Bmatrix} \text{manifolds} \\ C^\infty \text{maps} \end{Bmatrix}$$

is a covariant functor on the category which has manifolds as objects and C^∞ maps as morphisms. With the charts $\varphi : U \to V \subset \mathbb{R}^m, \psi : U' \to V' \subset \mathbb{R}^n, \omega : U'' \to V'' \subset \mathbb{R}^p$ the proof can also be thought of as the commutativity of the diagram

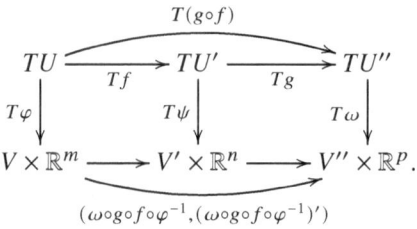

Fig. 1.10 The canonical Cartesian vector field $\frac{d}{dx}$ on $\mathbb{R}$.

Example 1.3.7 Let $I \subset \mathbb{R}$, $\gamma : I \to M$ be a curve on M with $\gamma(0) = p$ and set $\gamma'(0) := T_0\gamma(\frac{d}{dx}) = X \in T_pM$ where $\frac{d}{dx}$ is the Cartesian unit vector on $\mathbb{R}$ (Fig. 1.10; the notation $\frac{d}{dx}$ differs a bit from our other notation for derivatives). Then for any smooth map $f : M \to N$

$$T_pf(X) = (f \circ \gamma)'(0).$$

Example For $U \subset \mathbb{R}^{n+k}$, $f : U \to \mathbb{R}^k$ with $U \cap M^n = f^{-1}(0)$ and a local parametrization γ of M we have $f \circ \gamma = 0$, hence $Tf \circ T\gamma = 0$. So, for dimensional reasons, $TM \cong \ker f'$. For instance, for the sphere with $f(x) = \|x\|^2 - 1$, because $f'_{|x}(u) = 2x^t u$, one finds

$$T_xM \cong x^\perp \subset T_x\mathbb{R}^{n+1}.$$

Definition 1.3.8 *A **vector field** X on a manifold M is a C^∞ map $X : M \to TM$ with $X_{|p} \in T_pM$ for all $p \in M$. The set of vector fields is written as $\Gamma(M, TM)$.*

Exercises

Exercise 1.3.9 Consider the two maps φ, ψ

$$\varphi :]-3, 3[\times] - 5, 5[\to \mathbb{R}^2, (x, y) \mapsto (3x + 4y, 4y),$$
$$\psi :]1, 3[\times]1, 5[\to \mathbb{R}^2, (x, y) \mapsto (2x, x^2 + y).$$

Verify that φ, ψ with suitably restricted codomains are charts of $\mathbb{R}^2$, and compute when two tangent vectors $[(\varphi, v)]$, $[(\psi, w)]$ of $\mathbb{R}^2$ are equal.

Exercise* 1.3.10 A manifold M with dim $M \neq 0$ is said to be **orientable** if there exists an atlas such that the determinants of the Jacobi matrices of transition maps take only positive values. For dim $M = 0$, M is always said to be orientable. Show that TM is always orientable for any M.

Exercise 1.3.11 Show that Klein's bottle (with its C^∞ structure from Exercise 1.2.24) is not orientable. What is the maximal $n \in \mathbb{N}$ for which vector fields $X_1, \dots, X_n$ exist on Klein's bottle M such that at each point $p \in M$, the vectors $X_{1|p}, \dots, X_{n|p}$ are linearly independent?

Exercise 1.3.12 Construct a zero-free vector field on S^{2n-1} for $n \in \mathbb{Z}^+$. *Hint: Do not use the stereographic projections but the standard embedding $S^{2n-1} \subset \mathbb{C}^n$.*

Exercise 1.3.13 1) Consider the charts $\varphi_\pm$ of the sphere S^n from Exercise 1.1.6. Calculate precisely when two pairs (φ_+, u), (φ_-, v) with $u, v \in \mathbb{R}^n$ represent the same tangent vector.

2) Applying φ_+, φ_-, verify again that TS^1 and $S^1 \times \mathbb{R}$ are diffeomorphic. *Hint: When constructing the diffeomorphism, it may help to understand to which vector in $T\mathbb{R}$ a tangent vector of length 1 of the circle is mapped.*

Exercise 1.3.14 For an embedding $f : M \to N$, prove that $Tf : TM \to TN$ is also an embedding.

1.4 Second Description of the Tangent Bundle: Derivations

derivation	germs
direct image	jets
Lie bracket	

In this and the next section, smooth vector fields (as opposed to vectors at single points) are described in further ways. In terms of Exercise 1.4.13 this corresponds to further representations of the tangent bundle. The second description of vector fields is based on the following property, which characterizes first-order derivatives of $\mathbb{R}$-valued functions in a more algebraic way:

Definition 1.4.1 A *derivation* on $C^\infty(M)$ is an $\mathbb{R}$-*linear map* $\delta : C^\infty(M) \to C^\infty(M)$ *satisfying the requirement*

$$\delta(f \cdot g) = \delta(f) \cdot g + f \cdot \delta(g) \quad \text{for all } f, g \in C^\infty(M) \qquad \text{(Leibniz rule)}.$$

Example On $C^\infty(\mathbb{R})$, the map that maps $f \in C^\infty(\mathbb{R})$ to $x \mapsto f'(x) \cdot \sin x$ is a derivation.

Remark Given $f \equiv \text{const.}$ one gets $\delta(f) = 0$ as

$$\delta(f) = \delta(f \cdot 1) \overset{\text{Leibniz}}{=} \delta(f) \cdot 1 + f \cdot \delta(1) \overset{\mathbb{R}-\text{linear}}{=} 2\delta(f).$$

Proposition 1.4.2 *Consider* $f \in C^\infty(M)$, *a derivation* δ *and let* $U \subset M$ *be open. Then* $\delta(f)_{|U}$ *is uniquely determined by* $f_{|U}$ *("derivations are local operators").*

Proof. Let $f_{|U} = \widetilde{f}_{|U}$ and $p \in U$. Choose a C^∞ test function $\tau : M \to \mathbb{R}$ with $\tau(p) = 1, \tau_{|M \setminus U} \equiv 0$. With a chart $\varphi : \widetilde{U} \to V$ around p and a corresponding test function $\widetilde{\tau}$ on V, such a function can be obtained as $\tau := \widetilde{\tau} \circ \varphi$ (Fig. 1.11). Then $0 = (f - \widetilde{f}) \cdot \tau$ and hence

$$0 = \delta((f - \widetilde{f}) \cdot \tau)(p) = \delta(f - \widetilde{f})(p) \cdot \underbrace{\tau(p)}_{=1} + \underbrace{(f - \widetilde{f})(p)}_{=0} \cdot \delta\tau(p),$$

i.e. $\delta(f)(p) = \delta(\widetilde{f})(p)$ for all $p \in U$. $\square$

Let $\pi_2 : T\mathbb{R} = \mathbb{R} \times \mathbb{R} \twoheadrightarrow \mathbb{R}$ be the projection onto the second factor.

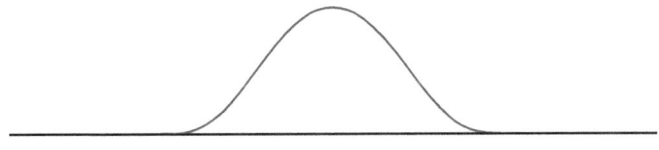

Fig. 1.11 Example of a test function $\tilde{\tau}$ on $\mathbb{R}$, in this case $e^{2-\frac{1}{(1+x)^2}-\frac{1}{(1-x)^2}}$ on $\,]-1,1[$.

Theorem 1.4.3 *There is a canonical isomorphism of* $\mathbb{R}$*-vector spaces*

$$L : \Gamma(M, TM) \to \{Derivations\ on\ C^\infty(M)\}$$
$$X \mapsto (f \mapsto L_X f := X.f := \pi_2(Tf(X))).$$

Remark This will be false if one replaces the C^∞ condition on transition maps by analytic, holomorphic or rational, or if one considers ∞-dimensional manifolds. For example, by Liouville's Theorem, all holomorphic $\mathbb{C}$-valued functions on $\mathbb{P}^1\mathbb{C}$ are constant, hence all derivatives are equal to 0. But there is a 3-dimensional vector space of holomorphic vector fields on $\mathbb{P}^1\mathbb{C}$.

For the sake of clarity, π_2 is usually not explicitly indicated in later calculations.

Proof. L is well-defined: Given a chart $\varphi : U \to V$ around $p \in M$, set $X_p = [(\varphi, u)] \in T_pM$. Then $L_X(gf)(p) = ((gf) \circ \varphi^{-1})'_{|\varphi(p)}(u) = (g \circ \varphi^{-1})'_{|\varphi(p)}(u) \cdot f(p) + g(p) \cdot (f \circ \varphi^{-1})'_{|\varphi(p)}(u) = (L_Xg \cdot f + g \cdot L_Xf)(p)$.

L is injective: Assume that $L_Xf = 0$ for all f and that there is a point $p \in M$ with $X_p \neq 0$. Choose $\tau \in C^\infty(V, \mathbb{R})$ such that $\tau = \left\{{1 \atop 0}\right.$ in a neighborhood of $\frac{\varphi(p)}{\partial V}$. Choose $g : V \to \mathbb{R}$ with $g(\varphi(p)) = 0, T_{\varphi(p)}g(u) \neq 0$ and $f := \left\{{(\tau \cdot g) \circ \varphi \atop 0}\right.$ on ${U \atop M\backslash U}$. Then $(X.f)(p) \neq 0 \,\xi$.

L is surjective: Let δ be a derivation, $f \in C^\infty(M, \mathbb{R})$, $\varphi : U \to V$ a chart and $g := f \circ \varphi^{-1}$. In a star-shaped neighborhood of $x_0 := \varphi(p)$ for $p \in U$ one finds

$$g(x) = g(x_0) + \int_0^1 \frac{\partial[g(t \cdot (x - x_0) + x_0)]}{\partial t} dt$$
$$= g(x_0) + \sum_{j=1}^n (x_j - x_{0,j}) \cdot \int_0^1 \left(\frac{\partial g}{\partial x_j}\right)(t \cdot (x - x_0) + x_0)\, dt$$

(Fig. 1.12). Thus because of Proposition 1.4.2,

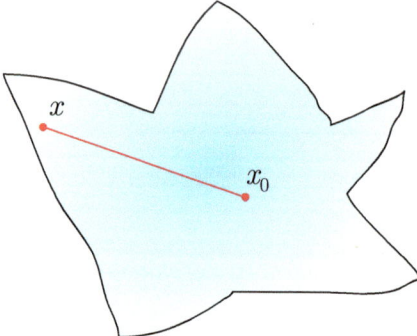

Fig. 1.12 Integration path to prove surjectivity of L.

$$\delta f(p) = \delta(g \circ \varphi)(p) = \underbrace{\delta(g(x_0))(p)}_{=0}$$

$$+ \sum \delta(x_j \circ \varphi - x_{0,j})(p) \cdot \int_0^1 \left(\frac{\partial g}{\partial x_j}\right)(t \cdot (x - x_0) + x_0)\, dt_{|x=x_0}$$

$$+ \sum \underbrace{(x_j \circ \varphi - x_{0,j})(p)}_{=0} \cdot \delta\left(\int_0^1 \left(\frac{\partial g}{\partial x_j}\right)(t \cdot (x - x_0) + x_0)\, dt \circ \varphi\right)(p)$$

$$= \sum \delta(x_j \circ \varphi)(p) \cdot \frac{\partial(f \circ \varphi^{-1})}{\partial x_j}\bigg|_{x=x_0} = T_p f(X)$$

for $X_p := [(\varphi, \begin{pmatrix} \delta(x_1 \circ \varphi)(p) \\ \vdots \\ \delta(x_n \circ \varphi)(p) \end{pmatrix})]$. This vector field depends smoothly on p. It is independent of the choice of chart: Because of injectivity, it follows from $\delta(\cdot)_{|U_1} = L_{X_1}(\cdot)_{|U_1}$, $\delta(\cdot)_{|U_2} = L_{X_2}(\cdot)_{|U_2}$ that $X_1 = X_2$ on $U_1 \cap U_2$. By construction, X satisfies $L_X = \delta$. $\quad\square$

The definition of derivatives is shorter and more elegant than that of vector fields, and it does not use charts. On the other hand, a single tangent vector at a point cannot be directly described with derivatives in this form. Some concepts related to vector fields are easier and more elegant to study with derivatives. The last part of this chapter shows some of these applications.

It is not surprising that an identification of manifolds by means of a diffeomorphism also transforms vector fields into each other. This leads to the following conceptualization (see Fig. 1.13):

Lemma and Definition 1.4.4 *Let $f : M \to N$ be a diffeomorphism. The **direct image** $f_* : \Gamma(M, TM) \to \Gamma(N, TN)$ of a vector field $X \in \Gamma(M, TM)$ is defined as*

$$L_{f_*X}g := (L_X(g \circ f)) \circ f^{-1} \qquad \text{for all } g \in C^\infty(N).$$

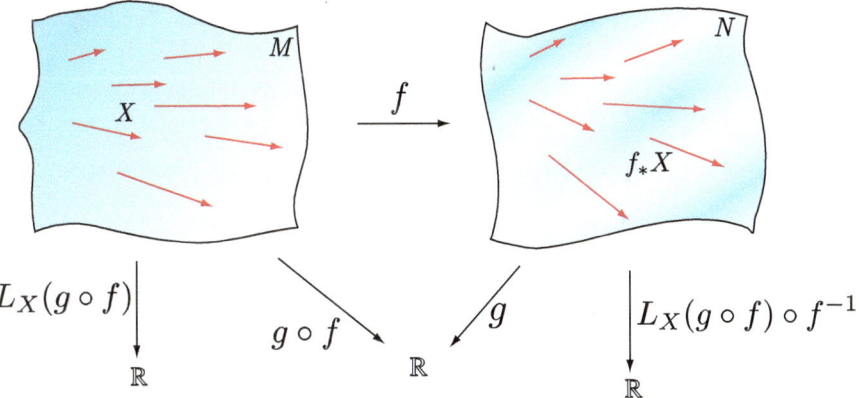

Fig. 1.13 Direct image of a vector field X under a diffeomorphism f.

Pointwise this can be written as

$$(f_*X)_{|p} = (T_{f^{-1}(p)}f)(X_{|f^{-1}(p)}).$$

The composition with f^{-1} is easy to forget in calculations, but it is of fundamental importance here, because without this composition one gets $L_X(g \circ f) \in C^\infty(M)$.

Proof. L_{f_*X} is a derivation as

$$
\begin{aligned}
L_{f_*X}(g \cdot h) &= [L_X((g \cdot h) \circ f)] \circ f^{-1} \\
&= [L_X(g \circ f) \circ f^{-1}] \cdot h + g \cdot [L_X(h \circ f) \circ f^{-1}] \\
&= L_{f_*X}g \cdot h + g \cdot L_{f_*X}h.
\end{aligned}
$$

Theorem 1.4.3 and the chain rule provide the pointwise formula:

$$
\begin{aligned}
L_{Tf(X_{|f^{-1}(p)})}g &= Tg(Tf(X_{|f^{-1}(p)})) = T(g \circ f)(X_{|f^{-1}(p)}) \\
&= (L_X(g \circ f) \circ f^{-1})(p). \qquad \square
\end{aligned}
$$

Lemma 1.4.5 *If f, g are diffeomorphisms, then $f_* \circ g_* = (f \circ g)_*$, i.e.*

$$
\left\{ \begin{array}{c} manifolds \\ diffeomorphisms \end{array} \right\} \rightarrow \left\{ \begin{array}{c} \mathbb{R}\text{-}vector\ spaces \\ vector\ space\ isomorphisms \end{array} \right\}
$$

$$M \mapsto \Gamma(M, TM)$$

$$f \mapsto f_*$$

is a covariant functor.

Proof. For any real-valued function h on the codomain manifold of f and any vector field X one gets

$$L_{f_*g_*X}h = (L_{g_*X}(h \circ f)) \circ f^{-1} = (L_X(h \circ f \circ g)) \circ g^{-1} \circ f^{-1} = L_{(f \circ g)_*X}h. \quad \square$$

This proof uses only the definition of the direct image and not the chain rule. This does not replace the corresponding proofs in the last section, since the chain rule was used in the proof of Theorem 1.4.3.

The composition of two derivation operators is a second-order differential operator. Remarkably, however, the following difference yields again a vector field:

Lemma and Definition 1.4.6 *Given two vector fields $X, Y \in \Gamma(M, TM)$, the **Lie bracket** $L_X \circ L_Y - L_Y \circ L_X$ is again a derivation and thus a vector field $[X, Y]$.*

Proof.

$$L_X L_Y(f \cdot g) = L_X(L_Y f \cdot g + f \cdot L_Y g)$$
$$= L_X L_Y f \cdot g + (L_Y f)(L_X g) + (L_X f)(L_Y g) + f \cdot L_X L_Y g,$$

and subtract from this the analog for the second term. $\quad \square$

Remark By this definition $[\cdot, \cdot]$ is skew-symmetric and $\mathbb{R}$-bilinear: $[X, Y] = -[Y, X]$ and for all $a, b \in \mathbb{R}$ the equation $[aX + bY, Z] = a[X, Z] + b[Y, Z]$ holds. Another basic property follows later in Lemma 1.5.7.

For more general maps f there is no direct image of vector fields. For example, for a curve $f : \mathbb{R} \to M$ there is in general no canonical extension of a vector field from $f(\mathbb{R})$ to M. And if the curve f intersects itself transversely at $p \in M$, the corresponding tangent spaces in $T_p M$ intersect only at 0. But in general one can impose an analogous compatibility condition that relates many properties of vector fields on the domain and codomain of f similarly well, as is the case for direct images. The following key lemma will be used many times in this book:

Lemma 1.4.7 *Let $f : M \to \widetilde{M}$ be a C^∞ map, let X, Y be vector fields on M and let $\widetilde{X}, \widetilde{Y}$ be vector fields on $\widetilde{M}$, such that $T_p f(X) = \widetilde{X}_{f(p)}, T_p f(Y) = \widetilde{Y}_{f(p)}$ for all $p \in M$. Then $T_p f([X, Y]) = [\widetilde{X}, \widetilde{Y}]_{f(p)}$.*

Proof. For any $g \in C^\infty(\widetilde{M})$ by assumption $L_X(g \circ f) = (L_{\widetilde{X}}g) \circ f$ holds and so

$$(L_{\widetilde{Y}}(L_{\widetilde{X}}g)) \circ f = L_Y((L_{\widetilde{X}}g) \circ f) = L_Y L_X(g \circ f).$$

Hence

$$[\widetilde{X}, \widetilde{Y}]_{|f(p)} . g = (([\widetilde{X}, \widetilde{Y}].g) \circ f)(p) = ([X, Y].(g \circ f))(p) = T_p f([X, Y]).g. \quad \square$$

For a diffeomorphism f these terms match well, as one would expect:

Corollary 1.4.8 *If $f : M \to N$ is diffeomorphism and $X, Y \in \Gamma(M, TM)$ are vector fields, one gets*

$$f_*[X, Y] = [f_*X, f_*Y].$$

Exercises

Exercise 1.4.9 Consider $M = \mathbb{R}^n$ and $A := (a_k)_{k=1}^n, B := (b_k)_{k=1}^n \in \Gamma(M, TM)$ with $a_j, b_j \in C^\infty(M, \mathbb{R})$ for all j. Show that

$$[A, B] = \left(\sum_{j=1}^n a_j \frac{\partial b_k}{\partial x_j} - \sum_{j=1}^n b_j \frac{\partial a_k}{\partial x_j} \right)_{k=1}^n .$$

For vector fields on $\mathbb{R}^n$ the notation $A = (a_k)_{k=1}^n = \sum_{k=1}^n a_k \frac{\partial}{\partial x_k}$ is useful, where the vectors of the Cartesian basis are written as $\frac{\partial}{\partial x_k}$.

Exercise 1.4.10 Consider the vector fields X, Y, Z on $\mathbb{R}^3$ given by

$$X := z\frac{\partial}{\partial y} - y\frac{\partial}{\partial z}, \quad Y := x\frac{\partial}{\partial z} - z\frac{\partial}{\partial x}, \quad Z := y\frac{\partial}{\partial x} - x\frac{\partial}{\partial y}.$$

Let V be the subspace of $\Gamma(\mathbb{R}^3, T\mathbb{R}^3)$ spanned by X, Y, Z and set

$$\varphi : V \to \mathbb{R}^3, aX + bY + cZ \mapsto (a, b, c).$$

1) Given an isometry $g \in O(3)$ of the Euclidean vector space for the canonical Euclidean metric, show that $g_*V = V$.
2) Prove that $\varphi([A, B]) = \varphi(A) \times \varphi(B)$ for $A, B \in V$ and the cross product $\times$ on $\mathbb{R}^3$ defined by $\langle u \times v, w \rangle = \det(u, v, w)$ for all $u, v, w \in \mathbb{R}^3$.

Exercise 1.4.11 Show that for $X \in \Gamma(M, TM)$ the map

$$L_X : \Gamma(M, TM) \to \Gamma(M, TM), Y \mapsto [X, Y]$$

is a derivation on $\Gamma(M, TM)$ with respect to the Lie bracket as product (i.e. it is $\mathbb{R}$-linear and it satisfies the Leibniz product rule).

Exercise 1.4.12 Let $M := \mathbb{R}^2/(2\pi\mathbb{Z})^2$ be a two-dimensional torus and consider $X, Y \in \Gamma(M, TM)$ with

$$X_{(x,y)} := \frac{\partial}{\partial x} + \cos(2y)\frac{\partial}{\partial y}, \quad Y_{(x,y)} := \sin(x)\frac{\partial}{\partial x} + \cos(x + y)^2\frac{\partial}{\partial y}$$

(or in the other notation, $X_{(x,y)} := \begin{pmatrix} 1 \\ \cos 2y \end{pmatrix}, Y_{(x,y)} := \begin{pmatrix} \sin x \\ \cos(x+y)^2 \end{pmatrix}$). Set $f : M \to M, (x, y) \mapsto (y, x)$ and $g : M \to \mathbb{R}, (x, y) \mapsto \cos(x + ny)$ with $n \in \mathbb{N}$. Compute $[X, Y], f_*X, f_*Y, L_Xg$ and L_Yg.

Exercise* 1.4.13 In this exercise we discuss yet another description of T_pM, more precisely of T_p^*M. Let M be a manifold and $p \in M$. In the $\mathbb{R}$-algebra of $\mathbb{R}$-valued functions $C^\infty(M, \mathbb{R})$, let I_p be the ideal of functions vanishing on an open neighborhood of p. The elements of $\mathcal{F}_p := C^\infty(M, \mathbb{R})/I_p$ are called **germs** of $\mathbb{R}$-valued functions. Show that:

1) $\mathcal{J}_p := \{[f] \in \mathcal{F}_p \mid f(p) = 0\}$ is well-defined and it is an ideal in $\mathcal{F}_p$.
2) $\mathcal{J}_p/(\mathcal{J}_p)^2 \to T_p^*M$, $[[f]] \mapsto T_pf$ is a vector space isomorphism.

(Analogously, k-**jets** are defined as elements of $\mathcal{J}_p/(\mathcal{J}_p)^{k+1}$. This definition can be applied to the analytic, holomorphic or rational case if we restrict ourselves to representatives defined on a neighborhood of p.)

1.5 Third Description of the Tangent Bundle: Flows

flow	one-parameter group
integral curve	Jacobi identity
trajectory	general linear group

In this section vector fields will be characterized as infinitesimal diffeomorphisms.

Theorem and Definition 1.5.1 *Given a manifold M, a vector field X and $p \in M$, there exist an $\varepsilon > 0$ and a neighborhood U of p such that there exists a unique (local) flow $\Phi^X_\cdot :] - \varepsilon, \varepsilon[\times U \to M$ with $\Phi^X_0(q) = q$, $\frac{\partial \Phi^X_t(q)}{\partial t} = X_{|\Phi^X_t(q)}$ (Fig. 1.14). The curves $\Phi^X_\cdot(q) :]\varepsilon, \varepsilon[\to M$ are called integral curves (or trajectories).*

As in Example 1.3.7, here $\frac{\partial \Phi^X_t(p)}{\partial t} = T\Phi^X_t(p)(\frac{\partial}{\partial t})$ holds with the cartesian vector field $\frac{\partial}{\partial t}$ on $\mathbb{R}$. The flow equation imposes that the canonical vector field $\frac{\partial}{\partial t}$ and X are related with respect to $\Phi^X(p)$ as in Lemma 1.4.7.

Proof. Via a chart $\varphi : U \to V$ and with $X = [(\varphi, u)], u \in C^\infty(V, \mathbb{R}^n)$, the condition on Φ^X reads

$$\frac{\partial}{\partial t}(\varphi(\Phi^X_t(p))) = u_{|\varphi(\Phi^X_t(p))}.$$

The Picard–Lindelöf Theorem on local existence and uniqueness of solutions of ordinary differential equations of 1st order (in this case an autonomous system) provides the assertion.   □

Corollary 1.5.2 *There is an open neighborhood $U_\Phi \subset \mathbb{R} \times M$ around $\{0\} \times M$, on which the flow $\Phi^X \in C^\infty(U_\Phi, M)$ is defined. It is uniquely determined by*

1) $\Phi^X_0 = \mathrm{id}_M$,
2) $\Phi^X_t \circ \Phi^X_s = \Phi^X_{t+s}$ (at those $p \in M$ where at least three of these values of Φ^X are defined),

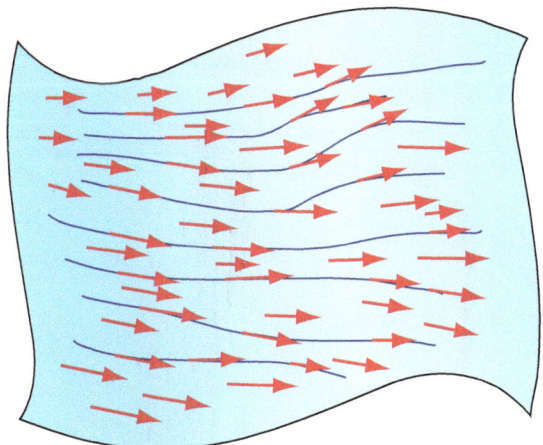

Fig. 1.14 Integral curves of a vector field

3) $\frac{\partial}{\partial t}|_{t=0} \Phi_t^X(p) = X_{|p}$.

Proof. Consider the union U_Φ of all neighborhoods $]-\varepsilon, \varepsilon[\times U$ as in Theorem 1.5.1. Since X is smooth, the solution Φ of the differential equation is also C^∞.

(1),(3) follow immediately from the definition of Φ.

(2) is valid for Φ because of the uniqueness of the solution of the differential equation:

$$\frac{\partial}{\partial t}\left(\Phi_t^X \circ \Phi_s^X(p)\right) = X_{|\Phi_t^X(\Phi_s^X(p))},$$

hence $\Phi_{t+s}^X(p) = \Phi_t^X(\Phi_s^X(p))$.

Conversely, (1)–(3) imply the definition of the integral curves.   □

By (2), if $\{t\} \times M \subset U_\Phi$, then Φ_t is a diffeomorphism since $\Phi_t \circ \Phi_{-t} = \mathrm{id}_M$. From that point of view, vector fields are infinitesimal diffeomorphisms of a manifold onto itself. Conversely, a main advantage of this relation is the construction of diffeomorphisms by vector fields.

Example Considering $X = y^2\frac{\partial}{\partial y}$ on $M = \mathbb{R}$ (Fig. 1.15) the flow differential equation implies that

$$t + C = \int dt = \int \frac{d\Phi_t^X(p)}{\Phi_t^X(p)^2} = -\frac{1}{\Phi_t^X(p)}.$$

Since $\Phi_0^X(p) = p$ implies $-\frac{1}{p} = C$, one finds $\Phi_t^X(p) = \frac{1}{\frac{1}{p}-t}$ on $U_\Phi = \{(t, p) \in \mathbb{R}^2 \mid tp < 1\}$.

Theorem 1.5.3 *If X has compact support* supp X *(e.g., if M is compact), then* Φ^X *is defined on all of* $\mathbb{R} \times M$.

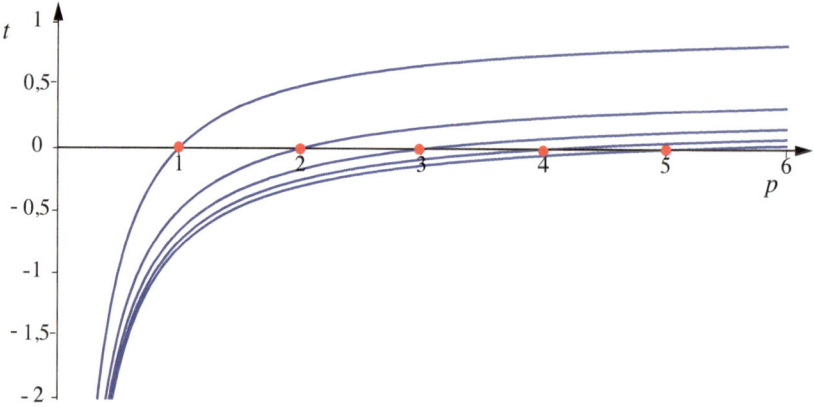

Fig. 1.15 The flow of the vector field $y^2 \frac{\partial}{\partial y}$ is not defined on all of $\mathbb{R} \times \mathbb{R}$.

Proof. Set $\varepsilon := \frac{1}{2} \min_{p \in \text{supp} X} \sup\{t \mid (t, p) \in U_\Phi\}$. Because supp X is compact, we have $\varepsilon > 0$. Then for $t > 0, t = k \cdot \varepsilon + r, k \in \mathbb{N}_0, 0 \leq r < \varepsilon$ set

$$\Phi_t^X := \Phi_r^X \circ \underbrace{\Phi_\varepsilon^X \circ \cdots \circ \Phi_\varepsilon^X}_{k-\text{mal}},$$

and analogously for $t < 0$. According to Corollary 1.5.2, Φ_t^X is the flow to X. $\qquad\square$

Thus, in the latter case $\Phi : (\mathbb{R}, +) \rightarrow (\text{Diff}(M, M), \circ)$ defines a group homomorphism to the diffeomorphisms of M, a so-called **one-parameter group** of diffeomorphisms.

Proposition 1.5.4 *Consider a diffeomorphism $f : M \rightarrow N$ and $X \in \Gamma(M, TM)$. Then the local flow of $f_* X$ is given by $f \circ \Phi^X \circ f^{-1}$.*

Proof. We find $f \circ \Phi_0^X \circ f^{-1} = \text{id}_N$ and

$$\frac{\partial}{\partial t}\Big|_{t=0} (f \circ \Phi_t^X \circ f^{-1})(p) = T_{f^{-1}(p)} f\big(\frac{\partial}{\partial t}\Big|_{t=0} (\Phi_t^X \circ f^{-1})(p)\big) = f_* X_{|p}. \qquad\square$$

Corollary 1.5.5 *Let M, N be manifolds, let $f : M \rightarrow N$ be a diffeomorphism and $X \in \Gamma(M, TM)$ with $f_* X = X$. Then for any choice of t we have $f \circ \Phi_t^X = \Phi_t^X \circ f$.*

Theorem 1.5.6 *For $X, Y \in \Gamma(M, TM)$ the formula $\frac{\partial}{\partial t}\Big|_{t=0}(\Phi_{t\,*}^Y X) = [X, Y]$ holds.*

This is meant to be a local formula, since Φ_t^Y does not need to be defined globally on M.

Proof. Since the formula is local, Y has compact support. This can always be achieved by multiplying by a suitable test function which is constant 1 on a neighborhood of a point p under consideration. For $f \in C^\infty(M)$, because $(\Phi_t^Y)^{-1} = \Phi_{-t}^Y$, one finds

$$\frac{\partial}{\partial t}_{|t=0}\left(X.(f \circ \Phi_t^Y) \circ (\Phi_t^Y)^{-1}\right)$$

$$= \frac{\partial}{\partial t}_{|t=0}\left(X.(f \circ \Phi_t^Y) \circ \underbrace{\Phi_0^Y}_{=\mathrm{id}}\right) + \frac{\partial}{\partial t}_{|t=0}\left(X.(f \circ \Phi_0^Y) \circ \Phi_{-t}^Y\right)$$

$$= X.(Y.f) - Y.(X.f). \qquad \square$$

Remark For the left-hand side in Theorem 1.5.6 the skew-symmetry of the Lie bracket is much less obvious. In contrast, it becomes immediately apparent that the Lie bracket yields a vector field. The following Jacobi identity is also much more transparent, because it only says that $[\cdot, X]$ is a derivation on $\Gamma(M, TM)$ with respect to the Lie bracket as product. The identity can also be shown elementarily with derivations on $C^\infty(M)$, but the reason why it holds remains more obscure.

Lemma 1.5.7 (Jacobi identity[7]) *Given vector fields X, Y, Z, the identity*

$$[[X,Y], Z] + [[Y, Z], X] + [[Z, X], Y] = 0$$

holds.

Proof. Applying Theorem 1.5.6 to Lemma 1.4.8 yields

$$[[Y, Z], X] = \frac{\partial}{\partial t}_{|t=0} \Phi_{t\,*}^X [Y, Z] = \frac{\partial}{\partial t}_{|t=0} [\Phi_{t\,*}^X Y, \Phi_{t\,*}^X Z]$$

$$= [[Y, X], Z] + [Y, [Z, X]]. \qquad \square$$

Proposition 1.5.8 *The flows of two vector fields X, Y commute locally around 0 if and only if the vector fields commute. More precisely, $[X, Y] = 0$ is equivalent to $\Phi_s^X \circ \Phi_t^Y = \Phi_t^Y \circ \Phi_s^X$ for s, t in a neighborhood of $0 \in \mathbb{R}$.*

Proof. Assume that $[X, Y] = 0$. Then $\Phi_{t\,*}^Y X = X$ holds, because for $t = 0$ this equation is true. And according to Corollary 1.5.2(2) and the chain rule 1.4.5 one obtains

$$\frac{\partial}{\partial t} \Phi_{t\,*}^Y X = \Phi_{t\,*}^Y \frac{\partial}{\partial \varepsilon}_{|\varepsilon=0} \Phi_{\varepsilon*}^Y X \overset{1.5.6}{=} \Phi_{t\,*}^Y \underbrace{[X, Y]}_{=0}.$$

---

[7] Carl Gustav Jacob Jacobi, 1804–1851 (estate).

The assertion follows by Corollary 1.5.5. The converse follows by differentiating the relation $\Phi_s^X = \Phi_{-t}^Y \circ \Phi_s^X \circ \Phi_t^Y$:

$$
0 \quad = \quad \frac{\partial^2}{\partial t \partial s}_{|s=t=0} \Phi_s^X \quad = \quad \frac{\partial^2}{\partial t \partial s}_{|s=t=0} \Phi_{-t}^Y \circ \Phi_s^X \circ \Phi_t^Y
$$

$$
\overset{\text{Proposition 1.5.4}}{=} \frac{\partial}{\partial t}_{|t=0} \Phi_{-t*}^Y X \overset{\text{Th. 1.5.6}}{=} -[X, Y]. \qquad \square
$$

The hypothesis of the following result will be further weakened in Frobenius' Theorem 2.3.10.

Theorem 1.5.9 *Suppose $X_1, \ldots, X_k$ are pairwise commuting vector fields on M which are linearly independent at $p \in M$. Then there exists a k-dimensional submanifold $N \subset M$ with $p \in N$ such that $X_{1|q}, \ldots, X_{k|q}$ form a basis of $T_q N$ for all $q \in N$.*

Proof. For a sufficiently small neighborhood $V \subset \mathbb{R}^k$ of 0, define

$$
\gamma : V \to M
$$
$$
(t_1, \ldots, t_k) \mapsto (\Phi_{t_1}^{X_1} \circ \cdots \circ \Phi_{t_k}^{X_k})(p).
$$

Then

$$
\frac{\partial \gamma}{\partial t_j} \overset{1.5.8}{=} \frac{\partial}{\partial t_j} \left[\Phi_{t_j}^{X_j} \circ \Phi_{t_1}^{X_1} \circ \cdots \circ \widehat{\Phi_{t_j}^{X_j}} \circ \cdots \circ \Phi_{t_k}^{X_k} \right](p)
$$

$$
= X_{j|\left[\Phi_{t_j}^{X_j} \circ \Phi_{t_1}^{X_1} \circ \cdots \circ \widehat{\Phi_{t_j}^{X_j}} \circ \cdots \circ \Phi_{t_k}^{X_k} \right](p)}
$$

$$
= X_{j|\gamma(t_1,\ldots,t_k)}.
$$

Because of the linear independence of the X_j at p, γ is a local parametrization for sufficiently small V according to the Implicit Function Theorem. $\qquad \square$

Exercises

Exercise 1.5.10 Compute the flow Φ of the vector field

$$
Y \in \Gamma(\mathbb{R}^+, T\mathbb{R}^+), \quad x \mapsto \frac{1}{3x^2} \frac{\partial}{\partial x}.
$$

What is the largest possible domain of definition of Φ?

Exercise 1.5.11 Let $G := \mathrm{GL}(n, \mathbb{R}) = \{A \in \mathbb{R}^{n \times n} \mid \det A \neq 0\}$ be the **general linear group**, $A \in \mathbb{R}^{n \times n}$ and set $X_g = g \cdot A$ for any $g \in G$. Why does this define a vector field on G? For the flow Φ^X of X show that

$$\Phi_t^X(g) = g \cdot \sum_{k=0}^{\infty} \frac{(tA)^k}{k!}.$$

Exercise* 1.5.12 Verify the Jacobi identity 1.5.7 by substituting the definition of the Lie bracket six times.

1.6 Lie Groups

Lie group	exponential map
Lie subgroup	representation
unitary group	standard representation
special unitary group	irreducible representation
Lie algebra	adjoint representation
left-invariant vector field	adjoint representation (of a Lie algebra)
trivialization	representation of a Lie algebra
Lie group homomorphism	symplectic group
Lie algebra homomorphism	ideal

Interesting and at the same time easy to study examples are obtained if one additionally demands a group structure on the manifolds. Thereby many properties can be traced back to properties of the tangent space at the neutral element, which of course is considerably easier to understand than the whole manifold. The resulting Lie groups are the basis for the investigations of homogeneous and symmetric spaces in the later chapters.

Definition 1.6.1 A **Lie group**[8] G is a group with a C^∞ structure such that the map $m_G : G^2 \to G, (g, h) \mapsto g \cdot h$ is C^∞. For $g \in G$, let

$$L_g : G \to G, \qquad R_g : G \to G$$
$$h \mapsto gh \qquad\qquad h \mapsto hg$$

be the left and right multiplication, respectively.

For a Lie group G, let e_G (and sometimes just e) be the neutral element.

Example $\mathbb{Z}/5\mathbb{Z}$ is a disconnected Lie group of dimension 0. The manifolds $\mathbb{R}^n$ and $\mathbb{R}^n/\mathbb{Z}^n$ are Lie groups. So are $GL(n, \mathbb{R})$ and $GL(n, \mathbb{C})$ as open subsets of $\mathbb{R}^{n \times n}$ and $\mathbb{C}^{n \times n}$, respectively.

Lemma 1.6.2 *For any Lie group G, the map $k_G : G \to G, g \mapsto g^{-1}$ is a C^∞ map.*

---

[8] 1884, Sophus Lie, 1842–1899.

Proof. The derivative of m_G at (g, h) with respect to the second variable is $T_h L_g$. The map L_g is a diffeomorphism, because the inverse map is $L_{g^{-1}} \in C^\infty(G, G)$. Thus $T_h L_g$ is invertible and according to the Implicit Function Theorem, the preimage $m_G^{-1}(\{e_G\}) = \{(g, g^{-1}) \mid g \in G\}$ of the neutral element is locally parametrized by a C^∞ function in g.   □

Lemma and Definition 1.6.3 *A **Lie subgroup** $H \subset G$ of a Lie group G is defined to be a subgroup which is also a submanifold of G. Then H is a Lie group and closed.*

Remark Sometimes in the literature Lie subgroups are not required to be submanifolds.

Proof. For $\iota : H \to G$, by Lemma 1.2.15 $\iota^{-1} : \iota(H) \to H$ is a C^∞ map. Hence

$$m_H : H \times H \to H,$$
$$(g, h) \mapsto gh = \iota^{-1}(\iota(gh)) = \iota^{-1}(\iota(g) \cdot \iota(h))$$

is smooth.

Let $(h_j)_{j \in \mathbb{N}} \in H^{\mathbb{N}}$ be a sequence convergent in G with limit g and let $U \subset G$ be a neighborhood of e on which a chart identifies $U \cap H$ with $\mathbb{R}^k \subset \mathbb{R}^n$ in a neighborhood of 0. In particular, $U \cap H$ is closed in U. Because of the continuity of m_G, k_G there exists a neighborhood $V \subset G$ of e_G with $\overline{V^{-1} \cdot V} \subset U$. Choose N such that $g^{-1} h_j \in V$ for $j > N$. For $j, k > N$ it follows that $h_j^{-1} h_k = (g^{-1} h_j)^{-1} g^{-1} h_k \in \overline{V^{-1} \cdot V} \cap H \subset U \cap H$. Thus for $j \to \infty$ the limit satisfies $h_k^{-1} g \in H$ and hence $g \in H$.   □

Example $SL(n, \mathbb{R})$, $SO(n)$, $O(n)$. For the **unitary group**

$$U(n) := \{A \in \mathbb{C}^{n \times n} \mid A^t \overline{A} = \mathrm{id}\}$$

and the **special unitary group**

$$SU(n) := \{A \in U(n) \mid \det A = 1\}$$

the proof proceeds as in the case $SO(n)$, e.g. by using the preimage $f^{-1}(0)$ with $f : \mathbb{C}^{n \times n} \to \mathbb{C}^{n \times n}_{\mathrm{Hermitian}}$, $A \mapsto A^t \overline{A} - \mathrm{id}$. This map has a surjective derivative as for any $Y \in \mathbb{C}^{n \times n}_{\mathrm{Hermitian}}$, it satisfies $T_A f(\frac{1}{2} A Y^t) = Y$.

Remark One can show that every closed subgroup of a Lie group is a Lie subgroup (Cartan's Theorem). But not every subgroup which is an image of a Lie group under a differentiable group homomorphism is also closed: Consider, for example, images of straight lines of irrational slope in the torus (Fig. 1.16).

Definition 1.6.4 *A **Lie algebra** $\mathfrak{g}$ is an $\mathbb{R}$-vector space together with a skew-symmetric bilinear form $[\cdot, \cdot] : \mathfrak{g}^2 \to \mathfrak{g}$ satisfying the Jacobi identity.*

Example By Lemma 1.5.7, $\Gamma(M, TM)$ is a Lie algebra. So are the skew-symmetric matrices with $[A, B] := AB - BA$, as can be seen by writing down this definition

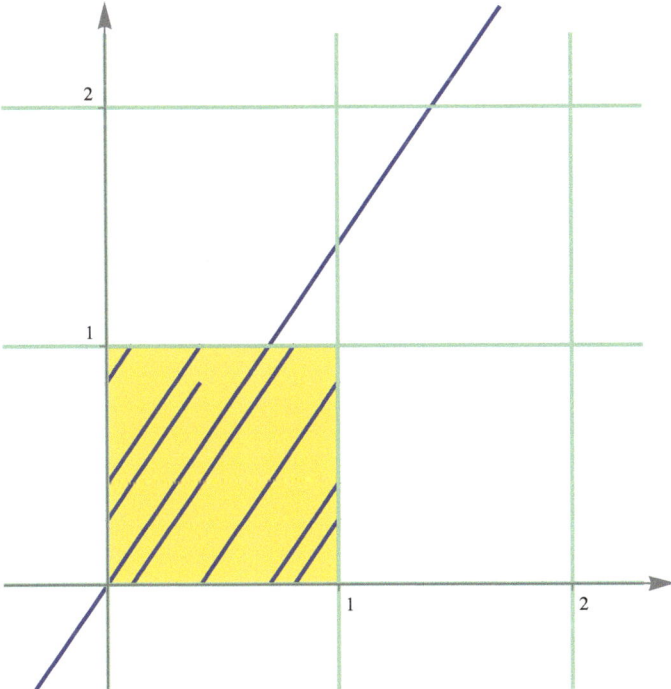

Fig. 1.16 Subgroup of a torus that is not a Lie subgroup.

six times and checking the mutual cancellation of the resulting twelve terms. We will learn about a more conceptual proof applying Lemma 1.6.12 and the preceding example.

Definition 1.6.5 *A vector field* $X \in \Gamma(G, TG)$ *is called* **left-invariant** *(or* **right-invariant***) if* $L_{g*}X = X$ *(or* $R_{g*}X = X$*, respectively) for all* $g \in G$.

Example The left-invariant vector fields on $\mathbb{R}^n$ are obtained by translating a vector. On $S^1 \subset \mathbb{C}$ left-invariant vector fields are obtained by rotation around the origin (Fig. 1.17).

Proposition 1.6.6 *The left-invariant vector fields on* G *form a Lie algebra.*

Proof. For X, Y left-invariant, $g \in G$ we have $L_{g*}[X, Y] = [L_{g*}X, L_{g*}Y] = [X, Y]$. Therefore the left-invariant vector fields form a Lie subalgebra of the vector fields $\Gamma(G, TG)$. $\square$

Lemma 1.6.7 *If* G *is a Lie group* G*, then the map*

$$\rho : T_{e_G}G \to \{\text{left-invariant vector fields on } G\}$$
$$X \mapsto T_{e_G}L_g(X) =: \widetilde{X}_{|g}$$

is a vector space isomorphism.

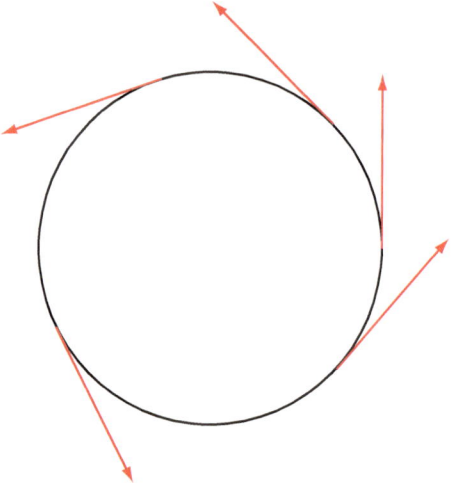

Fig. 1.17 Left-invariant vector field on S^1.

Thus, in contrast to the Lie algebra $\Gamma(G, TG)$, the subalgebra of left-invariant vector fields is finite-dimensional.

Proof. i) $\rho(X)$ is a left-invariant vector field, since for any $h, g \in G$ we get

$$(L_{h*}\rho(X))_{|g} = T_{h^{-1}g}L_h(T_{e_G}L_{h^{-1}g}X) \stackrel{\text{chainrule}}{=} T_{e_G}L_{hh^{-1}g}X = \rho(X)_{|g}.$$

ii) The inverse map is $\widetilde{X} \mapsto \widetilde{X}_{|e_G}$. $\qquad\qquad\qquad\qquad\qquad\qquad\qquad\square$

Corollary 1.6.8 *The tangent bundle TG has the **trivialization***

$$TG \to G \times T_e G$$
$$(g, X) \mapsto (g, T_g L_{g^{-1}} X).$$

Trivializations of bundles are discussed more extensively in the next chapter.

Definition 1.6.9 *The Lie algebra $\mathfrak{g}$ of a Lie group G is $\mathfrak{g} := T_{e_G} G$ with the Lie bracket $[\cdot, \cdot] : \mathfrak{g}^2 \to \mathfrak{g}$, $(X, Y) \mapsto [\rho(X), \rho(Y)]_{|e_G}$ induced by the left-invariant vector fields.*

Example 1.6.10 i) $G = \mathbb{R}^n$, $\mathfrak{g} = T_0\mathbb{R}^n \cong \mathbb{R}^n$, $L_g : \mathbb{R}^n \to \mathbb{R}^n$, $x \mapsto x + g$,

$$L_{g*}\begin{pmatrix} a_1 \\ \vdots \\ a_n \end{pmatrix} = \begin{pmatrix} a_1 \\ \vdots \\ a_n \end{pmatrix} \quad \text{und} \quad \left[\begin{pmatrix} a_1 \\ \vdots \\ a_n \end{pmatrix}, \begin{pmatrix} b_1 \\ \vdots \\ b_n \end{pmatrix} \right] = 0.$$

ii) In a neighborhood of 0, $\mathbb{R}^n$ and $\mathbb{R}^n/\mathbb{Z}^n$ are identical, hence the Lie algebra of the torus is also $\mathbb{R}^n$ with the trivial Lie bracket.

iii) Because $\mathrm{GL}(n,\mathbb{R}) \overset{\text{open}}{\subset} \mathbb{R}^{n\times n}$, we have $\mathfrak{gl}(n,\mathbb{R}) = \mathbb{R}^{n\times n}$. Concerning the Lie bracket, with $\widetilde{X}_{|g} := \rho(X)_{|g} = g \cdot X$ (where $\cdot$ is the matrix multiplication) and $f \in C^{\infty}(\mathrm{GL}(n,\mathbb{R}))$ it follows that

$$(L_{\widetilde{X}}L_{\widetilde{Y}} - L_{\widetilde{Y}}L_{\widetilde{X}})_{|\mathrm{id}_{\mathbb{R}^n}} f = \left(X.(T_g f(gY)) - Y.(T_g f(gX)) \right)_{|\mathrm{id}_{\mathbb{R}^n}}$$
$$= f"_{|\mathrm{id}_{\mathbb{R}^n}}(X,Y) + f'_{|\mathrm{id}_{\mathbb{R}^n}}(X \cdot Y)$$
$$- f"_{|\mathrm{id}_{\mathbb{R}^n}}(Y,X) - f'_{|\mathrm{id}_{\mathbb{R}^n}}(Y \cdot X)$$
$$= L_{X \cdot Y - Y \cdot X} f.$$

Thus $[X,Y] = X \cdot Y - Y \cdot X$. It is crucial for this calculation that on open subsets of $\mathbb{R}^{n\times n}$ we can calculate with second derivatives of f, which we do not yet have in this form for general manifolds, for which the Schwarz Lemma can be applied here.

Definition 1.6.11 *A **Lie group homomorphism** $f : G \to H$ shall be a smooth group homomorphism between Lie groups. A **Lie algebra homomorphism** $A : \mathfrak{g} \to \mathfrak{h}$ is a vector space homomorphism with $A[X,Y] = [AX, AY]$ for all $X,Y \in \mathfrak{g}$.*

Lemma 1.6.12

1) If $f : G \to H$ is a Lie group homomorphism, then $T_e f : \mathfrak{g} \subset \mathfrak{h}$ is a Lie algebra homomorphism.

2) For every $G \subset H$, $\mathfrak{g} \subset \mathfrak{h}$ is a Lie subalgebra.

Proof. 1) Given left-invariant $\widetilde{X}, \widetilde{Y}$ on G, let $\widehat{X}, \widehat{Y}$ be left-invariant on H corresponding to $T_{e_G} f(\widetilde{X}), T_{e_G} f(\widetilde{Y}) \in T_{e_H} H$. For any $g \in G$ one gets $f \circ L_g = L_{f(g)} \circ f$ on G and therefore

$$T_g f(\widetilde{X}) = T_g f(T_{e_G} L_g(\widetilde{X}_{|e_G})) = T_{e_H} L_{f(g)}(T_{e_G} f(\widetilde{X})) = \widehat{X}_{|f(g)}.$$

By Lemma 1.4.7 we obtain $T_g f([\widetilde{X}, \widetilde{Y}]_{\mathfrak{g}}) = [\widehat{X}, \widehat{Y}]_{\mathfrak{h}|f(g)}$, in particular at $g = e_G$.

2) By part (1) applied to $\iota : G \hookrightarrow H$, $T_{e_G}\iota$ identifies the Lie bracket on $\mathfrak{g}$ with that on $\mathfrak{h}$. □

Conversely, subgroups corresponding to Lie subalgebras are constructed in Theorem 7.4.1.

Example Lie subgroups $H \subset \mathrm{GL}(n,\mathbb{R})$:

i) $T_{\mathrm{id}_{\mathbb{R}^n}} \mathrm{SL}(n,\mathbb{R}) = \mathfrak{sl}_n$: Setting $f : \mathbb{R}^{n\times n} \to \mathbb{R}, A \mapsto \det A$, its derivative $f'_A(X) = \det A \cdot \mathrm{Tr}\, A^{-1}X$ is surjective on $A \in \mathrm{SL}(n,\mathbb{R})$ and $T_{\mathrm{id}_{\mathbb{R}^n}} \mathrm{SL}(n,\mathbb{R}) = \ker f'_{\mathrm{id}} = \{X \in \mathbb{R}^{n\times n} \mid \mathrm{Tr}\, X = 0\}$.

ii) $\mathfrak{so}(n) = \mathfrak{o}(n)$: For $f : \mathbb{R}^{n\times n} \to \mathbb{R}^{n\times n}_{\mathrm{symm}}, A \mapsto AA^t$ the derivative $f'_A(X) = AX^t + XA^t$ is surjective on $A \in \mathrm{SO}(n)$ (to see this, set $X = Y(A^{-1})^t$ with Y symmetrical). Hence $T_{\mathrm{id}_{\mathbb{R}^n}} \mathrm{SO}(n) = \ker f'_{\mathrm{id}_{\mathbb{R}^n}} = \{X \in \mathbb{R}^{n\times n} \mid X^t = -X\}$, the skew-symmetric matrices, and so $\dim \mathrm{SO}(n) = \frac{n(n-1)}{2}$.

iii) In the same way $\mathfrak{u}(n) = \{X \in \mathbb{C}^{n\times n} \mid X^t = -\overline{X}\}$, $\dim \mathrm{U}(n) = n^2$ follows, and $\mathfrak{su}(n) = \{X \in \mathbb{C}^{n\times n} \mid X^t = -\overline{X}, \mathrm{Tr}\, X = 0\}$. One finds $\dim \mathrm{SU}(n) = n^2 - 1$ since $\mathrm{Tr}\, X \in i\mathbb{R}$ for $X^t = -\overline{X}$.

Theorem 1.6.13 *Let G be a Lie group and let $X \in \Gamma(G, TG)$ be a left-invariant vector field. Then the flow of X is defined on $\mathbb{R} \times G$ and it satisfies $\Phi_t^X(g) = g \cdot \Phi_t^X(e_G)$, i.e.*

Proof. Assume that $\Phi_t^X(e_G)$ is well-defined for $|t| < \varepsilon$. Because $L_{g*}X = X$, by Corollary 1.5.5 $L_g \circ \Phi_t^X = \Phi_t^X \circ L_g$, in particular $g \cdot \Phi_t^X(e_G) = \Phi_t^X(g)$. Hence for all $g \in G$, $\Phi_t^X(g)$ is defined on $|t| < \varepsilon$, and by the same construction as in Theorem 1.5.3, it is also defined on all $\mathbb{R} \times G$.   □

Correspondingly, the flow of right-invariant vector fields is given by left multiplication. Since left and right multiplication commute due to the associative law, according to Proposition 1.5.8, right- and left-invariant vector fields commute as well.

Definition 1.6.14 *The **exponential map** is the map*

$$\exp_G : \mathfrak{g} \to G$$
$$X \mapsto \Phi_1^{\widetilde{X}}(e_G)$$

with the left-invariant vector field $\widetilde{X}$ associated to $X \in T_{e_G}G$.

Example i) For $G = (\mathbb{R}^n, +)$ one finds $\Phi_t^X(p) = p + t \cdot X$, $\exp_{\mathbb{R}^n} X = X$.

ii) For $G = \mathrm{GL}(n)$, by Exercise 1.5.11 one gets $\exp_G X = e^X := \sum_j \frac{X^j}{j!}$. In particular, for $n = 1$, $\exp_{(\mathbb{R}^\times, \cdot)} : T_1\mathbb{R}^\times \cong \mathbb{R} \to \mathbb{R}^\times, r \mapsto \sum \frac{r^j}{j!}$.

In later chapters $\exp_G X$ will be written as e^X as long as there is no risk of confusion.

Proposition 1.6.15 *Consider a manifold M, $X \in \Gamma(M, TM)$ and a point $p \in M$. Then $\Phi_{st}^X(p) = \Phi_s^{tX}(p)$ for sufficiently small s and t. In particular, for $M = G$ and X left-invariant, $\Phi_t^X(g) = g \cdot \exp(tX)$.*

Proof. $\Phi_0^{tX} = \mathrm{id}_M = \Phi_{0 \cdot t}^X$, $\Phi_{s_1 t}^X \circ \Phi_{s_2 t}^X = \Phi_{(s_1 + s_2)t}^X$ and $\frac{\partial}{\partial s}\big|_{s=0} \Phi_s^{tX} = tX = \frac{\partial}{\partial s}\big|_{s=0} \Phi_{ts}^X$.   □

Theorem 1.6.16 *If G is a Lie group, then the tangent of the exponential map at $0 \in \mathfrak{g}$ is given by $T_0 \exp = \mathrm{id}_{\mathfrak{g}}$. In particular, there is a neighborhood $V \subset \mathfrak{g}$ of 0 on which $\exp$ is a diffeomorphism and there $\exp^{-1}$ is a canonical chart around e_G.*

Proof. Using Proposition 1.6.15 one obtains

$$(T_0 \exp)(X) = \frac{\partial}{\partial t}\Big|_{t=0} \exp(tX) = \frac{\partial}{\partial t}\Big|_{t=0} \Phi_1^{t\widetilde{X}}(e) \overset{1.6.15}{=} \frac{\partial}{\partial t}\Big|_{t=0} \Phi_t^{\widetilde{X}}(e_G) = X. \quad □$$

Theorem 1.6.17 *Suppose that $f : G \to H$ is a Lie group homomorphism. Then for any $X \in \mathfrak{g}$,*

$$f(\exp_G X) = \exp_H(T_{e_G} f(X)).$$

For instance, for $\det : \mathrm{GL}(n, \mathbb{R}) \to (\mathbb{R}^\times, \cdot)$ it follows that $\det e^A = e^{\mathrm{Tr}\,A}$.

Proof. With the vector field $\widetilde{Y}_{|h} := (L_h)_* T_{e_G} f(X)$ on H and the left-invariant extension $\widetilde{X}$ of X, we get $f(\Phi_t^{\widetilde{X}}(e)) = \Phi_t^{\widetilde{Y}}(e_H)$, since $f(\Phi_0^{\widetilde{X}}(e_G)) = f(e_G) = e_H$ and

$$\frac{\partial}{\partial t}_{|t=t_0} f(\Phi_t^{\widetilde{X}}(e_G)) = \frac{\partial}{\partial t}_{|t=0} f(\Phi_t^{\widetilde{X}}(\Phi_{t_0}^{\widetilde{X}}(e_G))) = \frac{\partial}{\partial t}_{|t=0} f(\Phi_{t_0}^{\widetilde{X}}(e_G) \cdot \Phi_t^{\widetilde{X}}(e_G))$$

$$= \frac{\partial}{\partial t}_{|t=0} \left[f(\Phi_{t_0}^{\widetilde{X}}(e_G)) \cdot f(\Phi_t^{\widetilde{X}}(e_G)) \right]$$

$$= (L_{f(\Phi_{t_0}^{\widetilde{X}}(e_G))})_* T_e f(X) = \widetilde{Y}_{|f(\Phi_{t_0}^{\widetilde{X}}(e_G))}.$$

In particular, we find

$$f(\exp_G X) = f(\Phi_1^{\widetilde{X}}(e_G)) = \Phi_1^{\widetilde{Y}}(e_H) = \exp_H(T_{e_G} f(X)). \qquad \square$$

Corollary 1.6.18 *Given any Lie subgroup* $\iota : H \hookrightarrow G$, $X \in \mathfrak{h}$, *one obtains* $\exp_H(X) = \iota(\exp_H(X)) \overset{1.6.17}{=} \exp_G(T_{e_H}\iota(X)) = \exp_G X$.

Example Therefore, the calculation for $\mathrm{GL}(n, \mathbb{R})$ implies for all matrix Lie groups, say $G = \mathrm{O}(n)$, that $\exp_G X = \sum_{n=0}^\infty \frac{X^n}{n!}$.

Theorem 1.6.19 *If G is a connected Lie group, then any Lie group homomorphism* $f : G \to H$ *is uniquely determined by* $T_{e_G} f : \mathfrak{g} \to \mathfrak{h}$.

Proof. First, we show that any neighborhood $U \subset G$ of e_G generates the group G. For this, set recursively

$$A_0 := U, \quad A_{k+1} := \bigcup_{u \text{ or } u^{-1} \in U} L_u A_k \quad \text{and} \quad A := \bigcup_{k \in \mathbb{N}} A_k.$$

Thus A is a subgroup of G and it is open since $L_u^{-1}, L_{u^{-1}}^{-1}$ are continuous. So A is a Lie subgroup and closed by Lemma 1.6.3, hence $A = G$.

On U sufficiently small, f is given by $f(\exp X) = \exp T_e f(X)$ and consequently on all of G its values are determined by this. $\qquad \square$

Remark In fact, one can even show that every element of G can be written as a product of two elements of $\exp \mathfrak{g}$ ([Wü]).

Theorem 7.4.2 examines when there is a Lie group homomorphism associated to a given Lie algebra homomorphism. A **representation** of a Lie group G is a group homomorphism $\rho : G \to \mathrm{GL}(V)$ for a vector space V. For example, for every subgroup $H \subset \mathrm{GL}(n)$ the embedding into $\mathrm{GL}(n)$ provides the **standard representation**. Differentiating a representation induces a representation of the associated Lie algebras which uniquely determines ρ for connected G by Theorem 1.6.19. An **irreducible representation** is one which is not decomposable into nontrivial summands.

Example 1.6.20 The vector space V^q of homogeneous polynomials of degree $q \in \mathbb{N}_0$ in two variables s, t equipped with the action

$$SL(2) \times V^q \to V^q$$
$$\left(\begin{pmatrix} a & b \\ c & d \end{pmatrix}, P(s,t) \right) \mapsto P(as + ct, bs + dt),$$

i.e. $(A \cdot P)(s,t) := P\big((s,t) \cdot A\big)$, is an $SL(2)$ representation.

Definition 1.6.21 *The **adjoint representation** $\mathrm{Ad} : G \to \mathrm{Aut}(\mathfrak{g})$ of G is the derivative of the conjugation by $g \in G$,*

$$C_g : G \to G, \qquad \mathrm{Ad}_g := T_{e_G} C_g : \mathfrak{g} \to \mathfrak{g}$$
$$h \mapsto ghg^{-1} \qquad\qquad X \mapsto (R_{g^{-1}} \circ L_g)_* X.$$

Lemma 1.6.22 *The adjoint representation satisfies*

1) Ad is a G-representation, i.e. $\mathrm{Ad}_g \circ \mathrm{Ad}_h = \mathrm{Ad}_{gh}$. In particular, $\mathrm{Ad}_g^{-1} = \mathrm{Ad}_{g^{-1}}$.
2) Ad_g is a Lie algebra automorphism of $\mathfrak{g}$, i.e.

$$\mathrm{Ad}_g[X,Y] = [\mathrm{Ad}_g X, \mathrm{Ad}_g Y].$$

3) $\mathrm{Ad}_g(X) = \frac{\partial}{\partial t}_{|t=0} g \cdot \exp(tX) \cdot g^{-1}$.

Proof. 1) Differentiate $C_g \circ C_h = C_{gh}$.
2) holds, since C_g is a diffeomorphism and $[\cdot, \cdot]$ commutes with direct images.
3) $t \mapsto \exp tX$ is a path with $\frac{\partial}{\partial t}_{|t=0} \exp tX = X$. $\square$

Theorem 1.6.23 *For $X, Y \in \mathfrak{g}$ one gets*

$$[X,Y] = \frac{\partial}{\partial s}_{|s=0} \mathrm{Ad}_{\exp sX} Y \overset{1.6.22(3)}{=} \frac{\partial^2}{\partial s \partial t}_{|{s=0 \atop t=0}} \exp(sX) \cdot \exp(tY) \cdot \exp(-sX),$$

i.e. $T_{e_G}(\mathrm{Ad}.(Y))(X) = [X,Y]$.

Proof. According to Theorem 1.5.6, with $\widetilde{X}, \widetilde{Y}$ left-invariant associated to X, Y, one finds

$$
\begin{aligned}
[X,Y] &= \left(\frac{\partial}{\partial s}_{|s=0} \Phi_s^{-\widetilde{X}} {}_* \widetilde{Y} \right)_{|e_G} = \frac{\partial}{\partial s}_{|s=0} \left[T\Phi_s^{-\widetilde{X}} (\widetilde{Y}_{|\Phi_s^{\widetilde{X}}(e_G)}) \right] \\
&= \frac{\partial^2}{\partial s \partial t}_{|{s=0 \atop t=0}} (\Phi_s^{-\widetilde{X}} \circ \Phi_t^{\widetilde{Y}} \circ \Phi_s^{\widetilde{X}})(e_G) \\
&\overset{1.6.13}{=} \frac{\partial^2}{\partial s \partial t}_{|{s=0 \atop t=0}} R_{\exp(sX) \cdot \exp(tY) \cdot \exp(-sX)}(e) \\
&= \frac{\partial^2}{\partial s \partial t}_{|{s=0 \atop t=0}} \exp(sX) \cdot \exp(tY) \cdot \exp(-sX).
\end{aligned}
$$

$\square$

Example Consider $GL(n, \mathbb{R})$, $X, Y \in \mathbb{R}^{n \times n}$, then

$$[X, Y] = \frac{\partial}{\partial s}_{|s=0} e^{sX} \cdot Y \cdot e^{-sX} = X \cdot Y - Y \cdot X,$$

as has already been shown in Example 1.6.10(iii).

Differentiating the formula in Lemma 1.6.22 implies two variants of the Jacobi identity:

Corollary 1.6.24 *Let* $\mathrm{ad} := T_{e_G} \mathrm{Ad} : \mathfrak{g} \to \mathrm{End}(\mathfrak{g})$, $X \mapsto (Y \mapsto [X, Y])$ *be the adjoint (Lie algebra) representation. Then the following holds:*

1) ad *is a **Lie algebra representation**, i.e.* $\mathrm{ad}_{[X,Y]} = [\mathrm{ad}_X, \mathrm{ad}_Y]$.
2) Given $Z \in \mathfrak{g}$, ad_Z *is a derivation on* $\mathfrak{g}$, *i.e.*

$$\mathrm{ad}_Z[X, Y] = [\mathrm{ad}_Z X, Y] + [X, \mathrm{ad}_Z Y].$$

Applying Theorem 1.6.17, Theorem 1.6.23 implies:

Corollary 1.6.25 *For any* $X \in \mathfrak{g}$ *one obtains* $\mathrm{Ad}_{\exp X} = \exp(\mathrm{ad}_X)$.

Exercises

Exercise 1.6.26 Show that $G := \mathbb{R}^{\times} \times \mathbb{R}$ with the binary operation

$$(a, b) \cdot (a', b') := (aa', b + ab')$$

is a Lie group. Prove that G as a group is isomorphic to the group of affine transformations of the real line (i.e., of transformations of the form $x \mapsto ax + b$). Determine the left-invariant vector fields and thereby the Lie bracket (induced via ρ) on $T_{e_G} G = \mathbb{R}^2$.

Exercise 1.6.27 Set

$$V := \left\{ \begin{pmatrix} A & B \\ -\bar{B} & \bar{A} \end{pmatrix} \in \mathbb{C}^{2n \times 2n} \middle| A, B \in \mathbb{C}^{n \times n} \right\}.$$

The **symplectic group** $Sp(n)$ is defined as $Sp(n) := V \cap U(2n)$.

1) Prove that V is a subalgebra of $\mathbb{C}^{2n \times 2n}$ and that $V \to V$, $C \mapsto \bar{C}^t$ is well-defined.
2) Show that $Sp(n)$ is a Lie group.
3) Determine $T_{\mathrm{id}_{\mathbb{C}^{2n}}} Sp(n) \subset T_{\mathrm{id}_{\mathbb{C}^{2n}}} GL_{2n}(\mathbb{C}) = \mathbb{C}^{2n \times 2n}$ as a subset. How large is $\dim Sp(n)$?

Remark: $Sp(n)$ *can also be thought of as the (left-)$\mathbb{H}$-linear isometry group of* $\mathbb{H}^n$. *One can show that all compact Lie groups (up to covering) are products of* $SO(n)$, $SU(n)$, $Sp(n)$ *and 5 sporadic groups.*

Exercise 1.6.28 The **Heisenberg group** $H \subset \mathrm{GL}(3, \mathbb{R})$ consists of the upper tri-angular matrices with ones on the diagonal. Compute the Lie algebra $\mathfrak{h}$. Show that $\mathfrak{u} := [\mathfrak{h}, \mathfrak{h}] = \{[a, b] \mid a, b \in \mathfrak{h}\}$ is a real Lie subalgebra of $\mathfrak{h}$ and that

$$[[\mathfrak{h}, \mathfrak{h}], \mathfrak{h}] = 0.$$

Compute $\exp X$ for all $X \in \mathfrak{h}$ and conclude that $\exp$ is a diffeomorphism.

Exercise* 1.6.29 To a Lie group G associate the vector space $\mathfrak{g}^R$ of right-invariant vector fields. For $X \in \mathfrak{g}$, let $X^R \in \mathfrak{g}^R$ be the right-invariant vector field with $X_{|e_G} = X^R_{|e_G}$.

1) For $X, Y \in \mathfrak{g}$, show the equation

$$[X^R, Y^R] = -[X, Y]^R.$$

2) Prove that $\mathfrak{g}$, $\mathfrak{g}^R$ are isomorphic with respect to the Lie algebra structures induced by $\Gamma(G, TG)$.

The Lie algebra $\mathfrak{g}^R$ corresponds to the group structure $G \times G \to G$, $(g, h) \mapsto h \cdot g$, which is canonically isomorphic to the original one.

Exercise* 1.6.30 Consider the exponential map $\exp$ to $\mathrm{SL}(2)$.

1) Explicitly compute $\exp$ at a matrix $X = \begin{pmatrix} a & b \\ c & -a \end{pmatrix} \in \mathfrak{sl}(2)$ as a closed formula in a, b, c. Hint: What is X^2?
2) Show that $\exp$ is not surjective. Hint: What values does $\mathrm{Tr}\, \exp X$ take?
3) Show that $\exp$ is not injective.

Exercise 1.6.31 Show that the direct image under right multiplication $(R_g)_*$ for any $g \in G$ maps the set of left-invariant vector fields to itself and that it corresponds there to the action of $\mathrm{Ad}_{g^{-1}}$ on $\mathfrak{g}$. More precisely show that

$$\rho(\mathrm{Ad}_{g^{-1}} X) = (R_g)_* \rho(X).$$

Exercise* 1.6.32 (Representation theory of $\mathrm{SL}(2)$ **and** $\mathfrak{sl}(2)$**)** Let V be a finite-dimensional irreducible complex representation of $\mathrm{SL}(2)$, hence by differentiating V is also a representation of $\mathfrak{sl}(2)$. The elements

$$H := \begin{pmatrix} 1 & 0 \\ 0 & -1 \end{pmatrix}, \quad X := \begin{pmatrix} 0 & 1 \\ 0 & 0 \end{pmatrix}, \quad Y := \begin{pmatrix} 0 & 0 \\ 1 & 0 \end{pmatrix}$$

form a basis of $\mathfrak{sl}(2)$. Show that

1) $[H, X] = 2X$, $[H, Y] = -2Y$, $[X, Y] = H$.
2) For $\lambda \in \mathbb{C}$ set $V_\lambda := \{v \in V \mid Hv = \lambda v\} \subset V$. Then $X \cdot V_\lambda \subset V_{\lambda+2}$ and $Y \cdot V_\lambda \subset V_{\lambda-2}$.
3) Let $q \in \mathbb{C}$ be an eigenvalue of H with maximal real part, choose $v \in V_q \setminus \{0\}$ and let W be the subspace of V spanned by $\{Y^k v \mid k \in \mathbb{N}_0\}$. Then $X \cdot Y^k v = k(q - k + 1)Y^{k-1}v$, W is closed under the action of Y, H, X and $W = V$.

4) Using the formula for the operation of X from (3), deduce that $q \in \mathbb{N}_0$ and that q uniquely determines the representation of $\mathfrak{sl}(2)$ up to isomorphism.
5) Show that $q \in \mathbb{N}_0$ uniquely determines the representation of $\mathrm{SL}(2)$ up to isomorphism.
6) Example 1.6.20 yields for every $q \in \mathbb{N}_0$ a representation with largest H-eigenvalue q. By (5), these are exactly the finite-dimensional irreducible representations of $\mathrm{SL}(2)$ up to isomorphism.

Exercise 1.6.33 Consider a connected Lie group G, its center

$$Z(G) := \{g \in G \mid \forall h \in G : gh = hg\}$$

and $\mathfrak{z} := \ker \mathrm{ad}$.

1) Show that $Z(G) = \ker \mathrm{Ad}$.
2) Prove that the exponential map maps $\mathfrak{z}$ into the connected component Z_0 of $Z(G)$ at the neutral element.
3) Show that there is a neighborhood U of e such that $Z_0 \cap U = \exp \mathfrak{z} \cap U$.
4) Using Definition 1.1.1, deduce that $Z \subset G$ is a Lie subgroup with Lie algebra $\mathfrak{z}$.
5) Prove that $\exp_{|\mathfrak{z}} : \mathfrak{z} \to Z_0$ is a surjective group homomorphism.

Show that $Z(G) = \ker \mathrm{Ad}$ and prove that the exponential map maps $\mathfrak{z}$ onto $Z \cap \exp_G(\mathfrak{g})$. Conclude that $Z \cap \exp_G$ is a Lie subgroup with Lie algebra $\mathfrak{z}$.

Exercise 1.6.34 Let G be connected and let $\varphi : G \to \widetilde{G}$ be a Lie group homomorphism with discrete kernel Γ. Show that for any $a \in G, b \in \Gamma$ one gets $aba^{-1} = b$. Deduce that Γ lies in the center of G. In particular, it is abelian.

Exercise 1.6.35 A vector subspace $\mathfrak{h} \subset \mathfrak{g}$ of a Lie algebra $\mathfrak{g}$ is an **ideal** in $\mathfrak{g}$ if $[\mathfrak{h}, \mathfrak{g}] \subset \mathfrak{h}$. Show that $\mathfrak{sl}_n$ is an ideal in $\mathfrak{gl}_n$.

Exercise 1.6.36 Let $H \subset G$ be both a submanifold and a normal divisor. Show that $\mathfrak{h} \subset \mathfrak{g}$ is an ideal.

Chapter 2
Vector Bundles and Tensors

This chapter, like the previous one, belongs to the field of differential topology and not yet to Riemannian geometry. The tangent bundle will be generalized to arbitrary bundles of vector spaces, which very quickly become necessary for further constructions such as higher derivatives. In addition, some objects from linear algebra are provided: The algebra of tensor products of vectors and the finite-dimensional exterior algebra over a finite-dimensional vector space. The importance of these objects for differential geometry already becomes somewhat apparent in these sections by the fact that they allow the definition of further differential operators. In the penultimate section, the exterior algebra provides a topological tool for distinguishing manifolds, the de Rham cohomology. The exterior algebra generalizes the determinant. In the last section, the exterior algebra is used to define an integral on manifolds analogous to the notion of integration on an $\mathbb{R}^n$.

2.1 Vector Bundles

fiber bundle	vector (space) bundle
typical fiber	rank
local trivialization	zero section
base space	line bundle
total space	pullback of a vector bundle
trivial fiber bundle	pullback
fiber over x	pullback of a section
transition function	vector bundle homomorphism
Ehresmann's Fibration Theorem	direct sum of vector bundles
covering	normal bundle
section	tautological line bundle

One quickly encounters the need to generalize the construction of the tangent bundle. For example, given a function $f : M \to \mathbb{R}$ one can think of the derivative

K. Köhler, *Differential Geometry and Homogeneous Spaces*, Universitext,
https://doi.org/10.1007/978-3-662-69721-4_2

$T_p f : T_p M \to T_{f(p)} \mathbb{R} \overset{\cong}{\to} \mathbb{R}$ as a 1-form in the dual space $(T_p M)^*$; and then one would like to regard Tf as a 1-form at each point of M.

Definition 2.1.1 *Let M, B, Z be C^∞ manifolds. Let $\pi : M \to B$ be a C^∞ map, let $(U_j)_{j \in J}$ be an open cover of B, and let*

$$h_j : \pi^{-1}(U_j) \to U_j \times Z$$

be diffeomorphisms such that $\pi_{|\pi^{-1}(U_j)} = (\text{projection onto the 1st factor}) \circ h_j$., i.e. the diagram

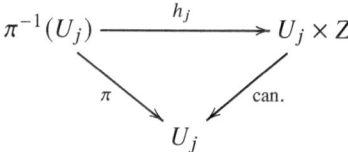

*commutes. Then π together with $(h_j)_{j \in J}$ is called a **fiber bundle** with **typical fiber** Z. The h_j are called **local trivializations**. B is called the **base (space)**, M the **total space** of the bundle.*

As in the definition of C^∞ structures on manifolds, fiber bundles with compatible local trivializations are considered as equivalent.

Example 2.1.2 i) By Lemma 1.3.4, the tangent bundle $TB \to B$ is a fiber bundle with local trivializations

$$h_\varphi : \quad TU \quad \to U \times \mathbb{R}^n$$
$$(x, [(\varphi, u)]) \longmapsto (x, u)$$

for each chart $\varphi : U \to V$.

ii) The **trivial bundle** is $M := B \times Z \overset{\text{proj}_1}{\to} B$.

iii) Setting $K := \mathbb{R}, \mathbb{C}$ or $\mathbb{H}$, the space $K^{n+1} \setminus \{0\}$ is a fiber bundle over $\mathbb{P}^n K$ with fiber $K \setminus \{0\}$ via

$$\pi : K^{n+1} \setminus \{0\} \to \mathbb{P}^n K$$
$$(x_0, \ldots, x_n) \mapsto (x_0 : \cdots : x_n).$$

Remark 2.1.3 i) As a composition of the submersion $U_j \times Z \to U_j$ and the diffeomorphism h_j, the map π is a submersion. It is surjective, since $(U_j)_{j \in J}$ is a cover.

ii) By Lemma 1.1.3(2), the fibers $\pi^{-1}(\{x\})$ are submanifolds. For $x \in U_j$, h_j induces a diffeomorphism $\pi^{-1}(\{x\}) \to \{x\} \times Z$. Therefore $M_x := \pi^{-1}(\{x\})$ is called a **fiber over** x (Fig. 2.1).

iii) For $j, k \in J$, $h_j \circ h_k^{-1} : (U_j \cap U_k) \times Z \to (U_j \cap U_k) \times Z$ is of the form $(\pi_{U_j \cap U_k}, g_{jk})$, where $\pi_{U_j \cap U_k}$ is the projection on the first factor and for all $x \in U_j \cap U_k$ the **transition function** $g_{jk|x}$ is a diffeomorphism of Z.

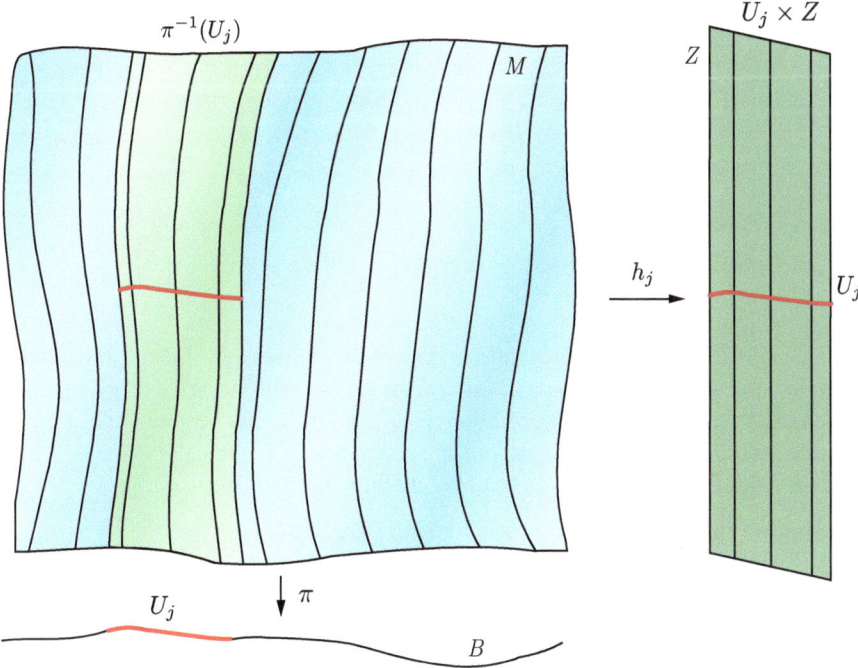

$\pi^{-1}(U_j)$

M

$U_j \times Z$

Z

h_j

U_j

U_j

π

B

Fig. 2.1 Fiber bundle.

iv) Ehresmann's Fibration Theorem [Du, ch. 9.5] states that every proper submersion is a fiber bundle (cf. also Hermann's Fibration Theorem 6.3.10).

Example 2.1.4 For the tangent bundle, $h_\psi \circ h_\varphi^{-1}(x, u) = (x, (\psi \circ \varphi^{-1})'_{|x} u)$ for any two charts ψ, φ and so $g_{\psi\varphi}(x) = (\psi \circ \varphi^{-1})'_{|x}$.

Definition 2.1.5 A *#Z-fold covering* is a fiber bundle with discrete fiber Z.

For example, $S^n \to \mathbb{P}^n\mathbb{R}$ is a twofold covering.

Definition 2.1.6 A *section* of a fiber bundle $\pi : M \to B$ is a C^∞ map $s : B \to M$ with $\pi \circ s = \mathrm{id}_B$. The set of all sections is denoted by $\Gamma^\infty(B, M) = \Gamma(B, M)$.

In particular, a section embeds B into M.

Definition 2.1.7 A fiber bundle E together with a choice of local trivializations is called a *K-vector (space) bundle* of $r \in \mathbb{N}_0$ over $K = \mathbb{R}$ or $\mathbb{C}$ if

1) Z is an r-dimensional K-vector space and
2) the transition functions $g_{jk|x} : Z \to Z$ are K-linear.

Again, vector bundles with compatible local trivializations are considered as equivalent. Then each fiber $E_x := \pi^{-1}(\{x\})$ is a K-vector space with vector space operations given by

$$\lambda \cdot h_j^{-1}((p, v)) + \mu \cdot h_j^{-1}((p, w)) = h_j^{-1}((p, \lambda \cdot v + \mu \cdot w))$$

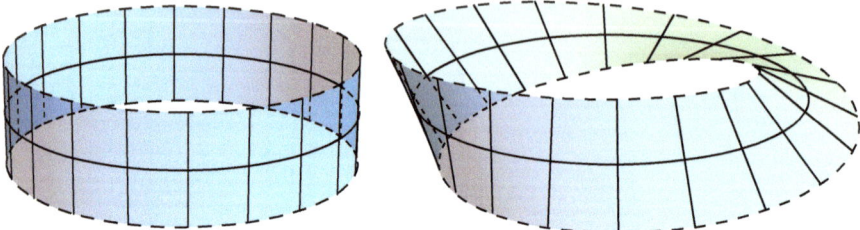

Fig. 2.2 Cylinder and Möbius strip as $\mathbb{R}$-line bundles over S^1.

for $v, w \in V, \lambda, \mu \in K$. These are independent of the choice of h_j. Indeed, for another trivialization h_k, $h_j^{-1}((p, v)) = h_k^{-1}((p, g_{kj}(v)))$, and the linearity of g_{kj} implies the equality of the vector space operations for h_j and h_k. Every vector bundle $E \to B$ has the **zero section** $s \equiv 0$ as a canonical section which canonically embeds B into E. A K-vector bundle of rank 1 is called a **K-line bundle**.

Example 1) The tangent bundle according to Example 2.1.4.
   2) Cylinder and Möbius strip $\to S^1$ form $\mathbb{R}$-line bundles (Fig. 2.2).

The following formal and comparatively simple construction is one of the most remarkable tools in the treatment of vector bundles.

Lemma and Definition 2.1.8 *Let $f : M \to N$ be a C^∞ map and let $\pi : E \to N$ be a fiber bundle. Then the **pullback** of E to M is the fiber bundle*

$$f^*E := \{(p, v) \in M \times E \mid \underbrace{\pi(v) = f(p)}_{v \in E_{f(p)}}\} \overset{\mathrm{proj}_1}{\to} M$$

equipped with the subspace topology induced by $M \times E$. If E is a vector bundle, then f^*E is likewise a vector bundle.

Remark Thus, the space $E_{f(p)}$ gets attached at the point $p \in M$. On the other hand, a reverse construction of a fiber bundle over N from one on M does not work this way; simply because the images $f(U_j)$ of open sets are not open in general.

Proof. If E has local trivializations $h_j : \pi^{-1}(U_j) \to U_j \times Z$, then f^*E has local trivializations

$$f^*h_j : (f^*E)_{|f^{-1}(U_j)} = \{(p, v) \mid p \in f^{-1}(U_j), v \in E_{f(p)}\} \overset{(\mathrm{proj}_1, \mathrm{proj}_2 \circ h_j)}{\to} f^{-1}(U_j) \times Z.$$

For these one finds

$$(f^*h_j) \circ (f^*h_k)^{-1}(p, v) = (f^*h_j)\big(p, h_k^{-1}(f(p), v)\big)$$
$$= \big(p, (h_j \circ h_k^{-1})_{|f(p)}(v)\big).$$

Hence the transition functions are $g_{jk|p}^{f^*E} = g_{jk|f(p)}$. This implies the second part.   □

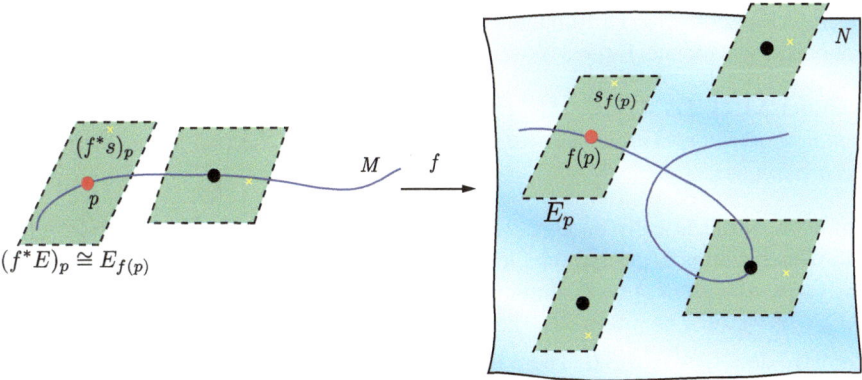

Fig. 2.3 Pullback of vector bundles and of sections.

This pullback induces a **pullback of sections** $f^* : \Gamma(N, E) \to \Gamma(M, f^*E)$, $s \mapsto s \circ f$ or $(f^*s)_p := s_{f(p)}$ (Fig. 2.3). In particular, for a basis $(s_\ell)_{\ell=1}^r$ of $E_{|U_j}$ a local basis of f^*E is given by $(f^*s_\ell)_{\ell=1}^r$. Therefore locally any section can be written as $\sum_{\ell=1}^r g_\ell \cdot f^*s_\ell$ with $g_\ell \in C^\infty(f^{-1}(U_j), \mathbb{R})$.

Definition 2.1.9 *Given vector bundles* $\pi : E \to M, \widetilde{\pi} : F \to M$, *a map* $f : E \to F$ *is called a **vector bundle homomorphism** if*

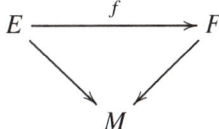

commutes and for all $p \in M$ *the map* $f_p : E_p \to F_p$ *is linear.*

Definition 2.1.10 *Let* $\pi : E \to M, \widetilde{\pi} : F \to M$ *be vector bundles with local trivializations* $h_j : \pi^{-1}(U_j) \to U_j \times V, \widetilde{h}_j : \widetilde{\pi}^{-1}(U_j) \to U_j \times W$ *(without restriction, the covers for* E, F *can be chosen to be the same by otherwise forming all intersections* $U_j \cap \widetilde{U}_k$*). Then the **direct sum** $E \oplus F \to M$ is the vector bundle* $\{(p, v, w) \mid v \in E_p, w \in F_p, p \in M\} \overset{\text{proj}_1}{\to} M$ *with local trivialization*

$$\widehat{h}_j : \{(p, v, w) \mid p \in U_j, v \in \pi^{-1}(U_j), w \in \widetilde{\pi}^{-1}(U_j)\} \to U_j \times (V \oplus W).$$

Analogously one defines the vector bundles $E^*, \mathrm{Hom}(E, F)$ and E/F for a subbundle $F \subset E$. That is, one sets $E^* := \{(p, \alpha) \mid \alpha \in E_p^*, p \in M\}$ with local trivializations given by

$$h_j^* : \{(p, \alpha) \in \mathrm{Hom}(E_p, \mathbb{R}) \mid p \in U_j\} \to U_j \times V^*$$
$$(p, \alpha) \mapsto (p, \alpha \circ h_j^{-1}(p, \cdot))$$

for $h_j : E_{|U_j} \to U_j \times V$. To bundle homomorphisms $f : E \to E', g : F \to F'$ we canonically associate $f \oplus g : E \oplus E' \to F \oplus F'$, $f \otimes g$ etc.

Lemma 2.1.11 *Let $E, F \to M$ be vector bundles and let $f : E \to F$ be a vector bundle homomorphism with constant (fiberwise) rank. Then im f and ker f form vector bundles.*

Proof. 1) im f: Locally on $U \subset M$ identify the vector bundles E, F with $\mathbb{R}^m$, $\mathbb{R}^k$ via trivializations h', h and thus identify $f_{|x}$ with $A_x \in \mathbb{R}^{k \times m}$ to $x \in U$. Denote the constant rank of A by ℓ. Without loss of generality at $x_0 \in U$ let the submatrix $(A_{x_0, rs})_{r,s=1}^{\ell}$ of A be invertible. Then there is a neighborhood $V \subset U$ of x_0 on which we have $\det(A_{rs})_{r,s=1}^{\ell} \neq 0$. There im $f_{|V} \to V \times \mathbb{R}^{\ell} \times \{0_{\mathbb{R}^{m-\ell}}\}$, $h^{-1}((x, A_x v)) \mapsto (x, v)$ is a trivialization of im f.

2) ker f: As the vanishing space of the image of $f^t : F^* \to E^*$, ker f is also a vector bundle.   □

Exercises

Exercise 2.1.12 Let $M^n \subset \mathbb{R}^{n+k}$ be an n-dimensional submanifold. Locally there exists an $f : U \to \mathbb{R}^k$ such that $f^{-1}(0) = M \cap U$. Thereby we identify $T_p M$ with $\ker T_p f \subset T_p \mathbb{R}^{n+k} \cong \mathbb{R}^{n+k}$. With respect to the canonical scalar product on $\mathbb{R}^{n+k}$, set

$$N_p := \{v \in T_p \mathbb{R}^{n+k} \mid v \perp T_p M\} \qquad (p \in M).$$

Show that the **normal bundle**

$$N := \{(p, N_p) \mid p \in M\} \to M, (p, N_p) \mapsto p$$

is a vector bundle. *Hint: Embed N as a submanifold.*

Exercise 2.1.13 Consider $M = S^n \subset \mathbb{R}^{n+1}$ and the normal bundle N from Exercise 2.1.12.

1) Prove that
$$N = \{(p, v) \in \mathbb{R}^{n+1} \times \mathbb{R}^{n+1} \mid v \in \mathbb{R} \cdot p, p \in S^n\}$$

and that this vector bundle is isomorphic to the trivial $\mathbb{R}$-line bundle O.
2) Show that $TS^n \oplus O$ is isomorphic to the trivial $\mathbb{R}^{n+1}$-bundle $O^{\oplus(n+1)}$.

Exercise 2.1.14 Let E, F be vector bundles over M with transition functions g_{jk}, h_{lm}. Describe the transition functions of E^*, $E \oplus F$ and $\mathrm{Hom}(E, F)$.

Exercise* 2.1.15 Set $K = \mathbb{R}$ or $\mathbb{C}$ and let L be a K-line bundle. Show that $\mathrm{Hom}_K(L, L)$ is canonically isomorphic to the trivial line bundle.

Exercise 2.1.16 Let $K = \mathbb{R}, \mathbb{C}$ or $\mathbb{H}$. Let L be the quotient of $K^{n+1} \setminus \{0\} \times K$ by the relation $(\mathbf{x}, \lambda) \sim (\mu\mathbf{x}, \lambda\mu^{-1})$ for each $\mu \in K^\times = K \setminus \{0\}$. Show that the projection map $\pi : L \to \mathbb{P}^n K$, $[(\mathbf{x}, \lambda)] \mapsto [\mathbf{x}]$ is a vector bundle (the **tautological line bundle** over $\mathbb{P}^n K$).

2.2 Tensors

tensor product	inner product
tensor	pullback of a covariant tensor
multilinear form	covariant tensor
tensor algebra	contravariant tensor
cotangent bundle	Lie derivative
contraction	tensorial
trace	differential

Tensor products were introduced by Graßmann in 1844 in his "Ausdehnungslehre" under the name "open product". Together with his student Tullio Levi-Civita, Gregorio Ricci-Curbastro developed the notion of tensors on manifolds ([RCLC]) from it in 1900 after some preliminary work. The tensor product solves, among others, the following two problems:

1) To represent spaces of the form $\operatorname{Hom}(\operatorname{End}(\operatorname{Bil}(V, W)), \operatorname{Bil}(V^*, V))$ in a uniform and concise way. For example, this space is canonically isomorphic to $\operatorname{End}(\operatorname{Hom}(W, \operatorname{End}V))$, which is not obvious at first sight.

2) Given K-vector spaces V, W, we are looking for a vector space U "large" enough to represent any bilinear map $\sigma : V \times W \to Z$ for any vector space Z by a linear map $f_\sigma : U \to Z$. Among all possible choices for U, we seek the "smallest" one.

Definition 2.2.1 *The **tensor product** $V \otimes W$ of two vector spaces V, W is a vector space with a bilinear map*

$$\kappa : V \times W \to V \otimes W, \qquad (v, w) \mapsto v \otimes w$$

which has the following universal property: for all bilinear maps $\sigma : V \times W \to Z$ there is exactly one linear map f_σ such that

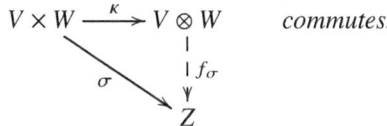

commutes.

That means $(V \otimes W)^* = \operatorname{Bil}(V, W)$ with the bilinear maps $V \times W \to \mathbb{R}$, because $Z := \mathbb{R}$ yields a monomorphism $\operatorname{Bil}(V, W) \hookrightarrow (V \otimes W)^*$, and composition with κ

gives a bilinear map for every element of $(V \otimes W)^*$. Thus, for finite-dimensional V, W, one gets $V \otimes W = \mathrm{Bil}(V, W)^*$ with $v \otimes w = (\sigma \mapsto \sigma(v, w))$. In particular, $\dim V \otimes W = \dim V \cdot \dim W$, and for bases $(v_1, \ldots, v_n)$, $(w_1, \ldots, w_m)$ of V, W, $(v_j \otimes w_k)_{\substack{1 \le j \le n \\ 1 \le k \le m}}$ is a basis of $V \otimes W$. For general $\sigma : V \times W \to Z$, one gets $f_\sigma : V \otimes W \to Z, \omega \mapsto \omega(\sigma)$. The elements of $V \otimes W$ are called **tensors**. So, every tensor has the form $v_1 \otimes w_1 + \cdots + v_m \otimes w_m$ with $v_1, \ldots, v_m \in V, w_1, \ldots, w_m \in W$. Not every tensor can be written as $v \otimes w$.

Remark In general, one constructs the tensor product as the vector space generated by $(v_j \otimes w_k)_{\substack{j \in J \\ k \in K}}$ for bases $(v_j)_j, (w_k)_k$ of V, W. Uniqueness holds because for a second tensor product $\widetilde{\kappa} : V \times W \to V \widetilde{\otimes} W$ the equality $\kappa = \widetilde{f_\kappa} \circ f_{\widetilde{\kappa}} \circ \kappa$ follows since

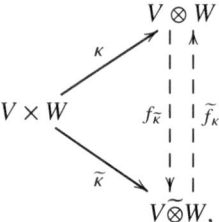

Because of the uniqueness in

$$
\begin{array}{ccc}
V \times W & \xrightarrow{\kappa} & V \otimes W \\
 & \searrow_{\kappa} & \downarrow \downarrow^{\mathrm{id} \; \widetilde{f_\kappa} \circ f_{\widetilde{\kappa}}} \\
 & & V \otimes W,
\end{array}
$$

one obtains $\widetilde{f_\kappa} \circ f_{\widetilde{\kappa}} = \mathrm{id}$ and hence $f_{\widetilde{\kappa}} : V \otimes W \xrightarrow{\cong} V \widetilde{\otimes} W$.

Lemma 2.2.2 *Given vector spaces U, V, W, one gets*

(1) $(U \oplus V) \otimes W \overset{\mathrm{can.}}{\cong} U \otimes W \oplus V \otimes W$,   (2) $V \otimes \mathbb{R} \overset{\mathrm{can.}}{\cong} V$,

(3) $(U \otimes V) \otimes W \overset{\mathrm{can.}}{\cong} U \otimes (V \otimes W)$,   (4) $V \otimes W \overset{\mathrm{can.}}{\cong} W \otimes V$,

(distributive, neutral element, associative, commutative). For finite-dimensional vector spaces V, W we find

(5) $(V \otimes W)^* \overset{\mathrm{can.}}{\cong} V^* \otimes W^*$,   (6) $\mathrm{Hom}(V, W) \overset{\mathrm{can.}}{\cong} V^* \otimes W$.

Remark 1) Although Bil satisfies (1) and (4), there is no canonical isomorphism $\mathrm{Bil}(\mathrm{Bil}(U, V), W) \cong \mathrm{Bil}(U, \mathrm{Bil}(V, W))$.

2) The identifications (1),(3),(5),(6) will be used without further comment, and constructions such as the Lie derivative in this chapter will be tacitly considered as compatible with these identifications. The identification (4) will be used very rarely,

and in such cases it will be specifically pointed out. Thus, the tensor product is considered to be associative, but not commutative.

3) For the associativity with more factors, one has to check that no contradictions arise due to different orderings of brackets (e.g. with four factors). For this purpose, the universal property for multilinear maps provides a suitable tool.

Proof. We show the last relation as an example. The map $f : V^* \otimes W \to \text{Hom}(V, W), \alpha \otimes w \mapsto (v \mapsto \alpha(v) \cdot w)$ is a monomorphism and for dimensional reasons it is bijective. $\square$

For $A \in \text{Hom}(\mathbb{R}^n, \mathbb{R}^m)$, we get more explicitly $f^{-1}(A) = \sum a_{kj} e^j \otimes e_k$.

Remark Thus any nested combination of Bil, Hom, $\otimes$, $*$ applied to finite-dimensional vector spaces $V_1, \dots, V_m$ can be written as a tensor product of $V_1, \dots, V_m, V_1^*, \dots, V_m^*$. For example, for vector space homomorphisms $f : V \to V', g : W \to W'$ one finds

$$f \otimes g \in \text{Hom}(V, V') \otimes \text{Hom}(W, W') = V^* \otimes V' \otimes W^* \otimes W' = \text{Hom}(V \otimes W, V' \otimes W').$$

Definition 2.2.3 *The space of q-**multi linear forms** on V is defined as*

$$V^{* \otimes q} = \underbrace{V^* \otimes \cdots \otimes V^*}_{q \text{ times}}$$

*for $q \in \mathbb{Z}^+$, and $V^{* \otimes 0} := \mathbb{R}$ for $q = 0$. The **tensor algebra** of V is $\bigotimes V^* := \bigoplus_{q \geq 0} V^{* \otimes q}$ with the ring structure associated to $(+, \otimes)$.*

Analogously to Definition 2.1.10, define $E \otimes F = \text{Hom}(E^*, F)$ and $f \otimes g$ for vector bundles E, F and vector bundle homomorphisms f, g, respectively.

Definition 2.2.4 *The vector bundle $T^*M := (TM)^*$ is called the **cotangent bundle** over M. The sections of $T_q^p M := TM^{\otimes p} \otimes T^*M^{\otimes q}$ are called (p, q)-**tensors or p-fold contravariant, q-fold covariant tensors**.*

Given $A, B \in T_0^\bullet M$ and $\omega, \eta \in T_\bullet^0 M$, set $(A \otimes \omega) \otimes (B \otimes \eta) := (A \otimes B) \otimes (\omega \otimes \eta)$. In this way, the tensors form a doubly graded $\mathbb{R}$-algebra, which is infinite-dimensional for $\dim M > 0$.

Definition 2.2.5 *For $1 \leq j \leq p, 1 \leq k \leq q$ the map $\text{Tr}_{jk} : T_q^p M \to T_{q-1}^{p-1} M$ is called a **contraction** (or **trace**)*

$$\text{Tr}_{jk}\left(\sum X_1 \otimes \cdots \otimes X_p \otimes \alpha_1 \otimes \cdots \otimes \alpha_q\right)$$
$$:= \sum \underbrace{\alpha_k(X_j)}_{\in C^\infty(M)} X_1 \otimes \cdots \otimes X_{j-1} \otimes X_{j+1} \otimes \cdots \otimes X_p$$
$$\otimes \alpha_1 \otimes \cdots \otimes \alpha_{k-1} \otimes \alpha_{k+1} \otimes \cdots \otimes \alpha_q.$$

The **inner product** of a vector field X with a $(p, q+1)$-tensor ω is the (p, q)-tensor $\iota_X\omega := \mathrm{Tr}_{11}(X \otimes \omega)$, i.e., for vector fields $Y_1, \ldots, Y_q$ one sets

$$(\iota_X\omega)(Y_1, \ldots, Y_q) = \omega(X, Y_1, \ldots, Y_q) \in T_0^P M$$

and 0 for $\omega \in T_0^P M$, respectively.

Contraction can be used to describe many simple vector space maps in a unified way, e.g. the composition of homomorphisms $\mathrm{Tr}_{12} : \mathrm{Hom}(V, W) \otimes \mathrm{Hom}(W, U) \to \mathrm{Hom}(V, U)$ or substitution maps like just the inner product.

Definition 2.2.6 Let $f : M \to N$ be C^∞ and let ω be a covariant tensor on N. The **pullback** $f^*\omega \in \Gamma(M, T^* M^{\otimes q})$ of ω is defined to be the covariant tensor on M given by

$$(f^*\omega)_x(X_1, \ldots, X_q) := \omega_{f(x)}(T_x f(X_1), \ldots, T_x f(X_q)) \qquad (X_1, \ldots, X_q \in T_x M).$$

In fundamental difference to the direct image of vectors, here f does **not** have to be a diffeomorphism.

Lemma 2.2.7 If $M \xrightarrow{f} N \xrightarrow{g} P$ are two C^∞ maps, then $(g \circ f)^* = f^* \circ g^*$ holds. Hence for all $q \in \mathbb{N}_0$

$$\left\{ \begin{array}{c} \text{manifolds} \\ C^\infty \text{maps} \end{array} \right\} \to \left\{ \begin{array}{c} \mathbb{R}\text{-vector spaces} \\ \text{linear maps} \end{array} \right\}$$
$$M \mapsto \Gamma(M, T_q^0 M)$$
$$f \mapsto f^*$$

is a contravariant functor.

Proof. Using the chain rule one obtains

$$\begin{aligned} [(g \circ f)^*\omega](X_1, \ldots, X_q) &= \omega(T(g \circ f)X_1, \ldots, T(g \circ f)X_q) \\ &= (g^*\omega)(Tf(X_1), \ldots, Tf(X_q)) \\ &= f^*(g^*\omega)(X_1, \ldots, X_q). \end{aligned} \qquad \square$$

Unfortunately, this does not correspond to the historical naming of the multilinear forms on TM as **covariant** tensors. Correspondingly, the sections in $T_0^P M$ are called contravariant tensors, although they are neither particularly covariant nor contravariant, because they can be transferred only by diffeomorphisms $f : M \to N$. For diffeomorphisms, f^* is defined on arbitrary tensors by setting $f^*X := (f^{-1})_*X$ for a vector field $X \in \Gamma(N, TN)$. Then $(g \circ f)^* = f^* \circ g^*$ still holds, and for any diffeomorphism f we have that $f^*\mathrm{Tr}_{jk}\omega = \mathrm{Tr}_{jk}f^*\omega$ since for $X \in \Gamma(M, TM), \alpha \in \Gamma(M, T^*M)$

$$\begin{aligned} \mathrm{Tr}\, f^*(\alpha \otimes X)_{|p} &= (f^*\alpha)(f_*^{-1}X)_{|p} = \alpha_{f(p)}(Tf \circ Tf^{-1}(X)) \\ &= \alpha(X)_{|f(p)} = f^*(\alpha(X))_{|p}. \end{aligned}$$

Applying the pullback, the previous derivations on real-valued maps and vector fields can be extended to arbitrary tensors:

Theorem 2.2.8 *Let X be a vector field on M. There exists uniquely an operator L_X on $\Gamma(M, T_\bullet^\bullet M)$, the **Lie derivative**, with the following properties:*

1) for all $f \in C^\infty(M)$ we have $L_X f = X.f$,
2) for all $Y \in \Gamma(M, TM)$ we have $L_X Y = [X, Y]$,
3) L_X is a derivation on the associative unitary algebra of tensors:

$$L_X(\omega \otimes \eta) = L_X \omega \otimes \eta + \omega \otimes L_X \eta,$$

4) L_X commutes with contractions: $L_X \mathrm{Tr}_{jk} \omega = \mathrm{Tr}_{jk}(L_X \omega)$.

Proof. Existence: Using

$$L_X : \Gamma(M, T_q^p M) \to \Gamma(M, T_q^p M)$$

$$\omega \mapsto \frac{\partial}{\partial t}_{|t=0} \Phi_t^{X^*} \omega$$

one finds:
1) $L_X f = \frac{\partial}{\partial t}_{|t=0} \Phi_t^{X^*} f = \frac{\partial}{\partial t}_{|t=0} f \circ \Phi_t^X = Tf(X).$
2) $L_X Y = \frac{\partial}{\partial t}_{|t=0} \Phi_t^{X^*} Y = \frac{\partial}{\partial t}_{|t=0} \Phi_{-t*}^X Y \overset{1.5.6}{=} -[Y, X] = [X, Y].$
3), 4) The pullback is $\mathbb{R}$-linear. Apply $\frac{\partial}{\partial t}_{|t=0}$ to

$$\Phi_t^{X^*}(\omega \otimes \eta) = \Phi_t^{X^*} \omega \otimes \Phi_t^{X^*} \eta \quad \text{and} \quad \Phi_t^{X^*} \mathrm{Tr}_{jk} \omega = \mathrm{Tr}_{jk} \Phi_t^{X^*} \omega,$$

respectively.

Uniqueness: Let K_X be a second operator of this kind. By (1),(2), $K_X = L_X$ holds on functions and vector fields. As in Proposition 1.4.2, K_X, L_X are local operators, therefore it suffices to compare them on a chart $\varphi : U \to V$ around any point $p \in M$. Now let $\alpha \in \Gamma(U, T^*U)$ and $Y_j := \varphi^* e_j$. Then for all j

$$(K_X \alpha)(Y_j) = \mathrm{Tr}\left((K_X \alpha) \otimes Y_j\right) \overset{(3)}{=} \mathrm{Tr}\left(K_X(\alpha \otimes Y_j) - \alpha \otimes K_X Y_j\right)$$
$$\overset{(4)}{=} K_X(\alpha(Y_j)) - \alpha(K_X Y_j) \overset{(1),(2)}{=} X.(\alpha(Y_j)) - \alpha([X, Y_j]). \quad (2.1)$$

Since $(Y_{j,p})$ is a basis of $T_p M$, it follows that $K_X \alpha_{|p} = L_X \alpha_{|p}$. By (3) it follows iteratively that $K_X = L_X$ holds on all tensors. □

It follows that L_X maps tensors of type (p, q) to just such.

Similar to Theorem 1.4.3, the uniqueness would be false if 'C^∞' were replaced by 'analytic' everywhere in the assumptions.

By iteration, for $\omega \in \Gamma(M, T_q^p M)$ in generalization of equation (2.1) it follows that

$$L_X(\omega(Y_1, \ldots, Y_q)) = (L_X\omega)(Y_1, \ldots, Y_q) + \omega([X, Y_1], Y_2, \ldots, Y_q)$$
$$+ \cdots + \omega(Y_1, \ldots, Y_{q-1}, [X, Y_q]).$$

Remark The derivative of real-valued functions can also be described pointwise as a derivative in the direction of a single vector. For vector fields, however, the Lie bracket yields only a derivative in the direction of a whole vector field, which depends not only on the value of a vector field X at a point p, but also on the first derivatives of this field at p. The Lie derivative has this flaw on all tensors except those of degree $(0, 0)$. In Chapter 3.2 we discuss a notion of derivation which fixes this problem, but it shall depend on additional choices.

Looking at the example from the previous proof, $L_X\alpha \in \Gamma(M, T^*M)$ with $(L_X\alpha)(Y) = X.\alpha(Y) - \alpha([X, Y])$, one can see that it is not always obvious whether an operator on tensors is a tensor. The following elegant characterization is often very useful for proving tensoriality.

Theorem 2.2.9 *Let E, F be vector bundles over M and consider a map P : $\Gamma(M, E) \to \Gamma(M, F)$. Then P is $C^\infty(M)$-linear if and only if $P \in \Gamma(M, \text{Hom}(E, F))$. In other words, $P(fs + g\widetilde{s}) = fP(s) + gP(\widetilde{s})$ holds for all $f, g \in C^\infty(M), s, \widetilde{s} \in \Gamma(M, E)$ if and only if P can be described fiberwise as $P_x \in \text{Hom}(E_x, F_x)$ for $x \in M$ such that the map $x \mapsto P_x$ is smooth.*

Example For $x_0 \in M$ fixed, the $\mathbb{R}$-linear map $P : \Gamma(M, \mathbb{R}) \to \Gamma(M, \mathbb{R}), f \mapsto f(x_0)$ does not satisfy this condition.

Proof. "$\Leftarrow$" Obvious.
"$\Rightarrow$" Let $x \in M, s, \widetilde{s} \in \Gamma(M, E)$ with $s_x = \widetilde{s}_x$. On a trivialization $h : \pi^{-1}(U) \to U \times \mathbb{R}^m$, choose a family $(s_j)_j$ of sections fiberwise forming a basis of $E_{|U}$ (say $s_{j|y} := h^{-1}(y, e_j)$). Then $(s - \widetilde{s})_{|U} = \sum_j f_j \cdot s_j$ with $f_j \in C^\infty(U)$. Let $\tau \in C^\infty(M)$, $\tau = \begin{cases} 1 & \text{on neighborhood } \widetilde{U} \text{ of } x \\ 0 & \text{on } M \backslash U \end{cases}$. Then

$$P(s)_x - P(\widetilde{s})_x = (\tau^2 \cdot P(s - \widetilde{s}))_x = P(\tau^2 \cdot (s - \widetilde{s}))_x$$
$$= P\left(\tau^2 \sum_j f_j s_j\right) = \sum_j \tau(x) \underbrace{f_j(x)}_{=0} \cdot P(\tau s_j)_x = 0.$$

Thus $\widetilde{P}_x : E_x \to F_x, \sum_j f_j \cdot s_j(x) \mapsto \sum_j f_j P(\tau s_j)_x$ (with $f_j \in \mathbb{R}$) defines a linear map independent of the choice of trivialization. On $\widetilde{U}$ this formula determines $\widetilde{P}$, and because $P(\tau s_j) \in \Gamma^\infty(M, F), y \mapsto \widetilde{P}_y$ on $\widetilde{U}$ depends smoothly on the base point y. $\square$

Corollary 2.2.10 *A map $P : \Gamma(M, T_q^p M) \to \Gamma(M, T_{q'}^{p'} M)$ is $C^\infty(M)$-linear if and only if $P \in \Gamma(M, T_{q'+p}^{p'+q} M)$.*

Therefore $C^\infty(M)$-linear maps between sections of vector bundles are called **tensorial**.

Example i) A q-fold covariant tensor is the same as a $C^\infty(M)$-linear map P : $\Gamma(M, TM^{\otimes q}) \to C^\infty(M)$, i.e. it is the same as a map

$$\widetilde{P} : \underbrace{\Gamma(M, TM) \times \cdots \times \Gamma(M, TM)}_{q-\text{mal}} \to C^\infty(M)$$

which is $C^\infty(M)$-linear in each variable.

ii) The map $\Gamma(M, TM \times TM) \to \Gamma(M, TM), (X, Y) \mapsto [X, Y]$ is not a tensor.

iii) If $f \in C^\infty(M)$, then $df : \Gamma(M, TM) \to C^\infty(M), X \mapsto L_X f$ is a tensor, since for all $g \in C^\infty(M)$ one finds $L_{gX} f = Tf(gX) = g \cdot Tf(X)$. For $X \in T_x M$, $df_x(X) := \mathrm{proj}_2(T_x f(X))$ defines the **differential** $df \in \Gamma(M, T^*M)$ of f. For example, for the Cartesian coordinate $x_j : \mathbb{R}^n \to \mathbb{R}$ it equals

$$dx_j = \left(\frac{\partial x_j}{\partial x_1}, \ldots, \frac{\partial x_j}{\partial x_n}\right) = (0, \ldots 0, 1, 0, \ldots, 0) \in (\mathbb{R}^n)^*.$$

Exercises

Exercise* 2.2.11 Given finite-dimensional vector spaces V, W, Z, represent the following maps as contractions:

1) The composition $\mathrm{Hom}(V, W) \otimes \mathrm{Hom}(W, Z) \to \mathrm{Hom}(V, Z), \varphi \otimes \psi \mapsto \psi \circ \varphi$.
2) The evaluation map $\mathrm{Hom}(V, W) \otimes V \to W, \varphi \otimes v \mapsto \varphi(v)$.

Exercise 2.2.12 Let

$$A = \begin{pmatrix} 0 & -2 \\ -1 & 1 \end{pmatrix} \in \mathrm{End}(\mathbb{C}^2).$$

Find a basis (v_1, v_2) of $\mathbb{C}^2$ with respect to which A takes the form $\lambda_1 v^1 \otimes v_1 + \lambda_2 v^2 \otimes v_2$ where $\lambda_{1,2} \in \mathbb{C}$. Here (v^1, v^2) is the dual basis to (v_1, v_2), i.e. $v^j(v_k) = \delta_{jk}$.

Exercise 2.2.13 1) Show that $\langle \cdot, \cdot \rangle : \mathfrak{su}(n) \times \mathfrak{su}(n) \to \mathbb{R}, (X, Y) \mapsto \mathrm{Tr}(X\overline{Y}^t)$ is a scalar product on $\mathfrak{su}(n)$. Prove that $\mathrm{Ad} : \mathrm{SU}(n) \to \mathrm{End}(\mathfrak{su}(n))$ takes values in the isometries associated to this scalar product.

2) For $n = 2$, interpret Ad as a map from $\mathrm{SU}(2)$ to $\mathrm{SO}(3)$. Determine the kernel of this map.

Exercise 2.2.14 Let $\iota : S^n \to \mathbb{R}^{n+1}$ be the canonical embedding and let $\varphi_\pm$ denote the stereographic projections. The Euclidean scalar product $\langle \cdot, \cdot \rangle$ is a bilinear form on $T\mathbb{R}^{n+1}$, hence $g := \iota^*\langle \cdot, \cdot \rangle \in T^*S^n \otimes T^*S^n$. Prove that

$$((\varphi_\pm^{-1})^* g)|_u = \frac{4}{(1 + \|u\|^2)^2} \langle \cdot, \cdot \rangle_{\mathrm{Eucl}}.$$

(Hint: This becomes a little easier to calculate if you remember that you only need to check the equality of the norms.)

Exercise 2.2.15 Consider M, X, Y, f as in Exercise 1.4.12 and $\omega \in \Gamma(M, \bigotimes T^*M)$ with $\omega_{(x,y)} := \sin(x)dx \otimes dy + \cos(y)(dx + 2dy) + 1$. Compute $\omega \otimes \omega$, $f^*\omega$, $\iota_X\omega$, $\iota_X(\omega \otimes \iota_Y\omega)$, $L_X\omega$ and $L_X(Y \otimes \omega)$.

2.3 Exterior Algebra

alternating	differential form
exterior power	exterior derivative
exterior product	de Rham operator
wedge product	naturality property
exterior algebra	Cartan Homotopy Formula
supercommutative	Frobenius' Theorem

The tensor algebra is an infinite-dimensional algebra into which T^*M is embedded. This infinite dimension often leads to difficulties. In this section we construct a finite-dimensional algebra ΛT^*M which contains T^*M and generalizes the construction of the determinant. This construction was introduced by Graßmann[1] in the same book in which he invented vector spaces.

Definition 2.3.1 *Consider an n-dimensional* $\mathbb{R}$*-vector space V. Given $q \in \mathbb{N}_0$, a q-form $\omega \in V^{*\otimes q}$ is called* **alternating**

$:\Leftrightarrow$*If of q vectors $v_1, \ldots, v_q$ two are equal, then $\omega(v_1, \ldots, v_q) = 0$*
$\Leftrightarrow$ *If q vectors $v_1, \ldots, v_q$ are linearly dependent, then $\omega(v_1, \ldots, v_q) = 0$.*

This is equivalent to the statement: For all distinct indices j, k one has

$$\omega(v_1, \ldots, v_j, \ldots, v_k, \ldots, v_q) = -\omega(v_1, \ldots, v_k, \ldots, v_j, \ldots, v_q),$$

as can be seen by substituting $v_j + v_k$ in place of v_j, v_k.

The q**th exterior power** of V^* is defined as the vector space

$$\Lambda^q V^* := \{\omega \in V^{*\otimes q} \mid \omega \text{ alternating}\}.$$

Then $\Lambda^q V^* = 0$ holds for $q > n$, as $q > n$ vectors are always linearly dependent. Furthermore $\Lambda^0 V^* = \mathbb{R}$, $\Lambda^1 V^* = V^*$, and $\Lambda^n V^*$ is generated by $(v_1, \ldots, v_n) \mapsto \det((e^j(v_k))_{jk})$ for any basis $(e^1, \ldots, e^n)$ of V^*. Given $\sigma \in \mathfrak{S}_q$ one has

$$(\omega \circ \sigma)(v_1, \ldots, v_q) := \omega(v_{\sigma(1)}, \ldots, v_{\sigma(q)}) = \text{sign}\,\sigma \cdot \omega(v_1, \ldots, v_q)$$

---

[1] 1844, Hermann Günther Graßmann, 1809–1877.

(to see this, write σ as a product of transpositions). Thus

$$\pi_\Lambda : V^{*\otimes q} \to \Lambda^q V^*$$

$$\omega \mapsto \frac{1}{q!} \sum_{\sigma \in \mathfrak{S}_q} \text{sign}\,\sigma \cdot \omega \circ \sigma$$

is a projection of vector spaces. Applying the Leibniz rule we get

$$\pi_\Lambda(\alpha_1 \otimes \cdots \otimes \alpha_q)(v_1, \ldots, v_q) = \frac{1}{q!} \det((\alpha_j(v_k))_{jk}).$$

Lemma and Definition 2.3.2 *Equipped with the* **exterior product** *(or* **wedge product***)*

$$\wedge : \Lambda^p V^* \times \Lambda^q V^* \to \Lambda^{p+q} V^*$$

$$(\alpha, \beta) \mapsto \frac{(p+q)!}{p!q!} \pi_\Lambda(\alpha \otimes \beta) = \sum_{[\sigma] \in \mathfrak{S}_{p+q}/\mathfrak{S}_p \times \mathfrak{S}_q} \text{sign}\,\sigma \cdot (\alpha \otimes \beta) \circ \sigma,$$

$\Lambda^\bullet V^* := \bigoplus_{q=0}^n \Lambda^q V^*$ *becomes an associative algebra, called the* **exterior algebra** *of V^*. The product $\wedge$ is* **supercommutative***, i.e. $\alpha \wedge \beta = (-1)^{pq} \beta \wedge \alpha$.*

Example One has $e^1 \wedge e^2 = 2\pi_\Lambda(e^1 \otimes e^2) = e^1 \otimes e^2 - e^2 \otimes e^1$.

Proof. As a projection of $\otimes$, $\wedge$ is bilinear. For a subspace $U \subset V$ denote by $U^0 \subset V^*$ the annihilator in the dual vector space. Then $\text{im}\,\pi_\Lambda = (V^*)^{\otimes q}/\ker \pi_\Lambda = ((\ker \pi_\Lambda)^0)^*$, hence $\ker \pi_\Lambda = ((\text{im}\,\pi_\Lambda)^*)^0$ or

$$\mathcal{J}_q := \ker \pi_{\Lambda|(V^*)^{\otimes q}} = \text{span}\{\alpha_1 \otimes \cdots \otimes \alpha_q \mid \exists j \neq k : \alpha_j = \alpha_k\}.$$

Therefore $\mathcal{J} := \ker \pi_{\Lambda| \otimes V^*}$ is an ideal in the algebra $\bigotimes V^*$. Thus the quotient $\bigotimes V^*/\mathcal{J} = \bigoplus_q (V^*)^{\otimes q}/\mathcal{J}_q$ is a ring with respect to the product $[\alpha] \wedge [\beta] := [\alpha \otimes \beta]$. According to the Fundamental Theorem on Homomorphisms, $\bigotimes V^*/\mathcal{J} \xrightarrow{\pi_\Lambda} \text{im}\,\pi_\Lambda = \Lambda^\bullet V^*$ gets a ring structure in this way.

Supercommutativity: Set $\tau := \begin{pmatrix} 1 & \cdots & p & p+1 & \cdots & p+q \\ q+1 & \cdots & p+q & 1 & \cdots & q \end{pmatrix} \in \mathfrak{S}_{p+q}$, then $\text{sign}\,\tau = (-1)^{pq}$ holds and so we get

$$\alpha \wedge \beta = \frac{1}{p!q!} \sum_{\sigma \in \mathfrak{S}_{p+q}} \text{sign}\,\sigma \cdot (\alpha \otimes \beta) \circ \sigma$$

$$= \frac{1}{p!q!} \sum_{\sigma \in \mathfrak{S}_{p+q}} \text{sign}\,(\sigma \circ \tau) \cdot (\alpha \otimes \beta) \circ (\sigma \circ \tau)$$

$$= \text{sign}\,\tau \cdot \frac{1}{p!q!} \sum_{\sigma \in \mathfrak{S}_{p+q}} \text{sign}\,\sigma \cdot (\beta \otimes \alpha) \circ \sigma = (-1)^{pq} \beta \wedge \alpha. \qquad \square$$

In this proof the detour via annihilators shows that $\ker \pi_\Lambda$ is an ideal, despite the fact that π_Λ is not a ring homomorphism. One can also check the associativity with the definition, which is a more cumbersome, yet more elementary method. In general for $\alpha_j \in \Lambda^{n_j} V^*$ $(1 \le j \le k)$ one obtains

$$
\alpha_1 \wedge \cdots \wedge \alpha_k = \binom{n_1 + \cdots + n_k}{n_1, \ldots, n_k} \pi_\Lambda(\alpha_1 \otimes \cdots \otimes \alpha_k)
$$

$$
= \sum_{[\sigma] \in \mathfrak{S}_{n_1 + \cdots + n_k}/\mathfrak{S}_{n_1} \times \cdots \times \mathfrak{S}_{n_k}} \operatorname{sign} \sigma \cdot (\alpha_1 \otimes \cdots \otimes \alpha_k) \circ \sigma.
$$

Thus, for a basis $(e^1, \ldots, e^n)$ of V^* the family $(e^{j_1} \wedge \cdots \wedge e^{j_q} \mid 1 \le j_1 < \cdots < j_q \le n)$ is a basis of $\Lambda^q V^*$, because for $1 \le k_1 \le \cdots \le k_q \le n$ one gets

$$
(e^{j_1} \wedge \cdots \wedge e^{j_q})(e_{k_1}, \ldots, e_{k_q}) = \begin{cases} 1 \text{ if for all } \ell \text{ we have } j_\ell = k_\ell, \\ 0 \text{ otherwise.} \end{cases}
$$

In particular one finds $\dim \Lambda^q V^* = \binom{n}{q}$ and $\dim \Lambda^\bullet V^* = 2^n$.

Notation: Let $\mathfrak{A}^\bullet(M) := \Gamma(M, \Lambda^\bullet T^* M)$ be the **space of differential forms**.

According to the rules for $\otimes$ one finds for $f : M \to N, \alpha, \beta \in \mathfrak{A}^\bullet(N), X \in \Gamma(N, TN)$ that

$$
f^*(\alpha \wedge \beta) = f^* \alpha \wedge f^* \beta, \quad L_X(\alpha \wedge \beta) = L_X \alpha \wedge \beta + \alpha \wedge L_X \beta.
$$

On the exterior algebra there is a fundamental new differential operator:

Theorem and Definition 2.3.3 *There exists uniquely an additive map $d : \mathfrak{A}^\bullet(M) \to \mathfrak{A}^\bullet(M)$ satisfying the axioms*

1) for all $\alpha \in \mathfrak{A}^q(M), \beta \in \mathfrak{A}^\bullet(M)$ one has $d(\alpha \wedge \beta) = d\alpha \wedge \beta + (-1)^q \alpha \wedge d\beta$,
2) $d : C^\infty(M) \to \Gamma(M, T^* M)$ is the differential on functions,
3) for any $f \in C^\infty(M)$ one has $d^2 f = 0$.

*The map d is called the **exterior derivative** or the **de Rham operator**.*[2]

Proof. Uniqueness: As in Proposition 1.4.2, an operator d satisfying (1) must be a local operator. Consider now a chart $\varphi : U \to V$ on M and $\omega \in \mathfrak{A}^q(M)$. Then $(\varphi^{-1})^* \omega_{|U} = \sum_{|I|=q} f_I \, dx_{j_1} \wedge \cdots \wedge dx_{j_q}$, where $f_I \in C^\infty(V)$ for each multi-index $I = \{j_1, \ldots, j_q\}, 1 \le j_1 < \ldots j_q \le n$, i.e.

$$
\omega_{|U} = \varphi^* \left(\sum_{|I|=q} f_I \, dx_{j_1} \wedge \cdots \wedge dx_{j_q} \right)
$$

$$
= \sum_{|I|=q} (f_I \circ \varphi) \, d(x_{j_1} \circ \varphi) \wedge \cdots \wedge d(x_{j_q} \circ \varphi).
$$

---

[2] 1899, Élie Cartan.

Let d^U be an operator defined on U, which satisfies (1)–(3). Then d^U is uniquely given by

$$d^U \omega_{|U} \overset{(1),(3)}{=} \sum_{|I|=q} \underbrace{d(f_I \circ \varphi)}_{\overset{(2)}{=} \varphi^* \sum_j \frac{\partial f_I}{\partial x_j} dx_j} \wedge d(x_{j_1} \circ \varphi) \wedge \cdots \wedge d(x_{j_q} \circ \varphi). \qquad (2.2)$$

Existence: The above d^U satisfies (2). (3) follows from the Schwarz Lemma combined with $dx_j \wedge dx_k = -dx_k \wedge dx_j$:

$$(d^U)^2(f \circ \varphi) = \varphi^* \sum_{j,k} \frac{\partial^2 f}{\partial x_j \partial x_k} dx_j \wedge dx_k = 0.$$

And (1) follows using the special case

$$\alpha = \varphi^*(f \, dx_{j_1} \wedge \cdots \wedge dx_{j_q}),$$
$$\beta = \varphi^*(g \, dx_{k_1} \wedge \cdots \wedge dx_{k_p}).$$

In particular, on any manifold there is at most one operator d with (1)–(3). Thus for any two charts $\varphi : U \to V, \psi : U' \to V'$ we have that $d^U_{|U\cap U'} = d^{U'}_{|U\cap U'}$, i.e. d^U is independent of the choice of U and d is thus globally defined. □

Example According to the chain rule, for $g \in C^\infty(N, \mathbb{R})$ and $f : M \to N$,

$$f^* dg = dg \circ Tf = d(g \circ f) = d(f^* g)$$

holds.

Lemma 2.3.4 *If $U, U' \subset \mathbb{R}^n$ are open subsets and $f : U \to U'$ is a smooth map, one obtains*
$$f^*(dx_1 \wedge \cdots \wedge dx_n) = \det f' \cdot dx_1 \wedge \cdots \wedge dx_n.$$

Proof. This follows from the unique axiomatic characterization of the determinant. Because $\Lambda^n(\mathbb{R}^n)^*$ is one-dimensional, there exists a function $g \in C^\infty(\mathbb{R}^n)$ with $f^*(dx_1 \wedge \cdots \wedge dx_n) = g \cdot dx_1 \wedge \cdots \wedge dx_n$. The function g is alternating in the n rows of f', so it equals $\det f'$ up to a constant factor. For $f = \mathrm{id}$ the factor is 1, hence one obtains $g = \det f'$. □

Lemma 2.3.5 *The de Rham operator satisfies:*

1) $d(\mathfrak{A}^q(M)) \subset \mathfrak{A}^{q+1}(M)$.
2) $d^2 = 0$, *or* $\mathrm{im}\, d \subset \ker d$, *i.e.*

$$0 \to \mathfrak{A}^0(M) \overset{d}{\to} \cdots \overset{d}{\to} \mathfrak{A}^n(M) \to 0$$

is a complex of $\mathbb{R}$-vector spaces.

3) *For all* $\varphi \in C^\infty(M, N), \omega \in \mathfrak{A}^\bullet(N)$ *one has* $d(\varphi^*\omega) = \varphi^* d\omega$.
4) *For all* $X \in \Gamma(M, TM), \omega \in \mathfrak{A}^\bullet(M)$ *one has* $L_X d\omega = dL_X \omega$.

Proof. All of these properties are local, thus without loss of generality one can assume that the forms are monomials.
(1) is a consequence of the local formula (2.2).
(2) follows from the second axiom for d and by induction on the degree as

$$d^2(\alpha \wedge \beta) = (-1)^{q+1} d\alpha \wedge d\beta + (-1)^q d\alpha \wedge d\beta = 0.$$

(3) holds for 1-forms, as the formula holds for $f \cdot dg$. The assertion follows via induction on the degree applying the first axiom.
(4) follows when applying $\frac{\partial}{\partial t}\big|_{t=0}$ to $\Phi_t^{X*} d\omega = d\Phi_t^{X*} \omega$. $\square$

Remark 2.3.6 1) The operator $\iota_X : \mathfrak{A}^q(M) \to \mathfrak{A}^{q-1}(M)$ has similar properties: One gets $\iota_X^2 = 0$ and $\iota_X(\alpha \wedge \beta) = \iota_X \alpha \wedge \beta + (-1)^q \alpha \wedge \iota_X \beta$ (Exercise 2.3.18).
2) One can show that the **naturality property** Lemma 2.3.5(3) also determines the de Rham operator in an essentially unique way. Palais proves in [Pal, p. 127] that on a compact manifold every linear operator on the differential forms satisfying (3) is a linear combination of d, id and the integral from Section 2.5. Kolář, Michor and Slovák show in [KMS, 25.4] that for $q > 0$ any (not necessarily linear) map satisfying (1) and (3) must be a scalar multiple of d.

Theorem 2.3.7 (Cartan Homotopy Formula[3]) *Given a vector field X and $\omega \in \mathfrak{A}^\bullet(M)$, the equality $L_X \omega = (d \circ \iota_X + \iota_X \circ d)\omega = (d + \iota_X)^2 \omega$ holds.*

In particular, L_X has the square root $d + \iota_X$ when acting on differential forms.

Proof. Together with d and ι_X, $K_X := (d + \iota_X)^2$ is also a local operator.
a) On functions f one gets $(d \circ \iota_X + \iota_X \circ d)f = df(X) = L_X f$.
b) The operator K_X acts as a derivation on $\mathfrak{A}^\bullet(M)$: Given $\alpha \in \mathfrak{A}^q(M), \beta \in \mathfrak{A}^\bullet(M)$ one finds

$$\begin{aligned}
K_X(\alpha \wedge \beta) &= (d + \iota_X)((d + \iota_X)\alpha \wedge \beta + (-1)^q \alpha \wedge (d + \iota_X)\beta) \\
&= K_X \alpha \wedge \beta + (-1)^{q-1}(d + \iota_X)\alpha \wedge (d + \iota_X)\beta \\
&\quad + (-1)^q (d + \iota_X)\alpha \wedge (d + \iota_X)\beta + \alpha \wedge K_X \beta \\
&= K_X \alpha \wedge \beta + \alpha \wedge K_X \beta.
\end{aligned}$$

c) On 1-forms one gets $L_X = K_X$, as $K_X df = d\iota_X df = dK_X f = dL_X f = L_X df$, and locally every 1-form can be written as $\sum f_j \, dx_j$.
Thus it follows that $K_X = L_X$ holds on all of $\mathfrak{A}^\bullet(M)$ as in the proof of Theorem 2.2.8. $\square$

Thus, this formula allows the computation of L_X in terms of d. Conversely, one can apply the Cartan Homotopy Formula to describe d in terms of L_X:

---

[3] Élie Cartan, 1869–1951.

Theorem 2.3.8 *If $X_0, \ldots, X_q$ are vector fields on M and $\omega \in \mathfrak{A}^q(M)$, one gets*

$$d\omega(X_0, \ldots, X_q) = \sum_{j=0}^{q} (-1)^j X_j . \left(\omega(X_0, \ldots, \widehat{X_j}, \ldots, X_q) \right)$$
$$+ \sum_{0 \le j < k \le q} (-1)^{j+k} \omega([X_j, X_k], X_0, \ldots, \widehat{X_j}, \ldots, \widehat{X_k}, \ldots, X_q).$$

Proof. Via induction on q: The assertion is obvious for $q = 0$. Now assume that the assertion is true for $(q-1)$-forms. According to the Leibniz rule for tensors one gets

$$L_{X_0}\left(\omega(X_1, \ldots, X_q) \right)$$
$$= \underbrace{(L_{X_0}\omega)}_{d\iota_{X_0}\omega + \iota_{X_0}d\omega} (X_1, \ldots, X_q)$$
$$+ \sum_{j=1}^{q} \omega(X_1, \ldots, [X_0, X_j], \ldots, X_q)$$
$$= \underbrace{(d\iota_{X_0}\omega)(X_1, \ldots, X_q)}_{\substack{=\Sigma_{1 \le j < k \le q}(-1)^{j+k}\omega(X_0, [X_j, X_k], X_1, \ldots, \widehat{X_j}, \ldots, \widehat{X_k}, \ldots, X_q) \\ -\Sigma_{j=1}^{q}(-1)^j X_j . \omega(X_0, \ldots, \widehat{X_j}, \ldots, X_q) \text{ by induction hypothesis}}} + d\omega(X_0, \ldots, X_q)$$
$$- \sum_{k=1}^{q} (-1)^k \omega([X_0, X_k], X_1, \ldots, \widehat{X_k}, \ldots, X_q). \qquad \square$$

Example 2.3.9 Given $\alpha \in \mathfrak{A}^1(M)$, one obtains $d\alpha(X, Y) = X.\alpha(Y) - Y.\alpha(X) - \alpha([X, Y])$.

Frobenius' Theorem 2.3.10 [4] *Let $H \subset TM^n$ be a vector subbundle of rank k such the Lie bracket $[X, Y]$ of any $X, Y \in \Gamma(M, H)$ is again a section of H. Then for every $p \in M$ there exists a k-dimensional submanifold $N \subset M$ with $p \in N$ and $TN = H_{|N}$.*

Proof. Let φ be a chart around p. Then the linear forms $\alpha_j := \varphi^* dx_j$ $(1 \le j \le n)$ form a basis of T^*M, so there is a subset $I \subset \{1, \ldots, n\}$ for which $(\alpha_{j|H_p})_{j \in I}$ is a basis of H_p^*. Let U be the open neighborhood of p on which $(\alpha_{j|H})_{j \in I}$ remains a basis of H^*, and let $(X_j)_{j \in I}$ be pointwise the dual basis of H. For $j, m, \ell \in I$ it follows that

$$0 = (\varphi^* d^2 x_j)(X_m, X_\ell) = d\alpha_j(X_m, X_\ell)$$
$$\overset{2.3.9}{=} X_m.\underbrace{(\alpha_j(X_\ell))}_{=0 \text{ or } 1} - X_\ell.(\alpha_j(X_m)) - \alpha_j([X_m, X_\ell]).$$

---

[4] 1875, Ferdinand Georg Frobenius, 1849–1917. Actually, in this article Frobenius states that the result was proved in 1840 by Heinrich Wilhelm Feodor Deahna, 1815–1844, and explains his proof.

By assumption, $[X_m, X_\ell] \in H$, hence it equals zero. Thus $N \subset U$ exists by Theorem 1.5.9.   □

Exercises

Exercise* 2.3.11 The characteristic polynomial of $f \in \operatorname{End} V$ satisfies

$$\chi_f = \sum_{q=0}^{n} (-X)^{n-q} \operatorname{Tr} f^*_{|\Lambda^q V^*}.$$

Exercise* 2.3.12 If $U, W \subset V$ are subspaces satisfying $U \oplus W = V$, check that for any natural number k,

$$\Lambda^k(U^* \oplus W^*) \stackrel{\text{can.}}{\cong} \bigoplus_{q=0}^{k} \Lambda^q U^* \otimes \Lambda^{k-q} W^*$$

holds.

Exercise 2.3.13 Suppose $M := \mathbb{R}^3$, let $\omega \in \mathfrak{A}^\bullet(M)$ be the differential form $\omega = y\, dx \wedge dz + (x+y)\, dy$ and let $X := x\frac{\partial}{\partial x} - y\frac{\partial}{\partial z}, Y := x\frac{\partial}{\partial x} + y\frac{\partial}{\partial y} + z\frac{\partial}{\partial z} \in \Gamma(M, TM)$. Let $f : \mathbb{R}^3 \to \mathbb{R}^3, (x, y, z) \mapsto (e^z, e^y, e^x)$. Compute

$$d(\iota_X \omega) - L_X \omega, \qquad \text{the flow } \Phi \text{ of } Y, \qquad f_* X, \qquad f^* \omega.$$

Exercise 2.3.14 If M, X, Y, f are given as in Exercise 1.4.12 and $\omega \in \Gamma(M, \Lambda^\bullet T^* M)$ is given by $\omega_{(x,y)} := \sin(x)dx \wedge dy + \cos(y)(dx + 2dy) + 1$, compute $\omega \wedge \omega, \iota_X \omega,$ $\iota_X(\omega \wedge \iota_Y \omega), L_X \omega$ and $d\omega$.

Exercise* 2.3.15 Let E be a vector bundle over M with transition functions g_{jk}. Describe the transition functions of $\det E := \Lambda^{\operatorname{rang} E} E$.

Exercise 2.3.16 Let $M^m \subset \mathbb{R}^{m+1}$ be an m-dimensional submanifold. Show that $\Lambda^{m+1} \mathbb{R}^{m+1}_{|M} \cong N \otimes \Lambda^m TM$ where N is the normal bundle.

Exercise* 2.3.17 Show in a direct way that the right-hand side in Example 2.3.9 is tensorial in X and Y.

Exercise 2.3.18 For the inner product $\iota_X : T^* M^{\otimes q} \to T^* M^{\otimes(q-1)}$ with a vector field X verify that:

1) Given $\alpha_0, \ldots, \alpha_m \in \mathfrak{A}^1(M)$, one gets

$$\iota_X(\alpha_0 \wedge \cdots \wedge \alpha_m) = \sum_{j=0}^{m} (-1)^j \alpha_j(X) \cdot \alpha_0 \wedge \cdots \wedge \widehat{\alpha_j} \wedge \cdots \wedge \alpha_m.$$

This provides an alternative definition of ι_X that is independent of the embedding of the exterior algebra into the tensor algebra.

2) ι_X maps $\mathfrak{A}^q(M)$ to $\mathfrak{A}^{q-1}(M)$.

3) For $\omega \in \mathfrak{A}^q(M)$ one finds $\iota_X^2 \omega = 0$.

4) For $\alpha \in \mathfrak{A}^k(M)$, $\beta \in \mathfrak{A}^q(M)$, the Leibniz rule

$$\iota_X(\alpha \wedge \beta) = (\iota_X \alpha) \wedge \beta + (-1)^k \alpha \wedge \iota_X \beta$$

holds.

5) If $\alpha \in \mathfrak{A}^1(M)$, $\omega \in \mathfrak{A}^q(M)$, then the equation

$$\iota_X(\alpha \wedge \omega) + \alpha \wedge \iota_X \omega = \alpha(X) \cdot \omega$$

holds (i.e. as an equation of operators on the differential forms,

$$\iota_X \cup (\alpha \wedge) + (\alpha \wedge) \circ \iota_X = \alpha(X)$$

holds).

Hint: This can be verified pointwise.

Exercise 2.3.19 Let X, Y be vector fields and let $\iota_Y : \mathfrak{A}^\bullet(M) \to \mathfrak{A}^\bullet(M)$ denote the inner product. Verify as an equation of operators on $\mathfrak{A}^\bullet(M)$ that

$$L_X \circ \iota_Y - \iota_Y \circ L_X = \iota_{[X,Y]}.$$

2.4 De Rham Cohomology

closed form	homotopic
exact form	homotopy
de Rham cohomology	homotopy type
Poincaré duality	homotopy equivalent
cup product	contractible
cohomology with compact support	retraction

Using the exterior algebra and the de Rham operator, it is comparatively easy to construct an important invariant of manifolds, a graded $\mathbb{R}$-algebra $H^\bullet(M)$ for any manifold M. This invariant is of crucial importance for Chapter 4.

Definition 2.4.1 *The forms in* $\ker d$ *and in* $\operatorname{im} d$ *are called **closed** and **exact** forms, respectively. The **de Rham cohomology**[5] ([deR]) of a manifold M is the family of $\mathbb{R}$-vector spaces*

$$H^q(M) := \ker d_{|\mathfrak{A}^q(M)} / \operatorname{im} d_{|\mathfrak{A}^{q-1}(M)} \qquad \text{for } q \in \mathbb{N}_0.$$

---

[5] 1931, Georges de Rham, 1903–1990.

Remark If M is a compact manifold, harmonic analysis can be used to prove that $\dim H^q(M) < \infty$ and to show the **Poincaré duality** $H^p(M) \cong H^{n-p}(M)$ ([War]). Vector bundles can essentially be interpreted as lattice points in $H^\bullet(M)$; a much trimmed down version of this result follows in Exercise 3.2.27 and Theorem 4.2.6.

An integer-valued invariant of manifolds derived from cohomology is the Euler characteristic $\chi(M) := \sum (-1)^q \dim H^q(M) \in \mathbb{Z} \cup \{\infty\}$, which is further studied in Chapter 4.

Lemma 2.4.2 *The cohomology has the following properties:*

1) $H^q(M) = 0$ *for any* $q > \dim M$.
2) $H^0(M) \overset{\text{can.}}{\cong} \mathbb{R}$ *if M is connected.*
3) *The wedge product on $\mathfrak{A}^\bullet(M)$ induces a ring structure on $H^\bullet(M)$ (called the* **cup product***). With this product $H^\bullet(M)$ becomes a supercommutative $\mathbb{Z}$-graded $\mathbb{R}$-algebra*
4) *Every $\varphi \in C^\infty(M, N)$ induces an $\mathbb{R}$-algebra homomorphism*

$$\varphi^* : H^\bullet(N) \to H^\bullet(M).$$

Proof. 1) $\mathfrak{A}^q(M) = 0$ for $q > \dim M$, because in this case for all $p \in M$ we get $\Lambda^q T_p^* M = 0$.

2) $H^0(M) = \{f \in C^\infty(M) \,|\, df = 0\} = \{f : M \to \mathbb{R} \text{ constant}\} \cong \mathbb{R}$.

3) Given $\alpha, \beta \in \ker d$, one gets $d(\alpha \wedge \beta) = 0$. Thus for $\alpha \in \ker d$ one obtains $d\beta \wedge \alpha = d(\beta \wedge \alpha)$, hence $\ker d \wedge \ker d \subset \ker d$, $\operatorname{im} d \wedge \ker d \subset \operatorname{im} d$.

4) Because $d \circ \varphi^* = \varphi^* \circ d$, $\varphi^* : \mathfrak{A}^\bullet(N) \to \mathfrak{A}^\bullet(M)$ maps $\ker d^N$ to $\ker d^M$ and $\operatorname{im} d^N$ to $\operatorname{im} d^M$.   □

In this way, as with the differential forms, one obtains a contravariant functor

$$\begin{Bmatrix} \text{manifolds} \\ C^\infty \text{ maps} \end{Bmatrix} \to \begin{Bmatrix} \mathbb{R} - \text{algebras} \\ \text{algebra homomorphisms} \end{Bmatrix},$$
$$M \mapsto H^\bullet(M),$$
$$\varphi \mapsto \varphi^*.$$

Example We verify $H^1(S^1) \cong \mathbb{R}$: On $S^1 \cong \mathbb{R}/\mathbb{Z}$, dx is not exact, because $\int_0^1 dx = 1 \neq 0$. Now consider any $\alpha \in \mathfrak{A}^1(S^1)$, i.e. $\alpha = f\,dx$, where $f \in C^\infty(S^1)$. Set $c := \int_0^1 f\,dx$ and $F(x) := \int_0^x (f(t) - c)\,dt$. Then $F(0) = F(1) = 0$, hence $F \in C^\infty(S^1)$, and $\alpha = c\,dx + dF$, hence $[\alpha] = c \cdot [dx]$ in $H^1(S^1)$. Therefore $H^\bullet(S^1) = \mathbb{R}[X]/(X^2)$ as an $\mathbb{R}$-algebra, where $X := [dx]$.

Remark Quite analogously, the **cohomology with compact support** $H_c^\bullet(M)$ is defined as the quotient of the closed forms with compact support by the image of d on the forms with compact support (this is also in de Rham's PhD thesis). Thus for compact M one finds $H^\bullet(M) = H_c^\bullet(M)$.

Theorem 2.4.3 *Consider the maps* $M \times \mathbb{R} \underset{\pi}{\overset{s}{\rightleftarrows}} M$, *where* $s(p) := (p, 0)$ *denotes the zero section. Then* s^*, π^* *are mutually inverse isomorphisms* $H^\bullet(M \times \mathbb{R}) \cong H^\bullet(M)$.

Proof. As a start one observes $s^* \pi^* = (\pi \circ s)^* = \mathrm{id}_M^*$. In the other direction, set $N := \mathbb{R} \times M$. We construct a map $K : \mathfrak{A}^q(N) \to \mathfrak{A}^{q-1}(N)$ satisfying $1 - \pi^* s^* = d^N \circ K + K \circ d^N$. Then the right-hand side maps closed to exact forms, so it acts as 0 on $H^\bullet(N)$. Hence $1 = \pi^* s^*$ holds on $H^\bullet(N)$.

Set $\alpha_t : N \to N, (p, u) \mapsto (p, tu)$ for $t \in \mathbb{R}$ and $X \in \Gamma(N, TN), X_{|(p,u)} = u\partial_u$. Then $\Phi_t(p, u) = (p, e^t u) = \alpha_{e^t}(p, u)$ is the flow of X. For any $\omega \in \mathfrak{A}^q(N), t > 0$ one finds

$$\frac{\partial}{\partial t}\alpha_t^* \omega = \frac{\partial}{\partial t}\Phi_{\log t}^* \omega = \frac{1}{t}\Phi_{\log t}^* L_X \omega = \frac{1}{t}\alpha_t^*(d^N \iota_X \omega + \iota_X d^N \omega)$$

$$= d^N(\frac{1}{t}\alpha_t^* \iota_X \omega) + \frac{1}{t}\alpha_t^* \iota_X d^N \omega.$$

This equation can be extended continuously to $t = 0$, since $X_{\alpha_t((p,u))} = tu\partial_u$ and thus

$$(\frac{1}{t}\alpha_t^* \iota_X \omega)_{|(p,u)} = \omega_{(p,tu)}(u\partial_u, T_{(p,u)}\alpha_t(\cdot), \ldots, T_{(p,u)}\alpha_t(\cdot))$$

holds. Now set $K : \mathfrak{A}^q(N) \to \mathfrak{A}^{q-1}(N), \omega \mapsto \int_0^1 \left(\frac{1}{t}\alpha_t^* \iota_X \omega\right) dt$. Then we obtain

$$(1 - \pi^* s^*)\omega = \underbrace{\alpha_1^*}_{=\mathrm{id}} \omega - \underbrace{\alpha_0^*}_{=s \circ \pi} \omega = \int_0^1 \left(\frac{\partial}{\partial t}\alpha_t^* \omega\right) dt$$

$$= \int_0^1 \left(d^N(\frac{1}{t}\alpha_t^* \iota_X \omega)\right) dt + \int_0^1 \left(\frac{1}{t}\alpha_t^* \iota_X d^N \omega\right) dt$$

$$= d^N K\omega + K d^N \omega. \qquad \square$$

Corollary 2.4.4 (Poincaré[6] Lemma) $H^q(\mathbb{R}^n) = H^q(\text{point}) = \begin{cases} \mathbb{R} \\ 0 \end{cases} \text{ if } \begin{matrix} q=0 \\ q\neq 0 \end{matrix}$.

Proof. By induction using $H^\bullet(\mathbb{R}^n \times \mathbb{R}) \cong H^\bullet(\mathbb{R}^n)$. $\qquad \square$

This lemma says that for $q > 0$ every closed q-form on $\mathbb{R}^n$ (or on manifolds diffeomorphic to $\mathbb{R}^n$, like star-shaped regions in $\mathbb{R}^n$) must be exact.

Example Consider $\eta := e^{x+y} dx \wedge dy \in \mathfrak{A}^2(\mathbb{R}^2)$ and $s : \mathbb{R} \to \mathbb{R}^2, x \mapsto (x, 0)$. Then $s^* \eta = 0$ because $s^* dy = 0$, $X = y\frac{\partial}{\partial y}$ and

$$K\eta = \int_0^1 (\alpha_t^* \iota_X \eta) \frac{dt}{t} = -\int_0^1 (e^{x+ty} ty \, dx) \frac{dt}{t} = (e^x - e^{x+y}) \, dx.$$

Hence one obtains $\eta = \eta - \pi^* s^* \eta = d(K\eta) + K d\eta = d\left((e^x - e^{x+y}) \, dx\right)$.

[6] 1899 (without proof), Jules Henri Poincaré, 1854–1912. As Samelson explains in [Sam], this result had already been proved by Volterra in 1889 ([Volt]).

Definition 2.4.5 *Two maps $f, g \in C^\infty(M, N)$ are called (C^∞)* **homotopic** *if there is a map $F \in C^\infty(M \times \mathbb{R}, N)$ such that $F(\cdot, 0) = g, F(\cdot, 1) = f$ hold. In this case the map F is called a (C^∞)* **homotopy**.

Remark Applying approximations one can show that f, g are C^∞-homotopic if and only if they are homotopic via a continuous map F ([Hi, ch. 2.2]). In the definition here $[0, 1]$ is avoided as it is a manifold with boundary.

Lemma 2.4.6 *Smooth homotopy is an equivalence relation.*

Proof. Let $\simeq$ denote homotopy. We have to show that if $f \simeq g$ and $g \simeq h$, then $f \simeq h$. Let F, G be suitable homotopies and choose $\lambda \in C^\infty(\mathbb{R}, [0, 1])$ with $\lambda_{[0,1/3]} \equiv 0$, $\lambda_{[2/3,1]} \equiv 1$. Then

$$H(x, t) := \begin{cases} F(x, \lambda(2t)) \\ G(x, \lambda(2t - 1)) \end{cases} \text{ for } \begin{matrix} t \le 1/2 \\ 1/2 < t \end{matrix}$$

is a smooth homotopy between f and h. $\qquad \square$

Corollary 2.4.7 *If $f, g \in C^\infty(M, N)$ are homotopic maps, then $f^* = g^*$ on $H^\bullet(N)$.*

Proof. Consider the maps $M \times \mathbb{R} \overset{s_0, s_1}{\underset{\pi}{\rightleftarrows}} M$, where $s_0(p) = (p, 0), s_1(p) = (p, 1)$. Then $f = F \circ s_1, g = F \circ s_0$ and therefore

$$f^* = s_1^* F^* \overset{2.4.3}{=} (\pi^*)^{-1} F^* \overset{2.4.3}{=} s_0^* F^* = g^*. \qquad \square$$

Two manifolds have **the same** (C^∞) **homotopy type** (are **homotopy equivalent**) if there are C^∞ maps $f : M \to N, g : N \to M$ such that $f \circ g, g \circ f$ are homotopic to id_N, id_M. M is called (C^∞) **contractible** if it has the homotopy type of a point. This means for $p_0 \in M$ that the map $f : M \to M, f \equiv p_0$ is homotopic to id_M.

Example Every star-shaped open subset $U \subset \mathbb{R}^n$ is contractible, as can be seen using $F : U \times \mathbb{R} \to U, (x, t) \mapsto tx$.

Corollary 2.4.8 *If M, N have the same homotopy type, then $H^\bullet(M) \cong H^\bullet(N)$. In particular, one gets $H^\bullet(M) \cong \mathbb{R}$ for contractible M.*

Proof. By Corollary 2.4.7, $g^* f^* = \text{id}_M^*$ holds on $H^\bullet(M)$. This implies the surjectivity of g^* and the injectivity of f^*. Likewise, $f^* g^* = \text{id}_N^*$ implies that f^* is surjective and g^* is injective. $\qquad \square$

Consequently, cohomology can distinguish manifolds only up to homotopy equivalence. In other words, two manifolds M, N with different cohomology rings cannot have the same homotopy type. For example, S^1 does not have the same homotopy type as an $\mathbb{R}^n$.

Remark See Exercise 6.5.14 and the following exercises for the behavior of cohomology in the case of a finite covering.

Exercises

Exercise 2.4.9 Determine the cohomology ring of the 2-dimensional torus $M :=$ $\mathbb{R}^2/\mathbb{Z}^2$ as $H^1(M) \cong \mathbb{R}^2$, $H^2(M) \cong \mathbb{R}$. *Hint: Proceed similarly to S^1.*

Exercise 2.4.10 Let M be a manifold and $X \in \Gamma(M, TM)$.

1) Show that L_X induces the zero map on cohomology.
2) Let Φ denote the global flow associated to X. Assume that there exists a closed 2-form ω such that for all $t \in \mathbb{R}$ we have $\Phi_t^*\omega = \omega$. Show that locally on M there exists a real-valued function f satisfying $\iota_X\omega = df$.

Remark: Part (2) arises in Hamiltonian mechanics with symplectic form $\omega = \sum dp_j \wedge dq_j$, Hamiltonian flow Φ_t and Hamiltonian function $f = H$.

Exercise* 2.4.11 Prove that $H_c^0(M) = 0$ if and only if M is not compact, and that $H_c^0(M) = \mathbb{R}$ otherwise.

Exercise 2.4.12 Let M be a manifold, $A \subset M$ a submanifold, let $\iota_A : A \hookrightarrow M$ be the embedding and consider a map $r : M \to A$ with $r_{|A} = \mathrm{id}_A$ (a **retraction**).

1) Show for the pullback maps on the cohomology that r^* is injective and ι_A^* is surjective.
2) In addition, assume that $\iota_A \circ r$ is homotopic to id_M. Deduce that then r^*, ι_A^* are bijective.

Exercise 2.4.13 Let M, N be manifolds and let π_M, π_N denote the canonical projections from $M \times N$ onto M, N.

1) Show that the map

$$k : H^p(M) \otimes H^q(N) \to H^{p+q}(M \times N),$$
$$\alpha \otimes \beta \mapsto \pi_M^*\alpha \wedge \pi_N^*\beta$$

is well-defined.
2) Show that k is not an isomorphism if $M = N = \mathbb{Z}$.

Exercise 2.4.14 On $S^1 \subset \mathbb{C}$, set $f : S^1 \to S^1$, $z \mapsto z^n$ for $n \in \mathbb{Z}$. Determine the induced map f^* on $H^\bullet(S^1)$.

Exercise 2.4.15 Let $M := \mathbb{R}^2 \setminus \{0\}$.

1) Compute the ring $H^\bullet(M)$.
2) Show that $\omega \in \mathfrak{A}^1(M)$, $\omega_{|\binom{x}{y}} := \frac{-y\,dx + x\,dy}{x^2 + y^2}$ represents a nontrivial element of $H^1(M)$.

2.5 Integration

Shrinking Lemma	oriented manifold
partition of unity	integral
volume form	transformation formula
orientable	normal vector field
orientation	antipodal map

While most previous considerations were essentially local and mainly used that manifolds are locally diffeomorphic to an $\mathbb{R}^n$, integration combines local objects into a global object. Here, for the first time, the second countability condition on manifolds becomes unavoidable.

Proposition 2.5.1 (Shrinking Lemma) *Let* $(\widetilde{U}_k)_{k \in K}$ *be a locally finite cover of a manifold M. Then there is a cover* $(W_k)_{k \in K}$ *satisfying* $\overline{W}_k \subset \widetilde{U}_k$.

The covering $(W_k)_{k \in K}$ is thus also locally finite.

Proof. Let $(V_m)_{m \in \mathbb{N}}$ be a base of the topology. For each $p \in M$ choose a neighborhood $V_{m(p)}$ and indices $k_{m(p)} \in K$ with $\overline{V}_{m(p)} \subset \widetilde{U}_{k_{m(p)}}$. Choose a locally finite refinement $(Z_\ell)_\ell$ of the cover $(V_{m(p)})_{p \in M}$. Let $W_k := \bigcup_{\overline{Z}_\ell \subset \widetilde{U}_k} Z_\ell$. This is again a cover. Then according to Proposition 1.2.4 we get

$$\overline{W}_k = \overline{\bigcup_{\overline{Z}_\ell \subset \widetilde{U}_k} Z_\ell} = \bigcup_{\overline{Z}_\ell \subset \widetilde{U}_k} \overline{Z}_\ell \subset \widetilde{U}_k. \qquad \Box$$

Theorem 2.5.2 (Partition of unity) *Consider an open cover* $(U_j)_{j \in J}$ *of a* C^∞ *manifold M. Then there exists a **partition of unity subordinate to*** $(U_j)_j$, *i.e. a family of* C^∞ *maps* $(\tau_k : M \to \mathbb{R}_0^+)_{k \in K}$ *such that*

1) *there is a map* $j : K \to J$ *such that* $\operatorname{supp} \tau_k \subset U_{j(k)}$,
2) $(\operatorname{supp} \tau_k)_k$ *is locally finite,*
3) $\sum_k \tau_k \equiv 1$.

Example Fig. 2.4 shows the covering $(U_j)_j = (]-\infty, 3[,]0, 4[,]1, 5[,]1, 7[,]3, \infty[)$ of $\mathbb{R}$.

Proof. Choose without loss of generality an atlas $(\varphi_\ell : U'_\ell \to V_\ell)_\ell$ such that for all ℓ the closure $\overline{V}_\ell$ is compact and there exists an index j such that $U'_\ell \subset U_j$. This can be achieved by intersecting an atlas with the U_j. Choose a locally finite refinement $(\widetilde{U}_k)_{k \in K}$ of $(U'_\ell)_\ell$ and $\overline{W}_k \subset \widetilde{U}_k$ as in Proposition 2.5.1.

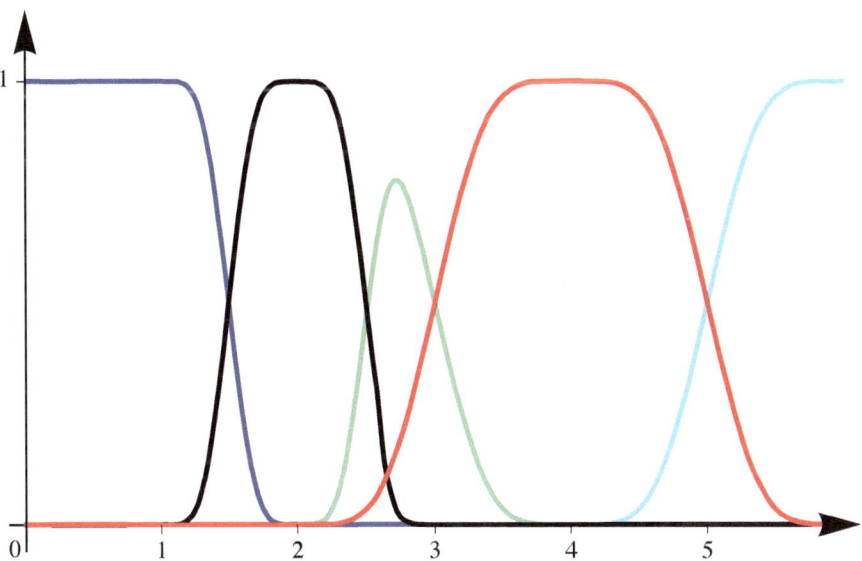

Fig. 2.4 A partition of unity on $\mathbb{R}$.

In particular, $\widetilde{\varphi}_k(W_k) \subset \mathbb{R}^n$ is bounded for each $k \in K$. Choose test functions $\lambda_k \in C_c^\infty(\mathbb{R}^n, \mathbb{R}_0^+)$ with $\lambda_{k|\widetilde{\varphi}_k(W_k)} > 0, \lambda_{k|\mathbb{R}^n \setminus \widetilde{\varphi}_k(\widetilde{U}_k)} \equiv 0$. Define $\mu_k : M \to \mathbb{R}_0^+$ as

$$\mu_k := \begin{cases} \lambda_k \circ \widetilde{\varphi}_k & \text{on} \quad \widetilde{U}_k \\ 0 & M \setminus \widetilde{U}_k \end{cases} .$$

Because $\bigcup_k \operatorname{supp} \mu_k \supset \bigcup \overline{W}_k = M$ and by the local finiteness of $(\widetilde{U}_k)_k$, $\tau_k := \frac{\mu_k}{\sum_m \mu_m}$ is well-defined. Then we find $\operatorname{supp} \tau_k \subset \widetilde{U}_k \subset U_{j(k)}$, $\tau_k \geq 0$ and $\sum \tau_k \equiv 1$.   □

Definition 2.5.3 *A **volume form** on an n-dimensional manifold M is an n-form ω with $\omega_p \neq 0$ for all $p \in M$. For $n > 0$, M is **orientable** if there exists an atlas whose transition maps all have positive Jacobi determinants (cf. Exercise 1.3.10). For $n = 0$ every manifold M is defined to be orientable.*

If $n = 0$, then each chart maps individual points to $\mathbb{R}^0$, and the chart is uniquely determined on each point. Thus, the transition maps are equal to the identity.

Theorem 2.5.4 *Given an n-dimensional manifold, the following are equivalent:*

1) There exists a volume form,
2) $\Lambda^n T^* M$ is a trivial bundle,
3) M is orientable.

Proof. $(1) \Rightarrow (2)$ The map $M \times \mathbb{R} \to \Lambda^n T^* M, (p, r) \mapsto (p, r \cdot \omega_p)$ associated to the volume form ω provides the trivialization.

(2)⇒(3) In the case $n = 0$ there is nothing to show. Assume now that $n > 0$. Without loss of generality assume that $\Lambda^n T^* M = M \times \mathbb{R}$ and let $(\varphi_j : U_j \to V_j)_j$ be an atlas of M with connected U_j. On each U_j one finds $v := \varphi_j^*(dx_1 \wedge \cdots \wedge dx_n) \neq 0$ everywhere, as φ_j^* is invertible with $(\varphi_j^*)^{-1} = (\varphi_j^{-1})^*$. So $v_{|p} \in \Lambda^n T_p^* M = \mathbb{R}$ is either positive everywhere on U_j or negative everywhere. Set

$$\psi_j := \begin{cases} \varphi_j & \text{if } v > 0 \\ \left(\begin{array}{c|c} \begin{array}{c|c} -1 & 0 \\ \hline 1 & 0 \end{array} & \\ \hline 0 & \begin{array}{cc} \ddots & \\ & 1 \end{array} \end{array} \right) \cdot \varphi_j & \text{if } v < 0. \end{cases}$$

Therefore there exists a function $f \in C^\infty(V_j \cap V_k, \mathbb{R}^+)$ such that $(\psi_j \circ \psi_k^{-1})^* dx_1 \wedge \cdots \wedge dx_n = f \cdot dx_1 \wedge \cdots \wedge dx_n$. On the other hand, according to Lemma 2.3.4 $f = \det T(\psi_j \circ \psi_k^{-1})$.

(3)⇒(1) For $n = 0$ every function $\omega : M \to \mathbb{R}^\times$ is a volume form. Now consider $n > 0$, let $(\varphi_j : U_j \to V_j)_j$ be an oriented atlas and let $(\tau_k)_k$ be a subordinate partition of unity. Set

$$\omega := \sum_k \tau_k \cdot \varphi_{j(k)}^* (dx_1 \wedge \cdots \wedge dx_n).$$

Then for all j,

$$(\varphi_j^{-1})^* \omega = \sum_k \underbrace{\tau_k \circ \varphi_j^{-1}}_{\geq 0} \cdot \overbrace{\underbrace{\det T(\varphi_{j(k)} \circ \varphi_j^{-1})}_{>0}}^{>0} dx_1 \wedge \cdots \wedge dx_n. \qquad \Box$$

Definition 2.5.5 *If* dim $M > 0$, *an* **orientation** *on a manifold M is the choice of an equivalence class of atlases with positive Jacobi determinant of all transition maps. If* dim $M = 0$, *let an orientation be a map $M \to \{\pm 1\}$. Together with such a choice, M is called an* **oriented manifold**.

Corollary 2.5.6 *A manifold M is connected and orientable if and only if M has exactly two distinct possible orientations.*

Proof. Fix a trivialization $\sigma : \Lambda^n T^* M \to M \times \mathbb{R}$. According to Theorem 2.5.4, the choice of an orientation corresponds to the choice of the sign of a section

$$\omega \in \Gamma(M, \sigma(\Lambda^n T^* M \setminus \{0\})) = \Gamma(M, M \times \mathbb{R} \setminus \{0\}) = C^\infty(M, \mathbb{R} \setminus \{0\}). \qquad \Box$$

Thus two volume forms ω_0, ω_1 induce the same orientation if and only if there is an $f \in C^\infty(M, \mathbb{R}^+)$ such that $\omega_0 = f \cdot \omega_1$. For $V \overset{\text{open}}{\subset} \mathbb{R}^n$ and $f \in C(V, \mathbb{R})$ define

$$\int_V f\, dx_1 \wedge \cdots \wedge dx_n := \int_V f\, d\lambda.$$

where $d\lambda$ is the Lebesgue measure.

Definition 2.5.7 *Let* M *be an oriented manifold. Given an oriented atlas* $(\varphi_j : U_j \to V_j)_j$ *of* M *and a subordinate partition of unity* $(\tau_k)_{k \in K}$, *define the* **integral of** $\omega \in \mathfrak{A}^q(M)$ **over** M *as*

$$\int_M \omega := \sum_{k \in K} \int_{V_{j(k)}} (\varphi_{j(k)}^{-1})^*(\tau_k \cdot \omega)$$

(i.e. 0 for $q < \dim M$).

In the zero-dimensional case, the integral is the sum over the values of ω, pointwise multiplied by the orientation. We restrict ourselves to integrals of forms with compact support in order to avoid discussing convergence problems.

Theorem 2.5.8 *The integral* $\int_M : \mathfrak{A}_c^\bullet(M) \to \mathbb{R}$ *depends on the orientation, but not on the choice of an atlas or of the partition of unity.*

In the proof we apply the **transformation formula** on $\mathbb{R}^n$ featured in lecture courses on Analysis III: Given $V \overset{\text{open}}{\subset} \mathbb{R}^n$, $f \in C_c(\mathbb{R}^n, \mathbb{R})$ and a diffeomorphism φ, then

$$\int_{\varphi(V)} f \, d\lambda = \int_V (f \circ \varphi) \cdot |\det T\varphi| \, d\lambda.$$

On the other hand according to Lemma 2.3.4, if $\omega = f \, dx_1 \wedge \cdots \wedge dx_n \in \mathfrak{A}_c^n(\mathbb{R}^n)$ is an n-form then

$$\varphi^*\omega = (f \circ \varphi)\varphi^*(dx_1 \wedge \cdots \wedge dx_n) = (f \circ \varphi) \cdot \det T\varphi \cdot dx_1 \wedge \cdots \wedge dx_n,$$

and thus $\int_{\varphi(V)} \omega = \text{sign} \det(T\varphi) \cdot \int_V \varphi^*\omega$. The possibility of a change of sign is the reason for the assumption of orientability.

Proof. If $(\psi_m : \tilde{U}_m \to \tilde{V}_m)_m$ is a second oriented atlas and if $(\sigma_\ell)_\ell$ is a matching partition of unity, then because $\det(T(\varphi_{j(k)} \circ \psi_{m(\ell)}^{-1})) > 0$, we obtain

$$\sum_k \int_{V_{j(k)}} (\varphi_{j(k)}^{-1})^*(\tau_k\omega) = \sum_{k,\ell} \int_{\varphi_{j(m)}(U_{j(k)} \cap \tilde{U}_{m(\ell)})} (\varphi_{j(k)}^{-1})^*(\tau_k\sigma_\ell\omega)$$

$$= \sum_{k,\ell} \int_{\psi_{m(\ell)}(U_{j(k)} \cap \tilde{U}_{m(\ell)})} (\varphi_{j(k)} \circ \psi_{m(\ell)}^{-1})^*(\varphi_{j(k)}^{-1})^*(\tau_k\sigma_\ell\omega)$$

$$= \sum_\ell \int_{\tilde{V}_{m(\ell)}} (\psi_{m(\ell)}^{-1})^*(\sigma_\ell\omega). \qquad \square$$

Remark 2.5.9 Every choice of a volume form ω induces a (signed) measure on M via

$$C_c(M) \to \mathbb{R}, f \mapsto \int_M f \cdot \omega.$$

Remark In an explicit calculation one will usually not calculate integrals by directly applying the definition, but first look for a subset $A \subset M$ of measure 0, such that $M \setminus A$ is disjoint union of the domains of definition of charts $\varphi_j : U_j \to V_j$. Using the partition of unity $\tau_j := \begin{cases} 1 & \text{on } U_j \\ 0 & \text{else} \end{cases}$ one gets $\int_M \omega = \sum_j \int_{V_j} (\varphi_j^{-1})^* \omega$.

Corollary 2.5.10 (Transformation formula) *Suppose $f : M \to N$ is an orientation preserving diffeomorphism, $U \subset M$ and $\omega \in \mathfrak{A}(N)$. Then $\int_{f(U)} \omega = \int_U f^* \omega$.*

Proof. Any atlas $(\varphi_j)_j$ of M induces an atlas $(\varphi_j \circ f^{-1})_j$ of N. $\qquad\square$

Theorem 2.5.11 (Special case of the Stokes–Cartan Theorem) *If M is orientable and $\omega \in \mathfrak{A}_c^\bullet(M)$, then $\int_M d\omega = 0$.*

Proof. If $(\varphi_j : U_j \to V_j)_j$ is an oriented atlas, applying a subordinate partition of unity $(\tau_k)_k$ one gets

$$
\begin{aligned}
\int_M d\omega &= \int_M d\left(\sum_k \tau_k \omega\right) = \sum_k \int_{V_{j(k)}} (\varphi_{j(k)}^{-1})^* \underbrace{d(\tau_k \omega)}_{\text{supp}\subset\subset U_{j(k)}} \\
&= \sum_k \int_{V_{j(k)}} d \underbrace{\left[(\varphi_{j(k)}^{-1})^* \tau_k \omega\right]}_{=:\sum_\ell f_{k,\ell}\, dx_1 \wedge \ldots \widehat{dx_\ell} \cdots \wedge dx_n,\, \text{supp}\subset\subset V_{j(k)}} \\
&= \int_{\mathbb{R}^n} \sum_{k,\ell} \frac{\partial f_{k,\ell}}{\partial x_\ell} dx_\ell \wedge dx_1 \wedge \ldots \widehat{dx_\ell} \cdots \wedge dx_n \\
&\overset{\text{Fubini}}{=} \int_{\mathbb{R}^{n-1}} \underbrace{\left(\int_\mathbb{R} \sum_{k,\ell} \frac{\partial f_{k,\ell}}{\partial x_\ell} dx_\ell\right)}_{=0} dx_1 \wedge \ldots \widehat{dx_\ell} \cdots \wedge dx_n. \qquad\square
\end{aligned}
$$

Corollary 2.5.12 *Given a compact oriented submanifold $N \subset\subset M$, $\int_N$ induces a map $\int_N : H^\bullet(M) \to \mathbb{R}$.*

Corollary 2.5.13 *If M^n is compact and oriented, then $H^n(M) \neq 0$.*

Proof. If ω is a volume form on M, then $[\omega] \in H^\bullet(M) \setminus \{0\}$ since $\int_M \omega \neq 0$. $\qquad\square$

Remark In particular, M with $n > 0$ is not contractible, since $H^n(M) \neq H^n(\text{point})$.

Corollary 2.5.14 *Suppose N is a compact, oriented manifold, $f, g : N \hookrightarrow M$ are two homotopic embeddings and $\omega \in \mathfrak{A}^\bullet(M)$. Then $\int_{f(N)} \omega = \int_{g(N)} \omega$.*

The homotopy does not have to consist of embeddings.

Proof. According to Corollary 2.4.7, $f^* \omega = g^* \omega$ holds in $H^\bullet(N)$ and so

$$
\int_{f(N)} \omega = \int_N f^* \omega \overset{\text{Cor. 2.5.12}}{=} \int_N g^* \omega = \int_{g(N)} \omega. \qquad\square
$$

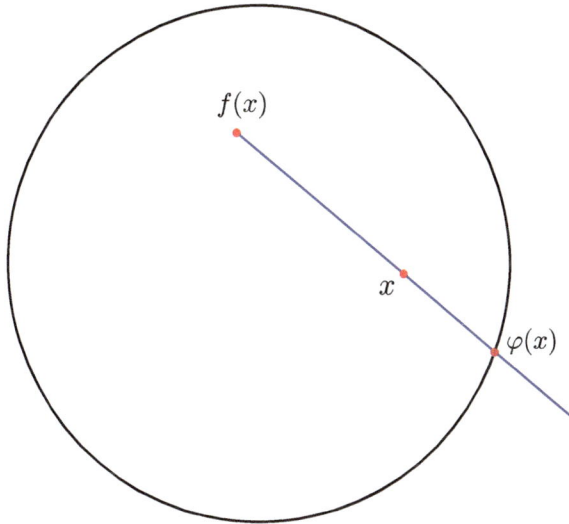

Fig. 2.5 The map $\varphi : \overline{B^n} \to S^{n-1}$.

Remark There is a notion of "manifold with boundary", for which quite similarly $\int_M d\omega = \int_{\partial M} \omega$ follows. In this case one obtains the following result:

Corollary 2.5.15 *Let M be a compact oriented (sub)manifold (of $\mathbb{R}^n$) with boundary $\partial M \neq 0$. Then there is no C^∞ map $\varphi : M \to \partial M$ such that $\varphi_{|\partial M} = $ id.*

Proof. Let ω be a volume form on ∂M. Then

$$0 = \int_M \varphi^* \underbrace{d\omega}_{=0} = \int_M d(\varphi^*\omega) \overset{\underset{\mathrm{Stokes-}}{\mathrm{Cartan}}}{=} \int_{\partial M} \varphi^*\omega = \int_{\partial M} \omega > 0. \, \text{\textlightning} \qquad \qquad \Box$$

Corollary 2.5.16 *Every C^∞ map $f : \overline{B^n} \to \overline{B^n}$ has at least one fixed point.*

Proof. Suppose f has no fixed point. Let $\varphi : \overline{B^n} \to \partial \overline{B^n} = S^{n-1}$ be the map that assigns to each $x \in \overline{B^n}$ the intersection of the ray $f(x) + \mathbb{R}^+ \cdot (x - f(x))$ with S^{n-1} (Fig. 2.5). Then $\varphi_{|S^{n-1}} = $ id. $\text{\textlightning}$ to Corollary 2.5.15. $\Box$

At the end of this section we can see for ourselves that the projection $\pi : TM \to M$, which locally looks like a projection $U \times \mathbb{R}^n \to U$ on every chart, does not always have this form globally.

Theorem 2.5.17 *Every vector field on S^{2m} has at least one zero ("hedgehogs can't be combed"). In particular, $\pi : TS^{2m} \to S^{2m}$ is not equal to $\tilde{\pi} : S^{2m} \times \mathbb{R}^{2m} \to S^{2m}$.*

Proof. Let X be a zero-free vector field and set $h : [0, \pi] \times S^{2m} \to S^{2m}$, $(t, p) \mapsto p \cdot \cos t + \frac{X_p}{\|X_p\|} \sin t$. The volume form $\omega := \iota_p dx_1 \wedge \cdots \wedge dx_{2m+1}$ on S^{2m} then satisfies

$$0 \neq \int_{S^{2m}} \omega = \int_{S^{2m}} h_0^* \omega = -\int_{S^{2m}} h_\pi^* \omega,$$

contradicting Corollary 2.4.7 or, alternatively,

$$0 = \int_{[0,\pi] \times S^{2m}} h^* \underbrace{d\omega}_{=0} = \int_{[0,\pi] \times S^{2m}} d(h^* \omega) \overset{\text{Stokes-}}{\underset{\text{Cartan}}{=}} \int_{S^{2m}} (h_\pi^* \omega - h_0^* \omega). \qquad \square$$

Exercises

Exercise* 2.5.18 Let $f : M \to N$ be a smooth map and let $E \to N$ be a vector bundle. Using a partition of unity, show that globally every section in f^*E can be written as a (locally finite) sum $\sum_{\ell \in \mathbb{N}} g_\ell \cdot f^* s_\ell$ with $g_\ell \in C^\infty(M, \mathbb{R})$, $s_\ell \in \Gamma(N, E)$. If N is compact, finitely many ℓ will suffice.

Exercise 2.5.19 Let $M^m \subset \mathbb{R}^{m+1}$ be an m-dimensional submanifold. A **normal vector field** $\mathfrak{n}$ on M is a section in N satisfying $\|\mathfrak{n}\|^2_{\mathbb{R}^{m+1}} \equiv 1$. Via Exercise 2.3.16, conclude that M is orientable if and only if M has a normal vector field. Show that in this case there are exactly two normal vector fields corresponding to the choice of an orientation.

Exercise 2.5.20 1) Prove that the torus $T^n = \mathbb{R}^n / \mathbb{Z}^n$ is orientable.
2) Show that $P^n\mathbb{R}$ is orientable if and only if n is odd. To do this, use Exercise 1.2.19 and the **antipodal map** $a : S^n \to S^n$, $u \mapsto -u$.

Exercise 2.5.21 Suppose M is an oriented connected manifold and ω is a volume form on M. If $f \in C_c^\infty(M, \mathbb{R}_0^+)$ is a function, then $f \equiv 0$ whenever $\int_M f \cdot \omega = 0$.

Exercise 2.5.22 Let $\omega \in \mathfrak{A}^2(\mathbb{R}^3 \setminus \{0\})$ be the form

$$\omega|_{(x,y,z)} := \frac{x \, dy \wedge dz + y \, dz \wedge dx + z \, dx \wedge dy}{\|(x, y, z)\|^3}.$$

Show that $d\omega = 0$, calculate $\int_{S^2} \omega$ and deduce that $[\omega] \neq 0$ in $H^2(S^2)$.

Chapter 3
Riemannian Manifolds

The previous sections dealt with some basic notions for C^∞ manifolds and thus belonged to differential topology. In this chapter another structure is added to manifolds, the Riemannian metric. This one structure will allow a remarkable number of geometric definitions: Angles and lengths of vectors, a canonical volume form, lengths of curves on manifolds, the distance between two points, curvatures, and n-fold directional derivatives of functions and tensors in general. On the other hand, every submanifold of the Euclidean $\mathbb{R}^n$ canonically carries such a Riemannian metric.

Bernhard Riemann developed this notion and essentially the contents of this chapter, including the notion of curvature, for his habilitation lecture in Göttingen on June 10, 1854 ([Rie]), building on ideas that Gauß published for surfaces in 1828 ([Gauß]). He also introduced the concept of a manifold.

In the middle sections, a derivative of vector fields and of more general sections in the direction of a single vector is defined. This leads to multiple derivatives by iteration, and a double derivative leads to the notion of curvature. Nicely, curvature turns out to not simply be a second-order differential operator, but a pointwise defined tensor. In the last section numerous symmetries are deduced for the curvature to the canonical connection of the tangent space.

3.1 Riemannian Metrics

metric

polarization identity

Riemannian metric

Riemannian manifold

Euclidean space

hyperbolic space

Lorentz metric

Minkowski form

musical isomorphisms

gradient

isometry

Riemannian submanifold

surface of revolution

canonical volume form

© The Editor(s) (if applicable) and The Author(s), under exclusive license to Springer-Verlag GmbH, DE, part of Springer Nature 2024
K. Köhler, *Differential Geometry and Homogeneous Spaces*, Universitext,
https://doi.org/10.1007/978-3-662-69721-4_3

volume	curve parameterized by arc length
(arc) length	pseudosphere
distance	ruled surface
helicoid	directrix
helical surface	striction line
upper half-plane	distribution parameter
biinvariant metric	Hodge $*$(star) operator
Pappus's Centroid Theorem	

Riemannian metrics are scalar products on the fibers of the tangent bundle. After the initial properties have been explained in this section, they are used to define volumes, in particular the lengths of curves. This is used to construct a distance metric on M.

Definition 3.1.1 *Let $E \to M$ be an $\mathbb{R}$-vector bundle. A **metric** on E is a section $h \in \Gamma(M, E^* \otimes E^*)$ which is pointwise a Euclidean scalar product, i.e. for all $p \in M$*

1) h_p is symmetric, i.e. for all $v, w \in E_p$ we have $h_p(v, w) = h_p(w, v)$,
2) h_p is positive, i.e. for all $v \in E_p \setminus \{0\}$ we have $h_p(v, v) > 0$.

Analogously one defines Hermitian metrics on $\mathbb{C}$-vector bundles.

Such a scalar product is uniquely determined by its norm because of the **polarization identity**

$$h(v, w) = \frac{1}{4}(\|v + w\|^2 - \|v - w\|^2).$$

This holds analogously for any other symmetric bilinear form.

Definition 3.1.2 *A **Riemannian metric** g on M is a metric on TM. The pair (M, g) is called a **Riemannian manifold**.*

So, for a chart $\varphi : U \to V \subset \mathbb{R}^n$ and $g \in \Gamma(M, T_2^0 M)$ one finds

$$(\varphi^{-1})^* g = \sum_{j,k=1}^{n} g_{jk} dx_j \otimes dx_k,$$

where $(g_{jk})_{j,k=1}^{n} : V \to \mathbb{R}^{n \times n}$ is a C^∞ map with values in the positive definite matrices. In particular, a Riemannian metric is quite different from the metrics one encounters in basic calculus lecture courses. To distinguish them from Riemannian metrics, we will call metrics in the latter sense distances.

Example 3.1.3 1) The **Euclidian space** is the manifold $\mathbb{R}^n$ equipped with the constant canonical metric $\langle \cdot, \cdot \rangle_{\mathbb{R}^n} = \langle \cdot, \cdot \rangle_{\text{can}} = \sum_{j=1}^{n} dx_j \otimes dx_j$ on $T_p \mathbb{R}^n \overset{\text{can}}{\cong} \mathbb{R}^n$, for which the Cartesian basis is an orthonormal basis.

2) The n-dimensional **hyperbolic space** is the Riemannian manifold $M := \{u \in \mathbb{R}^n \mid \|u\|_{\text{can}} < 1\}$ with $g_u := \frac{4}{(1 - \|u\|_{\text{can}}^2)^2} \langle \cdot, \cdot \rangle_{\text{can}}$.

Theorem 3.1.4 *On each vector bundle $\pi : E \to M$ there exists at least one metric.*

In particular, every manifold carries a Riemannian metric.

Proof. Suppose $(h_j : \pi^{-1}(U_j) \to U_j \times \mathbb{R}^m)_j$, $U_j \subset M$ is a family of trivializations of E. Choose a subordinate partition of unity $(\tau_k)_k$. Using the projection $\pi_2 : U_j \times \mathbb{R}^m \overset{\text{can.}}{\to} \mathbb{R}^m$ we construct

$$h(v, w) := \sum_k \tau_k \cdot \langle \pi_2 h_{j(k)}(v), \pi_2 h_{j(k)}(w) \rangle_{\text{can}}.$$

Then h is symmetric and $h(v, v) = \sum_k \tau_k \|\pi_2 h_{j(k)}(v)\|_{\text{can}}^2 > 0$ holds for all $v \in E_p \setminus \{0\}$. $\qquad\square$

Remark 3.1.5 More generally, one can consider non-degenerate bilinear forms g of arbitrary signature on a manifold. A **Lorentzian metric** g_L is defined in the same way as a Riemannian metric, except that we use **Minkowski forms** of signature $(1, -1, -1, \ldots, -1)$ in place of the scalar products. Many of the algebraically deduced statements in the following chapters apply to Lorentzian metrics, but Theorem 3.1.4 is false for Lorentzian metrics. For example, according to Theorem 2.5.17 there is no Lorentzian metric g_L on $M = S^{2n}$. This is because for a Riemannian metric g on S^{2n} and the endomorphism $g^{-1}g_L := A \in \text{End}(TM)$ with $g(\cdot, A\cdot) = g_L$, the eigenvectors corresponding to the single positive eigenvalue of $g^{-1}g_L$ form a line subbundle $\mathcal{L}$ of TM. By Exercise 3.1.19, there is a covering $\pi : N \to S^{2n}$ for which $\pi^*\mathcal{L}$ is trivial. But in Theorem 6.5.8, $N \cong S^{2n} \dot\cup S^{2n}$ is shown. Therefore $\mathcal{L}$ has a nowhere vanishing section X. Thus X is a zero-free vector field. See Corollary 4.2.16 for a more general criterion.

Corollary 3.1.6 *Suppose E is an $\mathbb{R}$-vector bundle, then $E \cong E^*$ holds (though in general not in a canonical way).*

Proof. Choose a metric h. Then the map $s \mapsto h(s, \cdot)$ is a vector bundle isomorphism. $\qquad\square$

In the case $E = TM$ these are the **musical isomorphisms** $TM \overset{\cong}{\to} T^*M$, $X \mapsto X^\flat := g(X, \cdot)$, $\alpha^\sharp \leftarrow\!\shortmid \alpha$. The **gradient** of $f \in C^\infty(M, \mathbb{R})$ is grad $f := (df)^\sharp$.

Definition 3.1.7 *An **isometry** of Riemannian manifolds (M, g), $(N, \widetilde{g})$ is a diffeomorphism $f : M \to N$ such that $f^*\widetilde{g} = g$, i.e. g satisfies $g_p(X, Y) = \widetilde{g}_{f(p)}(T_p f(X), T_p f(Y))$.*

Example Considering the transition map $f : \mathbb{R}^+ \times]0, 2\pi[\to \mathbb{R}^2 \setminus \mathbb{R}_0^+$, $\binom{r}{\vartheta} \mapsto \binom{r\cos\vartheta}{r\sin\vartheta}$ one gets $f^*(dx \otimes dx + dy \otimes dy) = dr \otimes dr + r^2 d\vartheta \otimes d\vartheta$.

Lemma 3.1.8 *If $(N, \widetilde{g})$ is a Riemannian manifold and $\iota : M \to N$ is an immersion, then $g := \iota^*\widetilde{g}$ is a Riemannian metric on M. If ι is an immersion, (M, g) is called a **Riemannian submanifold** of $(N, \widetilde{g})$.*

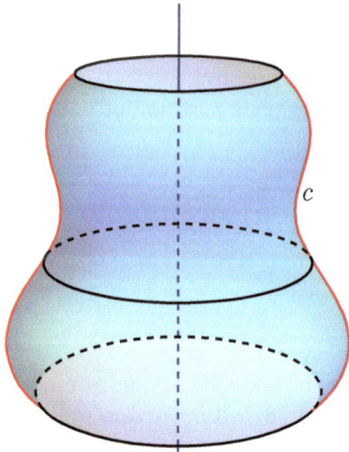

Fig. 3.1 A surface of revolution.

Proof. The form g is symmetric, and for $X \in T_p M \setminus \{0\}$ it follows that

$$g_p(X, X) = \widetilde{g}_{\iota(p)}(\underbrace{T\iota(X), T\iota(X)}_{\neq 0}) > 0$$

holds because of the injectivity of $T\iota$. $\square$

The metric g depends of course on the choice of ι. In particular, embedding a submanifold $\iota : M \hookrightarrow (\mathbb{R}^{n+k}, \langle \cdot, \cdot \rangle)$ canonically yields a Riemannian metric g on M by restricting the standard scalar product to $TM \cong \operatorname{im} T\iota$.

Example 3.1.9 The Riemannian metric induced by $\mathbb{R}^3$ on S^2, pulled back by the stereographic projections $\varphi_\pm$, equals $\left((\varphi_\pm^{-1})^* g\right)_{|u} = \frac{4}{(1+\|u\|^2)^2} \langle \cdot, \cdot \rangle_{\text{Eucl}}$ by Exercise 2.2.14.

Example 3.1.10 Let M be a **surface of revolution** in $\mathbb{R}^3$, i.e. M is the surface that results when a curve $c : I =]a, b[\to \mathbb{R}^+ \times \mathbb{R}^2, u \mapsto \begin{pmatrix} r(u) \\ 0 \\ z(u) \end{pmatrix}$ is rotated around the z-axis (Fig. 3.1). Thus with $J_0 :=] - \pi, \pi[$, $J_1 :=]0, 2\pi[$ one obtains the parametrizations

$$\iota_k : \underbrace{I \times J_k}_{=:V_k} \to \mathbb{R}^3, \quad (u, \vartheta) \mapsto \begin{pmatrix} r(u) \cos \vartheta \\ r(u) \sin \vartheta \\ z(u) \end{pmatrix} \qquad \text{for } k = 0, 1$$

of open subsets U_k of M. Thus $T\iota(TM)$ is spanned by

$$\frac{\partial}{\partial u}\iota = \begin{pmatrix} r'(u) \cos \vartheta \\ r'(u) \sin \vartheta \\ z'(u) \end{pmatrix} \qquad \text{and} \qquad \frac{\partial}{\partial \vartheta}\iota = \begin{pmatrix} -r(u) \sin \vartheta \\ r(u) \cos \vartheta \\ 0 \end{pmatrix}.$$

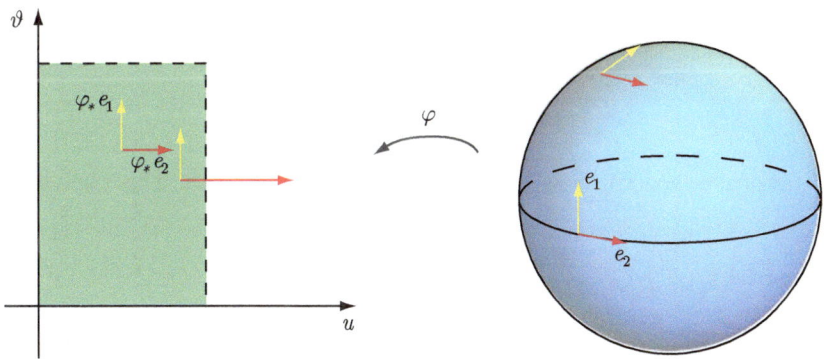

Fig. 3.2 Orthonormal basis.

Hence the induced metric g on M is uniquely determined by

$$g(\frac{\partial}{\partial u}, \frac{\partial}{\partial u}) = \langle \frac{\partial \iota}{\partial u}, \frac{\partial \iota}{\partial u} \rangle_{\mathbb{R}^3} = r'^2 + z'^2,$$

$$g(\frac{\partial}{\partial u}, \frac{\partial}{\partial \vartheta}) = \langle \frac{\partial \iota}{\partial u}, \frac{\partial \iota}{\partial \vartheta} \rangle_{\mathbb{R}^3} = 0, \qquad g(\frac{\partial}{\partial \vartheta}, \frac{\partial}{\partial \vartheta}) = \langle \frac{\partial \iota}{\partial \vartheta}, \frac{\partial \iota}{\partial \vartheta} \rangle_{\mathbb{R}^3} = r^2.$$

Written in another way, $g = (r'^2 + z'^2)du \otimes du + r^2 d\vartheta \otimes d\vartheta$ and

$$(g_{jk})_{jk} = \begin{pmatrix} r'^2 + z'^2 & 0 \\ 0 & r^2 \end{pmatrix} = (T\iota)^t T\iota.$$

For example, if $M = S^2$ is the sphere, then one obtains $g = du \otimes du + \sin^2 u \cdot d\vartheta \otimes d\vartheta$. By reparametrizing c one can always achieve that $r'^2 + z'^2 \equiv 1$.

Remark The Nash Embedding Theorem[1] states that any n-dimensional Riemannian manifold can be isometrically embedded into the Euclidean $\mathbb{R}^{n(n+1)(3n+11)/2}$.

If $\varphi : U \to V$ is a chart, let $G := (g_{jk}) \in C^\infty(V, \mathbb{R}^{n \times n})$, i.e. $(\varphi^{-1})^* g = \langle \cdot, G \cdot \rangle_{\mathbb{R}^n}$. As a positive definite matrix G has a positive definite root $\sqrt{G}$, we have $(\varphi^{-1})^* g = \langle \sqrt{G} \cdot, \sqrt{G} \cdot \rangle_{\mathbb{R}^n}$. Thus for the orthonormal basis $(\frac{\partial}{\partial x_j})_j$ of $\langle \cdot, \cdot \rangle_{\mathbb{R}^n}$, $((\varphi^{-1})_* \sqrt{G}^{-1} \frac{\partial}{\partial x_j}) =: (e_j)$ is an orthonormal basis for g on TU (Fig. 3.2).

Definition 3.1.11 *Suppose (M, g) is an oriented Riemannian manifold. Given an oriented orthonormal basis $(e^1, \ldots, e^n)$ of $T_p^* M$, let $d\mathrm{vol}_{g|p} := e^1 \wedge \cdots \wedge e^n$ be the* **canonical (or Riemannian) volume form.**

Theorem 3.1.12 *The canonical volume form is independent of the choice of an oriented orthonormal basis. In particular it is globally defined on M. For an oriented chart $\varphi : U \to V$, $(\varphi^{-1})^* d\mathrm{vol}_g = \sqrt{\det G}\, dx_1 \wedge \cdots \wedge dx_n$.*

[1] 1956, John Forbes Nash, 1928–2015.

Proof. Given a second oriented orthonormal basis $f^1, \ldots, f^n$ of $T_p^* M$ with associated isometry $A : e^j \mapsto f^j$, by Lemma 2.3.4 one gets $f^1 \wedge \cdots \wedge f^n = \underbrace{\det A}_{=1} \cdot e^1 \wedge \cdots \wedge e^n = d\text{vol}_g$. Considering the orthonormal basis $\left(\varphi^* \sqrt{G}^{-1} \frac{\partial}{\partial x_j}\right)_j$ of TU, the dual orthonormal basis of $T^* U$ equals $\left(\varphi^* (dx_j \circ \sqrt{G})\right)_j$. Thus one obtains

$$d\text{vol}_{g|U} = \varphi^* (\det \sqrt{G} \, dx_1 \wedge \cdots \wedge dx_n).$$

$\square$

Lemma 3.1.13 *If* $f : N \to M$ *is an orientation-preserving diffeomorphism, then the* **volume** $\text{vol}(M, g) := \int_M d\text{vol}$ *equals* $\text{vol}(N, f^* g)$.

Proof. If $(e^j)_j$ is an oriented orthonormal basis with respect to g, then $(f^* e^j)_j$ is an oriented orthonormal basis with respect to $f^* g$. Therefore $f^* d\text{vol}_g = d\text{vol}_{f^* g}$ and $\text{vol}(N, f^* g) = \int_N f^* d\text{vol}_g = \int_{f(N)} d\text{vol}_g = \text{vol}(M, g)$ hold. $\square$

Example 1) Suppose $A = (v_1, \ldots, v_n) \in \mathbb{R}^n$ is an oriented basis and Γ is the lattice generated by A over $\mathbb{Z}$ in $\mathbb{R}^n$ (Fig. 3.3). If $A := (v_1, \ldots, v_n) \in \mathbb{R}^{n \times n}$, then $\Gamma = A \cdot \mathbb{Z}^n$, and so for $\lambda \in \Gamma \setminus \{0\}$ one gets

$$1 \le \| \underbrace{A^{-1} \lambda}_{\in \mathbb{Z}^n \setminus \{0\}} \|_2 \le \|A^{-1}\| \cdot \|\lambda\|_2$$

with the operator norm associated to $\| \cdot \|_2$. Hence for all distinct $\lambda_1, \lambda_2 \in \Gamma$ one gets $\|\lambda_2 - \lambda_1\|_2 \ge 1/\|A^{-1}\|$ and Γ is discrete. Therefore $\mathbb{R}^n / \Gamma$ equipped with the quotient topology, the charts $\varphi_{B_r(x)} : B_r(x)/\Gamma \xrightarrow{\text{id}} B_r(x)$ for $r < 1/\|2A^{-1}\|$, $x \in \mathbb{R}^n$ and the standard metric on $\mathbb{R}^n$ is a Riemannian manifold.

The linear chart $\varphi := A^{-1}, \varphi(v_j) = \frac{\partial}{\partial x_j}$ shows that

$$\text{vol}(\mathbb{R}^n / \Gamma) = \int_{\mathbb{R}^n / \Gamma} dx_1 \wedge \cdots \wedge dx_n \overset{\substack{\text{transition} \\ \text{map}}}{=} \int_{\mathbb{R}^n / \mathbb{Z}^n} (\varphi^{-1})^* dx_1 \wedge \cdots \wedge dx_n$$

$$= \int_{\mathbb{R}^n / \mathbb{Z}^n} \det(A) \, dx_1 \wedge \cdots \wedge dx_n = \det A.$$

2) If $\iota : M \hookrightarrow \mathbb{R}^{n+k}$, $M = V \subset \mathbb{R}^n$ is a parametrized submanifold, then $g = \iota^* \langle \cdot, \cdot \rangle_{\mathbb{R}^{n+k}}$ and thus

$$d\text{vol}_g = \sqrt{\det(T\iota^t \cdot T\iota)} \, dx_1 \wedge \cdots \wedge dx_n.$$

Hence every manifold N immersed in a Riemannian manifold M has a (possibly infinite) volume by restricting the metric to N and integrating the canonical volume form there. In particular, one finds for the volume of paths on M:

Definition 3.1.14 *Suppose* $c :]a, b[\to M$ *is a smooth path on a Riemannian manifold* (M, g), *with* $c' \ne 0$ *everywhere. The* **(arc) length** *of* c *is the volume of* $]a, b[$ *with respect to the metric pulled back with* c,

$$L(c) := \text{vol}(]a, b[, c^* g) = \int_a^b \sqrt{(c^* g)_t \left(\frac{\partial}{\partial t}, \frac{\partial}{\partial t}\right)} \, dt = \int_a^b \sqrt{g_{c(t)}(c'(t), c'(t))} \, dt.$$

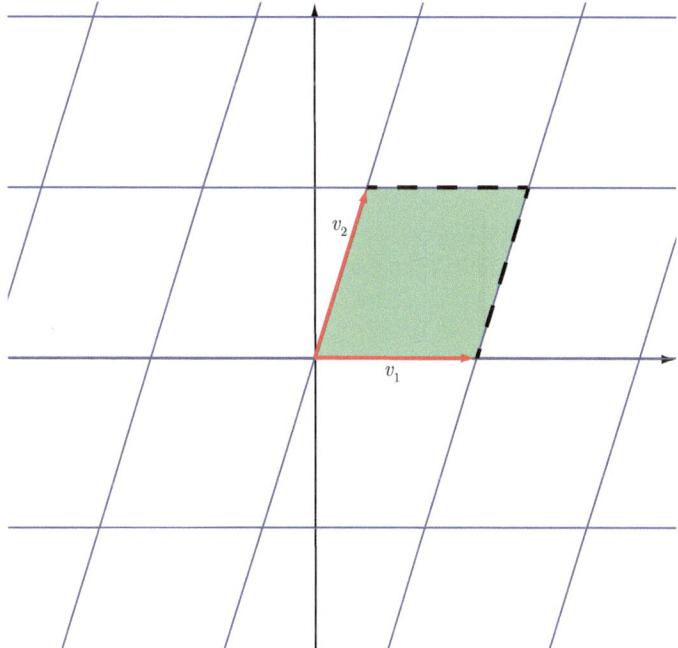

Fig. 3.3 Fundamental domain of a torus $\mathbb{R}^2/\Gamma$.

By Lemma 3.1.13 the length is independent of the parametrization of the curve. If c is an embedding, this is even more obvious for the volume of the submanifold $c(I) \subset M$.

Definition 3.1.15 *Let (M, g) be a (connected) Riemannian manifold. The **distance** $\mathrm{dist}(p, q)$ between points $p, q \in M$ is defined as the infimum of the lengths of all piecewise C^∞-paths from p to q.*

Remark This value is finite for all p, q. Indeed, for $p \in M$, let $N \subset M$ be the subset of all points which can be connected to p by a piecewise C^∞-path $c : [0, 1] \to M$. Then N is open, because any point $q \in N$ can be connected to all the points of a connected chart U around q, and $M \setminus N$ is open for the same reason. Hence $M = N$ because N is nonempty.

Proposition 3.1.16 *If $f : \mathbb{R} \to \mathbb{R}^n \setminus \{0\}$, then $\|f'\|_{\mathbb{R}^n} \geq \|f\|_{\mathbb{R}^n}'$ holds.*

Proof. Set $f_k := f(t_k)$. The difference quotient satisfies

$$\frac{\|f_1 - f_0\|^2}{(t_1 - t_0)^2} = \frac{\|f_1\|^2 + \|f_0\|^2 - 2\langle f_1, f_0\rangle}{(t_1 - t_0)^2}$$

$$\overset{\text{Cauchy--Schwarz}}{\geq} \frac{\|f_1\|^2 + \|f_0\|^2 - 2\|f_1\|\|f_0\|}{(t_1 - t_0)^2} = \frac{(\|f_1\| - \|f_0\|)^2}{(t_1 - t_0)^2}. \qquad \square$$

Theorem 3.1.17 *Suppose that M is connected. Then (M, dist) is a metric space, and the topology induced on M by* dist *is the original topology.*

Proof. 1) $\text{dist}(p, q) = \text{dist}(q, p)$, since each path from p to q yields a path of the same length from q to p.
2) $\text{dist}(p, q) \leq \text{dist}(p, r) + \text{dist}(r, q)$, since any path from p to r concatenated with one from r to q yields a piecewise C^∞ path from p to q.
3) $\text{dist}(p, q) = 0 \Leftrightarrow p = q$: Suppose $\varphi : U \to V$ is a chart around p with $\varphi(p) = 0$ and $\overline{B}_\varepsilon \subset V$ is a ball around 0. Then there exist numbers $\lambda_1, \lambda_2 \in \mathbb{R}^+$ such that for all $u \in T\overline{B}_\varepsilon$ we have

$$\lambda_1 \|u\|_{\mathbb{R}^n}^2 \leq (\varphi^{-1})^* g(u, u) \leq \lambda_2 \|u\|_{\mathbb{R}^n}^2,$$

as $((\varphi^{-1})^* g)_x(u, u)$ on $\overline{B}_\varepsilon \times S^{n-1} \ni (x, u)$ takes values in a compact interval $[\lambda_1, \lambda_2] \subset\subset \mathbb{R}^+$. Suppose $c : [0, 1] \to M$ is a path from p to q with $p \notin c(]0, 1[)$. Let q_0 be the first point where c meets $\varphi^{-1}(S_\varepsilon^{n-1})$, at a time t_0 (or $q_0 := q, t_0 := 1$ if $\varphi(c) \subset B_\varepsilon$). Then

$$L(c) \geq L(c_{|]0,t_0[}) = \int_0^{t_0} \sqrt{g_{c(t)}(c'(t), c'(t))} \, dt$$

$$= \int_0^{t_0} \sqrt{(\varphi^{-1*} g)_{|\varphi(c(t))}((\varphi \circ c)'(t), (\varphi \circ c)'(t))} \, dt$$

$$\geq \sqrt{\lambda_1} \int_0^{t_0} \|(\varphi \circ c)'(t)\| \, dt \overset{3.1.16}{\geq} \sqrt{\lambda_1} \int_0^{t_0} \|\varphi \circ c\|' \, dt$$

$$= \sqrt{\lambda_1} \|\varphi(q_0)\| > 0.$$

4) A subset $U \subset M$ is open with respect to the original topology if there is a chart φ around every point $p \in U$ and $r > 0$ with $\varphi^{-1}(B_r(\varphi(p))) \subset U$. With respect to the topology induced by dist, U is open if for every point $p \in U$ there is an $r > 0$ with $B_r^{\text{dist}}(p) \subset U$.

Thus, we have to show that for all $p \in M$ and a chart φ around p, every sufficiently small ball $\varphi^{-1}(B^{\text{Eucl}}(0))$ around p contains a ball B^{dist} with respect to dist and vice versa.

By (3), $B_{\varepsilon\sqrt{\lambda_1}}^{\text{dist}}(p) \subset \varphi^{-1}(B_\varepsilon^{\text{Eucl}}(0))$. Moreover, for $r < \varepsilon$, $q \in \varphi^{-1}(B_\varepsilon^{\text{Eucl}}(0))$, (3) shows that $\text{dist}(p, q) \geq \sqrt{\lambda_1} \|\varphi(q)\|$. Hence $B_{r\sqrt{\lambda_1}}^{\text{dist}}(p) \subset \varphi^{-1}(B_r^{\text{Eucl}}(0))$.

Conversely, for $r < \varepsilon$, $q \in B_r^{\text{dist}}(p)$ and the path $\varphi^{-1}(t\varphi(q))$ in $\varphi^{-1}(B_\varepsilon^{\text{Eucl}}(0))$, it follows that

$$\text{dist}(p, q) \leq L(c) = \int_0^1 \sqrt{(\varphi^{-1*} g)_{|t\varphi(q)}(\varphi(q), \varphi(q))} \, dt \leq \sqrt{\lambda_2} \|\varphi(q)\|_{\mathbb{R}^n}.$$

Therefore $\varphi^{-1}(B_r^{\text{Eucl}}(0)) \subset B_{r\sqrt{\lambda_2}}^{\text{dist}}(p)$ and the topology induced by dist is the standard topology. $\qquad\square$

Exercises

Exercise 3.1.18 Suppose $\pi : \mathcal{L} \to M$ is an $\mathbb{R}$-line bundle. Prove that $\mathcal{L} \otimes \mathcal{L}$ is isomorphic to the trivial line bundle.

Exercise 3.1.19 Suppose $\pi : \mathcal{L} \to M$ is an $\mathbb{R}$-line bundle with a metric h.

1) Let $N := \{s \in \mathcal{L}_p \,|\, p \in M, h(s, s) = 1\}$. Show that $\pi_1 := \pi_{|N}$ is a 2-fold covering.
2) Prove that $\pi_1^* \mathcal{L}$ is a trivial bundle.

Remark: This shows that if you cut a Möbius strip down the middle lengthwise, you get a cylinder.

Exercise 3.1.20 1) Equip the sphere S^n with the metric induced from the Euclidean $\mathbb{R}^{n+1}$. Determine the volume form $(\varphi_\perp^{-1})^* d\mathrm{vol}$ pulled back by the inverses of the stereographic projections $\varphi_\pm^{-1}$.
2) Compute the metric on Klein's bottle induced by the embedding into $\mathbb{R}^4$ given in Exercise 1.2.24.

Exercise 3.1.21 Let $a \in \mathbb{R}^+$ and let $h : \mathbb{R}^2 \to \mathbb{R}^3$, $(u, \vartheta) \mapsto (u \cos \vartheta, u \sin \vartheta, a\vartheta)$ be the **helicoid** (or **helical surface**). Show that h parametrizes a submanifold of $\mathbb{R}^3$ and compute the induced Riemannian metric and the canonical volume form.

Exercise 3.1.22 Let $H^2 := \{z \in \mathbb{C} \,|\, \mathrm{Im}\, z > 0\}$ be the **upper half-plane** with the Riemannian metric $g = \frac{\langle \cdot, \cdot \rangle_{\mathrm{Eucl.}}}{(\mathrm{Im}\, z)^2}$, where $\langle \cdot, \cdot \rangle_{\mathrm{Eucl.}}$ denotes the Euclidean metric on $\mathbb{R}^2$.

1) Show that H^2 is isometric to the hyperbolic plane (Example 3.1.3).
2) Prove that $\mathrm{SL}(2, \mathbb{R})$ acts isometrically on H^2 via $\begin{pmatrix} a & b \\ c & d \end{pmatrix} \cdot z := \frac{az+b}{cz+d}$.

Exercise 3.1.23 Consider the Lie group $G = \mathrm{SO}(k)$, $\mathfrak{g} = \mathfrak{so}(k)$ and $g_{\mathrm{id}}(A, B) := -\mathrm{Tr}\, AB$ with $A, B \in \mathfrak{g}$. Prove that:

1) g_{id} is a Euclidean scalar product on $\mathfrak{g}$ and it is invariant under conjugation by matrices in G.
2) $g_h := (L_{h^{-1}})^* g_{\mathrm{id}}$ (the pullback by left multiplication by $h^{-1} \in G$) is a Riemannian metric on G for which multiplication by $k \in G$ from the left or from the right is an isometry (i.e. g is a **biinvariant metric**).

Exercise* 3.1.24 Prove Pappus's Centroid Theorem[2]: Let $c : I \to \mathbb{R}^+ \times \mathbb{R}^2, u \mapsto (r(u), 0, z(u))$ be a **curve parametrized by arc length** (i.e. $\|c'\| = 1$). Then the associated surface of revolution M satisfies

$$\mathrm{vol}(M) = 2\pi \int_I r(u)\, du.$$

---

[2] Pappus of Alexandria, circa 320 AD.

Exercise 3.1.25 The **pseudosphere** M (or more precisely, half of it) is the surface of revolution with $r(u) = e^{-u}, z(u) = u + \log(1 + \sqrt{1 - e^{-2u}}) - \sqrt{1 - e^{-2u}}$ with $u \in \mathbb{R}^+$.

1) For a tangent of this surface at a point p that intersects the axis of rotation at a point q, determine the distance from p to q in $\mathbb{R}^3$.
2) Calculate the Riemannian metric of M in the coordinates u, ϑ.
3) Calculate the area of M.
4) Show that $f : M \setminus \{\vartheta = 0\} \to H, \left(\begin{smallmatrix} u \\ \vartheta \end{smallmatrix}\right) \mapsto \vartheta + ie^u$ is an isometry to a subset of the upper half-plane H defined in Exercise 3.1.22.

Exercise 3.1.26 Suppose M is the surface of revolution associated to $z(u) = u, r(u) = \frac{1}{u}$ for $u \in]1, \infty[$. Determine the area of M and the volume of the subset U of $\mathbb{R}^3$ bounded by M and a disk $\{(r \cos \vartheta, r \sin \vartheta, 1) \mid \vartheta \in \mathbb{R}, r \in [0, 1]\}$ (these may possibly be ∞).

Exercise* 3.1.27 Let I be an interval and $\alpha : I \to \mathbb{R}^3$, $w : I \to S^2$ be two C^∞ maps such that $w'(t) \neq 0$ for all $t \in I$. A **(noncylindrical) ruled surface** M in $\mathbb{R}^3$ is a surface with a parametrization

$$u : I \times \mathbb{R} \to \mathbb{R}^3, (t, s) \mapsto \alpha(t) + s \cdot w(t)$$

for such α, w. Thus M is represented as a family of straight lines. The curve α is called the **directrix** of M.

1) Show that for given (α, w) there exists a uniquely determined curve $\beta : I \to \mathbb{R}^3$ of the form

$$\beta(t) = \alpha(t) + s(t)w(t)$$

satisfying $\langle \beta'(t), w'(t) \rangle = 0$ everywhere. The curve β is called the **striction curve**.
2) Suppose $(\widetilde{\alpha}, w)$ is another ruled surface description of M with $\widetilde{\alpha}(t) = \alpha(t) + \widetilde{s}(t)w(t)$. Show that the corresponding striction curve equals β again. Thus the striction curve is a canonical choice for the directrix. Let

$$\lambda := \frac{\det(\beta', w, w')}{|w'|^2}$$

be the **distribution parameter**. Show that M has no tangent space at a point p if and only if p lies on the striction curve and λ has a zero there. *Hint: Choose α as the striction curve.*

See also Theorem 8.1.4 for the classification of surfaces carrying two families of straight lines.

Exercise 3.1.28 Let M be a ruled surface associated to (α, w) without singular points as in Exercise 3.1.27(3), where α is the striction curve. Determine the Riemannian metric on M induced by $\mathbb{R}^3$ and the corresponding volume form (or the pullback of both via u).

Exercise 3.1.29 Let (M, g) be a Riemannian manifold, $\omega := d\mathrm{vol}_g$ be the Riemannian volume form and let $X_1, \ldots, X_n, Y_1, \ldots, Y_n$ be vector fields. Show that:

1) $\omega(X_1, \ldots, X_n) \cdot \omega = X_1^\flat \wedge \cdots \wedge X_n^\flat$,
2) $\omega(X_1, \ldots, X_n) \cdot \omega(Y_1, \ldots, Y_n) = \det[g(X_j, Y_k)]_{j,k}$.

Exercise 3.1.30 Suppose M^n is an oriented Riemannian manifold and let locally (e^j) be an oriented orthonormal basis of $T_p^* M$. For

$$I := \{j_1, \ldots, j_k\} \subset \{1, \ldots, n\}, \quad j_1 < \cdots < j_k$$

define $e^I := e^{j_1} \wedge \cdots \wedge e^{j_k} \in \Lambda^k T_p^* M$. Let the **Hodge $*$(star) operator** $* : \Lambda^k T_p^* M \to \Lambda^{n-k} T_p^* M$ be the linear map given by

$$*e^I = \pm e^{\{1, \ldots, n\} \setminus I}.$$

Here the sign must be chosen such that $e^I \wedge *e^I = e^1 \wedge \cdots \wedge e^n$.

1) Using the metric for which e^I has norm 1, verify that for $\alpha, \beta \in \Lambda^q T^* M$

$$\alpha \wedge *\beta = \langle \alpha, \beta \rangle d\mathrm{vol}$$

   holds.
2) Prove that $*$ is well-defined independent of the choice of basis.
3) Check that $*^2 = (-1)^{k(n-k)}$ holds on $\mathfrak{A}^k(M)$.

3.2 Connections and Curvature

connection	pullback connection
covariant derivative	metric connection
form with coefficients in a vector bundle	parallel transport
curvature	torsion
parallel	torsion-free connection
second Bianchi equation	first chern class

So far no method has been introduced to define multiple derivatives of a function $f \in C^\infty(M)$ at a point $p \in M$. Analogous to derivatives on Euclidean space, such a q-fold derivative should be a q-fold covariant tensor. For example, there should exist a 2-form $\omega \in T_2^0 M$ which gives for $X, Y \in T_p M$ a twofold derivative $\omega(X, Y)$ of f in the direction X and Y. The Lie derivative, on the other hand, provides only $L_X L_Y f$ for a local vector field Y, not for a vector $Y \in T_p M$, and the result at a point p depends on the variation of Y at p. This also resembles what happens when the tangent functor is iterated to $TTf : TTM \to TT\mathbb{R}$. Thus, one would need a way to differentiate the 1-form df pointwise in the direction of a tangent vector. Dual to this (for example via musical isomorphisms) one would like to be able to differentiate vector fields pointwise. In general one obtains suitable differential operators on arbitrary vector bundles by the following assumption analogous to derivations of $\mathbb{R}$-valued functions:

Definition 3.2.1 *Suppose $E \to M$ is a vector bundle. A **connection** ∇ on E is an additive map $\nabla : \Gamma(M, E) \to \Gamma(M, T^*M \otimes E)$ satisfying the Leibniz rule:*

$$\text{for all } f \in C^\infty(M), s \in \Gamma(M, E) \text{ we have } \nabla(f \cdot s) = df \otimes s + f \nabla s.$$

According to Theorem 2.2.9, ∇ can also be interpreted as a map

$$\nabla : \Gamma(M, TM) \times \Gamma(M, E) \to \Gamma(M, E)$$
$$(X, s) \mapsto \nabla_X s$$

which is $C^\infty(M)$-linear in the first factor and satisfies the Leibniz rule in the second. Thus $\nabla_X s$ is a derivative of s in the direction $X \in T_p M$ at the point $p \in M$, a **covariant derivative**.

Example If $E := M \times \mathbb{R}^k \to M$ is the trivial bundle, then the componentwise applied de Rham operator

$$d : \Gamma(M, E) = C^\infty(M, \mathbb{R})^k \to \Gamma(M, T^*M)^k = \Gamma(M, T^*M \otimes E)$$

is a canonical connection on E.

Notation: Let $\mathfrak{A}^q(M, E) := \Gamma(M, \Lambda^q T^*M \otimes E)$ denote the **space of q-forms with coefficients in E.** Products $\mathfrak{A}^p(M, E) \otimes \mathfrak{A}^q(M, F) \to \mathfrak{A}^{p+q}(M, E \otimes F)$ are defined as $(\alpha \otimes s) \wedge (\beta \otimes \widetilde{s}) := (\alpha \wedge \beta) \otimes (s \otimes \widetilde{s})$.

Lemma 3.2.2 *The difference between two connections ∇^0, ∇^1 is an element of $\mathfrak{A}^1(M, \text{End}(E))$.*

Proof. If $f \in C^\infty(M)$, $s \in \Gamma(M, E)$, then

$$\nabla^0(f \cdot s) - \nabla^1(f \cdot s) = f \cdot \nabla^0 s - f \cdot \nabla^1 s + df \otimes s - df \otimes s = f \cdot (\nabla^0 - \nabla^1)s.$$

Hence $\nabla^0 - \nabla^1$ is tensorial by Theorem 2.2.9.   □

Conversely, if ∇^0 is a connection and $\vartheta \in \mathfrak{A}^1(M, \text{End}(E))$, then $\nabla^0 + \vartheta$ is again a connection. Thus, choosing a connection ∇^0 maps the set of all connections on E bijectively onto $\mathfrak{A}^1(M, \text{End}(E))$. In general, among all these, there is no distinguished canonical connection on a bundle E. However, in the next section, a canonical connection associated to a Riemannian metric is constructed on the tensor bundle.

Locally, any connection can be described by a matrix of 1-forms:

Corollary 3.2.3 *On a local trivialization $h : E_{|U} \to U \times \mathbb{R}^k$, every connection ∇ has the form $\nabla s = h^{-1}((d + \vartheta)h(s))$ with $\vartheta \in \mathfrak{A}^1(U, \mathbb{R}^{k \times k})$.*

Proof. The operator $h^{-1} \circ d \circ h$ is a connection on $E_{|U}$ because

$$h^{-1}(d(h(fs))) = h^{-1}(df \circ h(s) + f d(h(s))).$$

Hence $h^{-1} \circ d \circ h - \nabla$ is a tensor. $\qquad\qquad\qquad\qquad\qquad\qquad\qquad\square$

In particular, ∇ is a first-order differential operator. Thus, the definition forces ∇ to be locally equal to d except for a summand of order 0. Conversely, this further motivates the use of connections: In general, the local operators d do not form a global operator on $\Gamma(M, E)$, but according to the next corollary there are always global connections which then correspond to the local de Rham operators up to terms of 0th order.

As in the construction of the integral, a partition of unity enables us to prove the existence of a global object:

Corollary 3.2.4 *On every vector bundle $E \rightarrow M$ there exists at least one connection.*

Proof. Let $(h_j)_j$ be a family of trivializations of E covering M and let $(\tau_k)_k$ be a subordinate partition of unity. Set $\nabla := \sum_k \tau_k \cdot h_{j(k)}^{-1} \circ d \circ h_{j(k)}$. If $f \in C^\infty(M)$, $s \in \Gamma(M, E)$, then one obtains

$$\nabla(f \cdot s) = \sum_k \tau_k h_{j(k)}^{-1}(df \otimes h_{j(k)}(s) + f \cdot d(h_{j(k)}(s)))$$

$$= \sum_k (df \otimes s + f \cdot \tau_k h_{j(k)}^{-1} d(h_{j(k)}(s))) \cdot \tau_k = df \otimes s + f \nabla s. \qquad \square$$

The analogy to the de Rham operator can be extended even further:

Definition 3.2.5 *A connection ∇^E on E is extended to an operator $\nabla^E : \mathfrak{A}^q(M, E) \rightarrow \mathfrak{A}^{q+1}(M, E)$ by the rule*

$$\nabla^E(\alpha \otimes s) = d\alpha \otimes s + (-1)^{\deg \alpha} \alpha \wedge \nabla^E s$$

for any $\alpha \in \mathfrak{A}^q(M), s \in \Gamma(M, E)$.

Thus powers of ∇ can be taken in the sequence

$$0 \rightarrow \Gamma(M, E) \xrightarrow{\nabla^E} \mathfrak{A}^1(M, E) \xrightarrow{\nabla^E} \cdots \xrightarrow{\nabla^E} \mathfrak{A}^n(M, E) \rightarrow 0.$$

Henceforth the notation $\nabla^{\Lambda^q T^* M \otimes E}$, as opposed to ∇^E, will continue to indicate a connection

$$\nabla^{\Lambda^q T^* M \otimes E} : \mathfrak{A}^q(M, E) \rightarrow \Gamma(M, T^* M \otimes \Lambda^q T^* M \otimes E)$$

on the bundle $\Lambda^q T^* M \otimes E$.

Lemma 3.2.6 *For each $\alpha \in \mathfrak{A}^q(M), \beta \in \mathfrak{A}^\bullet(M, E)$ one gets*

$$\nabla^E(\alpha \wedge \beta) = d\alpha \wedge \beta + (-1)^{\deg \alpha} \alpha \wedge \nabla^E \beta.$$

Proof. If $\omega \in \mathfrak{A}^p(M)$, $s \in \Gamma(M, E)$ and $\beta = \omega \otimes s$, then one obtains

$$\nabla^E(\alpha \wedge \beta) = d(\alpha \wedge \omega) \otimes s + (-1)^{q+p}\alpha \wedge \omega \wedge \nabla^E s$$
$$= d\alpha \wedge \omega \otimes s + (-1)^q \alpha \wedge (d\omega \otimes s + (-1)^p \omega \wedge \nabla^E s)$$
$$= d\alpha \wedge \beta + (-1)^q \alpha \wedge \nabla^E \beta. \qquad \square$$

Example 3.2.7 As a generalization of the Example 2.3.9, the derivative of $\omega = \alpha \otimes \tilde{s} \in \Gamma(M, T^*M \otimes E)$ by a connection ∇^E on E is given by

$$(\nabla^E \omega)(X, Y) = d\alpha(X, Y)\tilde{s} - (\alpha \wedge \nabla^E \tilde{s})(X, Y)$$
$$= X.\alpha(Y)\tilde{s} - Y.\alpha(X)\tilde{s} - \alpha([X, Y])\tilde{s} - \alpha(X)\nabla^E_Y \tilde{s} + \alpha(Y)\nabla^E_X \tilde{s}$$
$$= \nabla^E_X(\omega(Y)) - \nabla^E_Y(\omega(X)) - \omega([X, Y]).$$

Surprisingly, the square of ∇^E is a differential operator of lower than 2nd order, namely of 0th order. This is a generalization of the behavior of the de Rham operator, which satisfies $d^2 = 0$. It yields pointwise (tensor-valued) invariants of the connection.

Theorem 3.2.8 *The* **curvature** $\Omega^E := (\nabla^E)^2$ *of a connection* ∇^E *on* E *is tensorial. More precisely,* $\Omega^E \in \mathfrak{A}^2(M, \text{End}(E))$ *holds.*

Proof.

$$\nabla^E(\nabla^E(fs)) = \nabla^E(df \otimes s + f\nabla^E s) = -df \wedge \nabla^E s + df \wedge \nabla^E s + f(\nabla^E)^2 s. \quad \square$$

In a local trivialization h of $E_{|U}$, Corollary 3.2.3 shows that $\nabla^E s = h^{-1}(d+\vartheta)h(s)$ with $\vartheta = (\vartheta_{j\ell})^k_{j,\ell=1}$, $\vartheta_{j\ell} \in \mathfrak{A}^1(U)$. More generally one gets $h \circ \nabla^E \circ h^{-1} = d + \vartheta \wedge$ on forms with coefficients in E. Thus one finds

$$h \circ \Omega^E \circ h^{-1} = d\vartheta + \vartheta \wedge \vartheta = (d\vartheta_{j\ell} + \sum_m \vartheta_{jm} \wedge \vartheta_{m\ell})^k_{j,\ell=1}.$$

Lemma 3.2.9 *If* $X, Y \in \Gamma(M, TM)$, *then*

$$\Omega^E(X, Y)s = \nabla^E_X \nabla^E_Y s - \nabla^E_Y \nabla^E_X s - \nabla^E_{[X,Y]} s.$$

In this formula, the left-hand side depends only on the values of X_p, Y_p at each point $p \in M$. The summands of the right-hand side, on the other hand, are defined only for vector fields in a neighborhood of p.

Proof. Substituting $\omega := \nabla^E s$ into the formula given in Example 3.2.7 provides the assertion. $\qquad \square$

All methods discussed so far for constructing new vector bundles from given vector bundles yield for connections on the bundles under consideration a connection on the new bundle. Connections ∇^E, ∇^F on bundles E, F induce connections $E \oplus F, E \otimes F$ via

$$\nabla^{E \oplus F}(s, s') := (\nabla^E s, \nabla^F s'),$$
$$\nabla^{E \otimes F}(s \otimes s') := \nabla^E s \otimes s' + s \otimes \nabla^F s' \qquad (s \in \Gamma(M, E), s' \in \Gamma(M, F)).$$

The latter product rule follows from this also for $\mathfrak{A}^\bullet(M, E) \otimes \mathfrak{A}^\bullet(M, F)$: Given $\alpha \otimes s \in \mathfrak{A}^q(M, E), \widetilde{s} \in \Gamma(M, F)$ one gets

$$\begin{aligned}
\nabla^{E \otimes F}((\alpha \otimes s) \otimes \widetilde{s}) &= d\alpha \otimes s \otimes \widetilde{s} + (-1)^q \alpha \wedge \nabla^{E \otimes F}(s \otimes \widetilde{s}) \\
&= d\alpha \otimes s \otimes \widetilde{s} + (-1)^q \alpha \wedge (\nabla^E s) \otimes \widetilde{s} + (-1)^q \alpha \wedge (s \otimes \nabla^F \widetilde{s}) \\
&= (\nabla^E(\alpha \otimes s)) \otimes \widetilde{s} + (-1)^q (\alpha \otimes s) \wedge \nabla^F \widetilde{s}
\end{aligned}$$

and analogously for $d(s \otimes (\alpha \otimes \widetilde{s}))$.

Lemma 3.2.10 *The curvatures of $E \oplus F$, $E \otimes F$ satisfy*

$$\Omega^{E \oplus F} = \begin{pmatrix} \Omega^E & 0 \\ 0 & \Omega^F \end{pmatrix}, \qquad \Omega^{E \otimes F} = \Omega^E \otimes \mathrm{id}_F + \mathrm{id}_E \otimes \Omega^F.$$

Proof. We show only the 2nd equation as an example: One finds

$$\begin{aligned}
(\nabla^{E \otimes F})^2(s \otimes s') &= \nabla^{E \otimes F}(\nabla^E s \otimes s' + s \otimes \nabla^F s') \\
&= (\nabla^E)^2 s \otimes s' - \nabla^E s \wedge \nabla^F s' + \nabla^E s \wedge \nabla^F s' + s \otimes (\nabla^F)^2 s'.
\end{aligned}$$

$\square$

Lemma 3.2.11 *A connection ∇^E on a bundle E induces a connection ∇^{E^*} on E^* such that for all $s \in \Gamma(M, E), \sigma \in \Gamma(M, E^*)$ we have*

$$d(\sigma(s)) = (\nabla^{E^*} \sigma)(s) + \sigma(\nabla^E s).$$

Applying the canonical isomorphism $^t : \mathrm{End}(E) \to \mathrm{End}(E^*)$, the curvature satisfies $\Omega^{E^*} = -(\Omega^E)^t$, i.e. $(\Omega^{E^*} \sigma)(s) = -\sigma(\Omega^E s)$.

Proof. If $f \in C^\infty(M)$ is a function, then

$$(\nabla^{E^*}(f\sigma))(s) = d(f\sigma(s)) - f\sigma(\nabla^E s) = df \otimes \sigma(s) + f \cdot (\nabla^{E^*} \sigma)(s).$$

Furthermore as in Lemma 3.2.10 one gets

$$\begin{aligned}
0 = d^2(\sigma(s)) &= d(\nabla^{E^*} \sigma(s) + \sigma(\nabla^E s)) \\
&= (\nabla^{E^*})^2 \sigma(s) - \underbrace{(\nabla^{E^*} \sigma)(\nabla^E s) + (\nabla^{E^*} \sigma)(\nabla^E s)}_{\text{using } \wedge \text{ on the 1-form factor}} + \sigma((\nabla^E)^2 s).
\end{aligned}$$

$\square$

Confusingly, this leads to the situation that a connection of the tangent bundle induces connections on $\mathfrak{A}^q(M, TM)$ in at least two ways which are in fact

different: First, the $\nabla : \Gamma(M, TM \otimes \Lambda^q T^*M) \to \Gamma(M, TM \otimes \Lambda^{q+1}T^*M)$ defined via the supercommutative Leibniz rule, and second, $\nabla^{\otimes T^*M \otimes \otimes T^*M}$:
$\Gamma(M, TM \otimes \Lambda^q T^*M) \to \Gamma(M, T^*M \otimes TM \otimes \Lambda^q T^*M)$ via the subbundle of the
tensor algebra $\Lambda T^*M \subset \bigotimes T^*M$. The relation between these two is determined in
Corollary 3.3.6.

The principle underlying the definition of $\nabla^{E \otimes F}$ and Lemma 3.2.11 is an extension
of the Leibniz rule from Definition 3.2.1: The derivative of a term that is multilinear
in several variables is to be a sum with one summand for each of the derivatives of
the individual variables.

Sections with vanishing covariant derivative are called **parallel**. For example, the
curvature of the connection induced by ∇^E on End E is parallel in the following
sense:

Corollary 3.2.12 (2nd Bianchi[3] identity) $\nabla^{\mathrm{End}\, E} \Omega = 0$.

Proof. If $s \in \Gamma(M, E)$ is a section, then

$$(\nabla \Omega)(s) = \nabla(\Omega s) - \Omega(\nabla s) = \nabla(\nabla^2 s) - \nabla^2(\nabla s) = 0. \qquad \square$$

If $X, Y, Z \in T_p M$ are vectors and $s \in E_p$, then this implies

$$0 = (\nabla \Omega)(X, Y, Z)s = (\nabla_X \Omega)(Y, Z)s + (\nabla_Y \Omega)(Z, X)s + (\nabla_Z \Omega)(X, Y)s.$$

If ∇^{TM} is a connection on TM, then in general Ω is not parallel with respect to the
connection $\nabla^{\Lambda^2 T^*M \otimes \mathrm{End}E}$ (see Lemma 7.2.5).

Definition 3.2.13 *Given a C^∞ map $\varphi : M \to N$ and a vector bundle $E \to N$ with
connection ∇^E, let the **pullback connection** ∇^{φ^*E} on φ^*E be the uniquely determined
connection such that for all $f \in C^\infty(M), s \in \Gamma(N, E)$ we have*

$$\nabla^{\varphi^*E}(f \cdot \varphi^*s) = df \otimes \varphi^*s + f\varphi^*\nabla^E s.$$

Here $\varphi^* : \mathfrak{A}^\bullet(N, E) \to \mathfrak{A}^\bullet(M, \varphi^*E)$ acts on the differential form factor as the
pullback of covariant tensors by Definition 2.2.6 and on E it acts as the pullback
by Definition 2.1.8. Since the connection acts locally, it is sufficient to apply the
defining rule to local sections of φ^*E; but the sections are even global sums of terms
of the form $f\varphi^*s$, cf. Exercise 2.5.18. Thus for any $X \in T_p M$ one gets

$$\left(\nabla_X^{\varphi^*E}(f \cdot \varphi^*s)\right)_{|p} = (X.f)(p) \cdot s(\varphi(p)) + f(p) \left(\nabla_{T_p \varphi(X)}^E s\right)_{|\varphi(p)},$$

where the right-hand side is interpreted as an element of $(\varphi^*E)_p$. The definition
implies $\Omega^{\varphi^*E} = \varphi^*\Omega^E$.

Example Given a curve $c : I \to M$ and a connection ∇ on a bundle $E \to M$, one
obtains a connection ∇^{c^*E} on the bundle $c^*E \to I$. In the literature, this connection
is often written as $\frac{\nabla}{dt}$. Sometimes it is combined with choosing a trivialization

---

[3] 1880, Ricci-Curbastro; 1902, Luigi Bianchi, 1856–1928 independently.

of $c^*E \cong \mathbb{R}^k$ so that it can be represented in the form $\nabla_{\partial_t}^{c^*E} = \frac{\partial}{\partial t} + A_t$ with $A_t = \vartheta(\frac{\partial}{\partial t}) : I \to \mathbb{R}^{k \times k}$. To emphasize which bundle the connection is defined on, we use the notation ∇^{c^*E} in this book.

Combined with the metrics on vector bundles discussed in the last section, we obtain the following refinement of the notion of a connection:

Definition 3.2.14 *Suppose h is a metric on a vector bundle $E \to M$. A connection ∇ on E is called* **metric** *if $\nabla h = 0$, i.e. if it satisfies that for all $s_1, s_2 \in \Gamma(M, E), X \in \Gamma(M, TM)$ we have*

$$X.(h(s_1, s_2)) = h(\nabla_X s_1, s_2) + h(s_1, \nabla_X s_2).$$

Then the following variant of Lemma 3.2.2 holds:

Lemma 3.2.15 *Suppose $E \to M$ is a vector bundle with metric h, ∇, ∇' are connections on E and ∇ is a metric. Then $S := \nabla - \nabla'$ is a 1-form with coefficients in the skew-symmetric endomorphisms of E with respect to h if and only if ∇' is metric.*

For the existence of metric connections, see also Exercise 3.2.24.

Proof. If s, s' are section in E, then

$$d(h(s, s')) - h(\nabla' s, s') - h(s, \nabla' s')$$
$$= h(\nabla s, s') - h(\nabla' s, s') + h(s, \nabla s') - h(s, \nabla' s')$$
$$= h(Ss, s') + h(s, Ss'). \qquad \square$$

Correspondingly, the curvature of a metric connection takes a special form.

Lemma 3.2.16 *The curvature of a metric connection is a 2-form with coefficients in the skew-symmetric endomorphisms of E with respect to h.*

Proof. If s_1, s_2 are sections of a bundle E equipped with a metric h and a metric connection ∇, then as in Lemma 3.2.10 one finds

$$0 = d^2(h(s_1, s_2)) = d(h(\nabla s_1, s_2) + h(s_1, \nabla s_2))$$
$$= h(\Omega s_1, s_2) - h(\nabla s_1, \nabla s_2) + h(\nabla s_1, \nabla s_2) + h(s_1, \Omega s_2).$$

Hence the $\mathrm{End}(E)$-factor of Ω is skew-symmetric, $\Omega^* = -\Omega$. $\qquad \square$

Lemma and Definition 3.2.17 *Suppose $c : I \to M$ with $I \subset \mathbb{R}$ is a* **curve**, *i.e. a C^∞ map satisfying $\dot{c}(t) \neq 0$ for all t, i.e. c is an immersion of I in M. Let ∇ be a connection on a vector bundle $E \to M$. The* **parallel transport** *of $s_0 \in E_{c(t_0)}$ along c is defined as the section $s \in \Gamma(I, c^*E)$ such that $\nabla^{c^*E} s = 0$ and $s_{|t_0} = s_0$ (Fig. 3.4). The parallel transport is a bijective element of $\mathrm{Hom}(E_{c(t_0)}, E_{c(t)})$ and it depends smoothly on the initial data.*

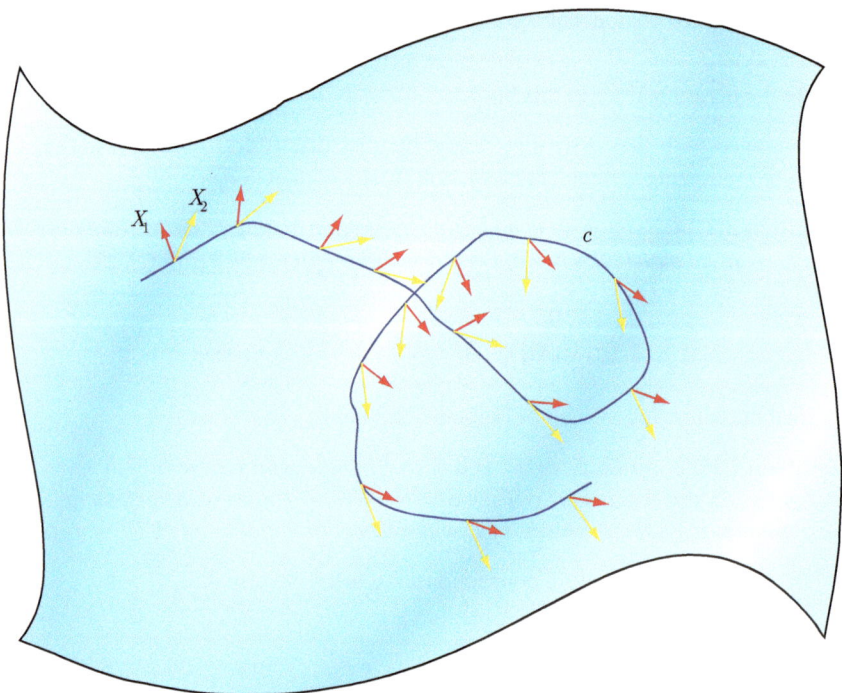

Fig. 3.4 Parallel transport of two vectors X_1, X_2.

Thus, locally at every $c(\widetilde{t})$, s has an extension $\widetilde{s} \in \Gamma(U, E)$ to $U \subset M$ such that $\nabla_{\dot{c}(t)} \widetilde{s}|_{c(t)} = 0$ for all t in a neighborhood of $\widetilde{t}$.

Proof. (existence and uniqueness) Given a local trivialization of $c^* E \to J$ on $J \subset I$, $\nabla^{c^* E}_{\partial/\partial t}$ takes the form $\frac{\partial}{\partial t} + A_t$ with $A_t \in \mathbb{R}^{k \times k}$, hence the condition above means

$$0 = \nabla^{c^* E}_{\partial/\partial t} s(t) = \frac{\partial s(t)}{\partial t} + A_t s(t). \tag{3.1}$$

This is an ordinary linear differential equation of 1st order, so globally on J it has a uniquely determined solution. Because of the uniqueness, the solutions on all trivializations form a global solution, and the inverse is given by parallel transport in the reverse direction. The coefficients of the equation are smooth, thus the solution depends smoothly on the initial data.   □

Similarly, when $E \to M$ is parametrized by an additional manifold N, the parallel transport depends smoothly on the parameters.

Remark 1) If $\nabla^E = d + \vartheta$ holds in a local trivialization of E, then $A_t = (c^* \vartheta)_t(\frac{\partial}{\partial t}) = \vartheta_{c(t)}(\dot{c})$ because $c^* ds = d(c^* s)$. Eq. (3.1) then reads $0 = \dot{s}(t) + \vartheta_{c(t)}(\dot{c}(t)) s(t)$.
2) See Remark 5.3.13 for an intuitive interpretation on surfaces.

Example 3.2.18 On the Euclidean $\mathbb{R}^n$ the differential equation becomes $s' \equiv 0$, hence $s \equiv$ const. and the parallel transport is independent of the curve.

Lemma 3.2.19 *If ∇ is a metric connection, then the parallel transport from $E_{c(t_0)}$ to $E_{c(t)}$ is an isometry of Euclidean vector spaces.*

Proof. If $s, \tilde{s} \in \Gamma(I, c^*E)$ are parallel sections, then one obtains

$$\frac{\partial}{\partial t}\langle s, \tilde{s}\rangle = \big\langle \underbrace{\nabla^{c^*E}_{\partial/\partial t} s}_{=0}, \tilde{s}\big\rangle + \big\langle s, \underbrace{\nabla^{c^*E}_{\partial/\partial t} \tilde{s}}_{=0}\big\rangle = 0. \qquad \square$$

In general, the parallel transport depends on the curve. However, it is independent of the parametrization of c, because for a diffeomorphism $\varphi : \tilde{I} \to I$, $\tilde{c} := c \circ \varphi$ one gets $\nabla^{\tilde{c}^*E}(s \circ \varphi) = \nabla^{\varphi^*c^*E}(\varphi^*s) = \varphi^*(\nabla^{c^*E}s)$. Thus with s the pullback φ^*s is also parallel.

Theorem 3.2.20 *If M, N are manifolds, $E \to N$ is a vector bundle and $f, g : M \to N$ are C^∞-homotopic, then $f^*E, g^*E \to M$ are isomorphic vector bundles.*

Proof. Choose a connection on E and a homotopy $F \in C^\infty(M \times \mathbb{R}, N)$ from f (at 0) to g (at 1). Let $s_t : M \hookrightarrow M \times \mathbb{R}$, $p \mapsto (p, t)$ for any $t \in \mathbb{R}$. Identify the fibers $f^*E = s_0^*F^*E$, $g^*E = s_1^*F^*E$ of $F^*E \to M \times \mathbb{R}$ applying the parallel transport associated to ∇^{F^*E}. $\qquad \square$

Corollary 3.2.21 *If M is C^∞ contractible, then every vector bundle $E \to M$ is trivializable.*

Proof. Let $p_0 \in M$. Apply Theorem 3.2.20 to a C^∞ homotopy $F : M \times \mathbb{R} \to M$ satisfying $F_0 = \mathrm{id}_M$, $F_1 \equiv p_0$. $\qquad \square$

The last two results can be generalized directly to general fiber bundles using a notion of connection for fiber bundles $P \to M$.

Exercises

Exercise* 3.2.22 Suppose ∇ is any connection on $TM \to M$ and X, Y are two vector fields. The **torsion** of ∇ is defined as

$$T(X, Y) := \nabla_X Y - \nabla_Y X - [X, Y].$$

Prove that T is a tensor and that it is an element of $\Lambda^2 T^*M \otimes TM$.

Exercise* 3.2.23 Let ∇ be a connection on TM and let T be its torsion. Show that the connection $\nabla' := \nabla - \frac{1}{2}T$ is **torsion free**, i.e. that its torsion vanishes.

Exercise 3.2.24 Show that every vector bundle E with metric h carries at least one metric connection by choosing any connection ∇ and setting $\nabla' := \nabla + \frac{1}{2}h^{-1}\nabla h$ (where $h^{-1}\nabla h \in \mathrm{End}\,E$ is defined by $h(\cdot, (h^{-1}\nabla_X h)\cdot) := \nabla_X h$ for any $X \in TM$).

Exercise 3.2.25 Let $\mathcal{L} \to M$ be an $\mathbb{R}$-line bundle equipped with a metric h. Show that there is a uniquely determined metric connection ∇ on $\mathcal{L}$.

Exercise 3.2.26 Applying Theorem 2.3.8, show that for any connection ∇ on $E \to M$, $\alpha \in \mathfrak{A}^\ell(M, E)$ and vector fields $X_0, \dots, X_\ell$ the equation

$$(\nabla \alpha)(X_0, \dots, X_\ell) = \sum_{j=0}^{\ell}(-1)^j \nabla_{X_j}\left[\alpha(X_0, \dots, \widehat{X_j}, \dots, X_\ell)\right]$$
$$+ \sum_{j<m}(-1)^{j+m}\alpha([X_j, X_m], X_0, \dots, \widehat{X_j}, \dots, \widehat{X_m}, \dots, X_\ell)$$

holds.

Exercise 3.2.27 Consider a k-line bundle $\mathcal{L}$ over a manifold M (where $k = \mathbb{R}$ or $\mathbb{C}$) and let ∇ be a (k-linear) connection on $\mathcal{L}$.

1) Show that $\nabla^{\mathrm{End}(\mathcal{L})} = h^{-1} \circ d \circ h$ holds by using the canonical identification

$$h : \mathrm{End}(\mathcal{L}) \xrightarrow{\cong} M \times k, \quad \sigma \otimes s \mapsto \sigma(s).$$

   We will not emphasize this isomorphism anymore in the rest of the exercise.
2) Show that the curvature $\Omega \in \mathfrak{A}^2(M)$ of ∇ is a closed form.
3) Let $\widetilde{\nabla}$ be another connection on $\mathcal{L}$ with curvature $\widetilde{\Omega}$. Prove that $\Omega - \widetilde{\Omega}$ is exact. Thus $\mathcal{L}$ canonically induces an element $c_1(\mathcal{L}) := [\frac{-1}{2\pi i}\Omega] \in H^2(M) \otimes_{\mathbb{R}} \mathbb{C}$, the **first Chern class**.
4) Let $\mathrm{Pic}(M)$ be the set of isomorphism classes of k-line bundles on M. Then $\mathrm{Pic}(M)$ with $\otimes$ as product and the dualization * as inverse (by Exercise 2.1.15) forms an abelian group. Prove that $c_1 : \mathrm{Pic}(M) \to H^2(M) \otimes_{\mathbb{R}} \mathbb{C}$ is a group homomorphism.
5) Deduce $c_1(\mathcal{L}) = 0$ if $k = \mathbb{R}$.
6) In general, for a vector bundle E one sets $c_1(E) := c_1(\Lambda^{\mathrm{rk}\,E} E)$. Compute this in terms of the curvature Ω^E of a connection ∇^E on E.

Remark: Such classes can also be used to obtain number-valued invariants, e.g. the degree $\deg \mathcal{L} := \int_M c_1(\mathcal{L})^{\dim M/2} \in \mathbb{R}$ *if M is even-dimensional.*

Exercise 3.2.28 Given any oriented Riemann surface M, TM can be regarded as a complex line bundle as follows: For any local oriented orthonormal basis (e_1, e_2) of TM, multiplying by i is defined as

$$i \cdot (a_1 e_1 + a_2 e_2) := -a_2 e_1 + a_1 e_2 \qquad (a_1, a_2 \in C^\infty(M, \mathbb{R})).$$

Calculate $\int_M c_1(TM)$ for this complex structure on TM and (a) $M = S^2$, (b) $M = \mathbb{R}^2/\mathbb{Z}^2$.

3.3 The Levi-Civita Connection

Levi-Civita connection	Killing vector field
metric connection	Hesse form
torsion-free connection	critical point
Koszul formula	nondegenerate critical point
Christoffel symbol	index

In the last section we saw that on every vector bundle there are uncountably many connections. But for the tangent bundle, and thus on all tensors, there is a canonical connection associated to any Riemannian metric.

Theorem 3.3.1 *Let (M, g) be a Riemannian manifold. Then there is a uniquely determined connection ∇ on TM, the* **Levi-Civita connection**[4] *([Levi]) such that*

*1) ∇ is **metric**: $\nabla g = 0$,*
*2) ∇ is **torsion free**: $X, Y \in \Gamma(M, TM) : \nabla_X Y - \nabla_Y X = [X, Y]$.*

This connection is uniquely determined by the **Koszul formula**[5]

$$2g(\nabla_X Y, Z) = X.g(Y, Z) + Y.g(Z, X) - Z.g(X, Y)$$
$$-g(X, [Y, Z]) + g(Y, [Z, X]) + g(Z, [X, Y]).$$

The first condition can be formulated for any metric vector bundle, but the second condition can only be formulated for the tangent bundle. The standard derivative on the Euclidean $\mathbb{R}^n$ satisfies both conditions. Substituting vector fields into the first condition yields the equivalent formulation

$$d(g(X, Y)) = g(\nabla X, Y) + g(X, \nabla Y).$$

Given $f \in C^\infty(M)$, the second condition implies

$$0 = d^2 f(X, Y) \overset{\text{example 2.3.9}}{=} X.df(Y) - Y.df(X) - df([X, Y])$$
$$\overset{(2)}{=} X.df(Y) - Y.df(X) - df(\nabla_X Y) + df(\nabla_Y X)$$
$$= (\nabla df)(X, Y) - (\nabla df)(Y, X),$$

i.e., this second derivative of f in the directions X, Y is independent of the order of the vectors, analogous to Schwarz's theorem on the Euclidean $\mathbb{R}^n$. Likewise, if α is a closed 1-form, then $\nabla \alpha$ is symmetric. Conversely, with local coordinates $f := x_j$, this is equivalent to torsion freeness.

---

[4] 1917, Tullio Levi-Civita, 1873–1941.
[5] 1950, Jean-Louis Koszul, 1921–2018.

Proof. If X, Y, Z are vector fields on M, then the two conditions imply that

$$X.g(Y, Z) + Y.g(X, Z) - Z.g(X, Y)$$
$$= g(\nabla_X Y, Z) + g(Y, \nabla_X Z) + g(\nabla_Y X, Z)$$
$$+ g(X, \nabla_Y Z) - g(\nabla_Z X, Y) - g(X, \nabla_Z Y)$$
$$= 2g(\nabla_X Y, Z) + g(X, [Y, Z]) + g(Y, [X, Z]) - g(Z, [X, Y]),$$

which is the Koszul formula. This formula is tensorial in X and Z: One gets

$$2g(\nabla_{fX} Y, Z) - 2fg(\nabla_X Y, Z) = Y.f \cdot g(X, Z) - Z.f \cdot g(X, Y)$$
$$- g(Y, [fX, Z]) + g(Z, [fX, Y]) + fg(Y, [X, Z]) - fg(Z, [X, Y]) = 0$$

and analogously for Z. For Y, however, this calculation yields the Leibniz rule.

(1) follows in the form $2g(\nabla_X Y, Z) + 2g(Y, \nabla_X Z) = 2X.g(Y, Z)$, since the five other terms cancel out when we swap Y and Z. In the same way, one gets (2). □

In the Koszul formula, one can see the Lie bracket $[X, Y]$ on the far right. Because of torsion freeness, it provides the skew-symmetric part of $(X, Y) \mapsto \nabla_X Y$ in X, Y, if the latter is understood as an $\mathbb{R}$-bilinear form on the $\mathbb{R}$-vector space $\Gamma(M, TM)$. Consequently, the remaining five terms yield an expression symmetric in X and Y.

Fortunately, for two typical choices of local bases, three of the six summands in the Koszul formula for $\nabla_X Y$ vanish: First, if X, Y, Z are from a local orthonormal basis. Then the derivatives of their scalar products on the left-hand side vanish. On the other hand, if X, Y, Z are basis vectors $\frac{\partial}{\partial x_j}$ of a local coordinate system, the three terms with Lie brackets on the right-hand side vanish and the whole formula necessarily becomes symmetric in X and Y. This case is described in more detail in the next paragraph:

Remark Consider a chart $\varphi : U \to V$. Then $\nabla^{\varphi^{-1*}TM}$ is a connection on $(\varphi^{-1})^* TM \overset{T\varphi}{\cong} TV \to V$. By Corollary 3.2.3, $\nabla^{\varphi^{-1*}TM} = d + \Gamma$ holds with a 2-fold covariant, 1-fold contravariant tensor $\Gamma \in \Gamma(V, T^*V \otimes T^*V \otimes TV)$, called the **Christoffel symbol**[6] ([Chr]). Of course, Γ is defined only on $V \subset \mathbb{R}^n$, not on M itself. Γ depends strongly on the choice of φ and cannot be defined independently of φ, just as in the last section the connection form ϑ depended on the choice of trivialization. As already explained, torsion freeness implies the symmetry $\Gamma(\frac{\partial}{\partial x_j}, \frac{\partial}{\partial x_k}) = \Gamma(\frac{\partial}{\partial x_k}, \frac{\partial}{\partial x_j})$. If $G = (g_{jk})$, $\varphi^{-1*}g = \langle \cdot, G \cdot \rangle_{\mathbb{R}^n}$ and $X := \frac{\partial}{\partial x_j}$, $Y := \frac{\partial}{\partial x_k}$, $Z := \frac{\partial}{\partial x_\ell}$, then Theorem 3.3.1 implies

$$\frac{\partial}{\partial x_j} g_{k\ell} + \frac{\partial}{\partial x_k} g_{j\ell} - \frac{\partial}{\partial x_\ell} g_{jk} = 2 dx_\ell \left(G \cdot \Gamma \left(\frac{\partial}{\partial x_j}, \frac{\partial}{\partial x_k} \right) \right).$$

---

[6] Elwin Bruno Christoffel, 1829–1900.

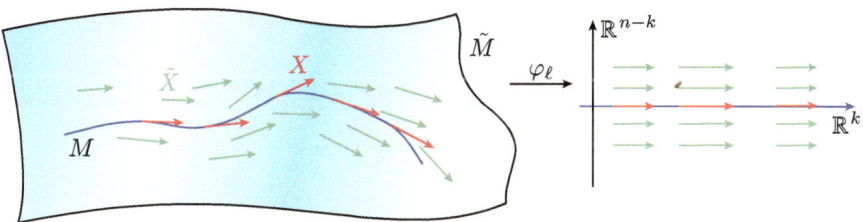

Fig. 3.5 Extension of a vector field.

Hence for $G^{-1} = (g^{jk})_{j,k}$ one obtains

$$\Gamma\left(\frac{\partial}{\partial x_j}, \frac{\partial}{\partial x_k}\right) = \frac{1}{2}G^{-1} \cdot \left(\frac{\partial g_{k\ell}}{\partial x_j} + \frac{\partial g_{j\ell}}{\partial x_k} - \frac{\partial g_{jk}}{\partial x_\ell}\right)^n_{\ell=1}$$

$$= \left(\frac{1}{2}\sum_{\ell=1}^n g^{\ell m} \cdot \left(\frac{\partial g_{k\ell}}{\partial x_j} + \frac{\partial g_{j\ell}}{\partial x_k} - \frac{\partial g_{jk}}{\partial x_\ell}\right)\right)^n_{m=1}.$$

Example If $G = \begin{pmatrix} 1 & 0 \\ 0 & r(u)^2 \end{pmatrix}$, in particular $g(\varphi_*^{-1}\frac{\partial}{\partial u}, \varphi_*^{-1}\frac{\partial}{\partial u}) = 1$, then $\nabla := \nabla^{\varphi^{-1*}TM}$
satisfies

$$\nabla_{\frac{\partial}{\partial u}}\frac{\partial}{\partial u} = 0, \quad \nabla_{\frac{\partial}{\partial \vartheta}}\frac{\partial}{\partial \vartheta} = -r'r\frac{\partial}{\partial u}, \quad \nabla_{\frac{\partial}{\partial u}}\frac{\partial}{\partial \vartheta} = \nabla_{\frac{\partial}{\partial \vartheta}}\frac{\partial}{\partial u} = \frac{r'}{r}\frac{\partial}{\partial \vartheta}.$$

The original motivation for the construction of the Levi-Civita connection was the following result for submanifolds (originally of the Euclidean $\mathbb{R}^m$). This result makes it easy to compute the connection on the submanifold from the one on the surrounding manifold, but the differentiated vector field has to be extended to the surrounding manifold. To this end, let $\iota : M \hookrightarrow \tilde{M}$ be an embedding, let X, Y be vector fields on M and let $\tilde{X}, \tilde{Y}$ be vector fields on $\tilde{M}$ such that $\tilde{X}_{|M} = X, \tilde{Y}_{|M} = Y$ (Fig. 3.5). Such vector fields $\tilde{X}, \tilde{Y}$ exist for any X, Y: Locally using charts φ_ℓ as in Lemma 1.1.3, $\iota : M \hookrightarrow N$ takes the form of the embedding $(\mathbb{R}^k \times \{0_{\mathbb{R}^{n-k}}\}) \cap \Omega \to \Omega \subset \mathbb{R}^n$. Extend vector fields X from $\mathbb{R}^k$ to $\mathbb{R}^n$ in parallel and combine these with a partition of unity to a vector field on $\tilde{M}$:

$$X^\ell_{|\binom{y}{z}} := (T\varphi_\ell(X))_y, \qquad \tilde{X} := \sum_j \tau_j \cdot T\varphi_{\ell(j)}^{-1}(X^{\ell(j)}).$$

To formulate the following theorem, however, we need $\tilde{X}$ only locally.

Theorem 3.3.2 *Let $\tilde{g}$ be a Riemannian metric on $\tilde{M}$, let $g := \iota^*\tilde{g}$ be the induced metric on $\iota : M \hookrightarrow \tilde{M}$, and let $\tilde{\nabla}, \nabla$ be the associated Levi-Civita connections (e.g. $(\tilde{M}, \tilde{g}) = (\mathbb{R}^n, \langle \cdot, \cdot \rangle_{\text{Eucl}}), \tilde{\nabla} = d$). If $\pi : T\tilde{M}_{|M} \to TM, X \mapsto X^{TM}$ is the orthogonal projection, then one gets*

$$\nabla_X Y = (\tilde{\nabla}_{\tilde{X}}\tilde{Y})^{TM} = (\nabla^{\iota^*T\tilde{M}}_X Y)^{TM}.$$

Proof. According to Lemma 1.4.7, $[\widetilde{X}, \widetilde{Y}]_{|M} = [X, Y]$ holds. Thus one finds

$$\pi(\widetilde{\nabla}_{\widetilde{X}}\widetilde{Y} - \widetilde{\nabla}_{\widetilde{Y}}\widetilde{X})_{|M} = \pi([\widetilde{X}, \widetilde{Y}])_{|M} = \pi([X, Y]) = [X, Y].$$

Furthermore for every $Z \in \Gamma(M, TM)$ and its extension $\widetilde{Z}$, one finds

$$g(\pi\widetilde{\nabla}_{\widetilde{Z}}\widetilde{X}, Y) + g(X, \pi\widetilde{\nabla}_{\widetilde{Z}}\widetilde{Y}) = \widetilde{g}(\widetilde{\nabla}_{\widetilde{Z}}\widetilde{X}, \widetilde{Y}) + \widetilde{g}(\widetilde{X}, \widetilde{\nabla}_{\widetilde{Z}}\widetilde{Y})$$
$$= \widetilde{Z}.\widetilde{g}(\widetilde{X}, \widetilde{Y}) = Z.g(X, Y).$$

Substituting in the Koszul formulas for ∇ and $\widetilde{\nabla}$ proves the assertion. □

Remark Typically, $\widetilde{\nabla}_{\widetilde{X}}\widetilde{Y}_{|M}$ is not tangent to M and thus in that case $\widetilde{\nabla}_{\widetilde{X}}\widetilde{Y}_{|M} \neq \nabla_X Y$.

Example Suppose $M := S^2 \subset \mathbb{R}^3 =: \widetilde{M}, \varphi : S^2 \setminus \text{poles} \to \mathbb{R}^2, \begin{pmatrix} \cos u \cos \vartheta \\ \cos u \sin \vartheta \\ \sin u \end{pmatrix} \mapsto \begin{pmatrix} u \\ \vartheta \end{pmatrix}.$
Then one finds

$$X := \varphi^* \frac{\partial}{\partial u} = \begin{pmatrix} -\sin u \cos \vartheta \\ -\sin u \sin \vartheta \\ \cos u \end{pmatrix}, \quad Y := \varphi^* \frac{\partial}{\partial \vartheta} = \begin{pmatrix} -\cos u \sin \vartheta \\ \cos u \cos \vartheta \\ 0 \end{pmatrix}.$$

If $\widetilde{Y}_{\begin{pmatrix} x \\ y \\ z \end{pmatrix}} := \begin{pmatrix} -y \\ x \\ 0 \end{pmatrix}$, then one gets

$$X.\widetilde{Y} = -\sin u \sin \vartheta \cdot \begin{pmatrix} 0 \\ 1 \\ 0 \end{pmatrix} - \sin u \cos \vartheta \cdot \begin{pmatrix} -1 \\ 0 \\ 0 \end{pmatrix} = -\tan u \cdot Y,$$

hence $\nabla_{\frac{\partial}{\partial u}} \frac{\partial}{\partial \vartheta} = -\tan u \cdot \frac{\partial}{\partial \vartheta}.$

Analogous to the formula for the Lie derivative of forms in terms of the de Rham operator, a corresponding formula holds for any torsion free connection.

Theorem 3.3.3 *Let ∇ be a torsion free connection on TM and let $\omega \in T_q^0 M$ be a q-fold covariant tensor. Given any vector fields $X_1, \ldots, X_q, X$,*

$$(L_X\omega)(X_1, \ldots, X_q) = (\nabla_X^{T_q^0 M}\omega)(X_1, \ldots, X_q) + \sum_{j=1}^{q} \omega(X_1, \ldots, \nabla_{X_j}^{TM}X, \ldots, X_q)$$

holds (in particular "$L_X = \nabla_X + \iota_{\nabla X}$" on $\mathfrak{A}^\bullet(M)$).

This formula reflects clearly that in contrast to $L_X\omega$, $\nabla_X\omega$ does not depend on the 1st derivatives of X. Therefore there must be a summand containing derivatives of X on the right-hand side.

Proof. According to Theorem 2.2.8 one finds

$$(L_X\omega)(X_1, \ldots, X_q)$$

$$= X.[\omega(X_1, \ldots, X_q)] - \sum_{j=1}^{q} \omega(X_1, \ldots, [X, X_j], \ldots, X_q)$$

$$\overset{\nabla \text{ torsion free}}{=} (\nabla_X\omega)(X_1, \ldots, X_q) + \sum_{j=1}^{q} \omega(X_1, \ldots, \nabla_X X_j, \ldots, X_q)$$

$$- \sum_{j=1}^{q} \omega(X_1, \ldots, \nabla_X X_j - \nabla_{X_j} X, \ldots, X_q). \qquad \square$$

This formula can be seen as a generalization of the formula concerning the de Rham operator, because the de Rham operator on forms is the alternating part of a torsion free covariant derivative in the following sense:

Theorem 3.3.4 *Suppose ∇ is a torsion free connection on TM and $\omega \in \mathfrak{A}^q(M)$ is a q-form. If $X_0, \ldots, X_q$ are vector fields, then*

$$d\omega(X_0, \ldots, X_q) = \sum_{j=0}^{q} (-1)^j (\nabla_{X_j}^{\Lambda^q T^* M} \omega)(X_0, \ldots, \widehat{X_j}, \ldots, X_q),$$

i.e. $d = \varepsilon \circ \nabla$ on $\Gamma(M, \Lambda^q T^* M)$, where ε is the map $\varepsilon : T^* M \otimes \Lambda^q T^* M \to \Lambda^{q+1} T^* M$, $\alpha \otimes \omega \mapsto \alpha \wedge \omega$.

In particular one obtains $\nabla\omega = 0 \Rightarrow d\omega = 0$.

Proof. According to Theorem 2.3.8, one gets

$$d\omega(X_0, \ldots, X_q) = \sum_{j=0}^{q} (-1)^j X_j.\omega(X_0, \ldots, \widehat{X_j}, \ldots, X_q)$$

$$+ \sum_{j<k} (-1)^{j+k} \omega([X_j, X_k], X_0, \ldots, \widehat{X_j}, \ldots, \widehat{X_k}, \ldots, X_q)$$

$$= \sum_{j \neq k} (-1)^j \omega(X_0, \ldots, \nabla_{X_j} X_k, \ldots, \widehat{X_j}, \ldots, X_q)$$

$$+ \sum (-1)^j (\nabla_{X_j}\omega)(X_0, \ldots, \widehat{X_j}, \ldots, X_q)$$

$$- \sum_{j<k} (-1)^j \omega(X_0, \ldots, \widehat{X_j}, \ldots, \nabla_{X_j} X_k, \ldots, X_q)$$

$$- \sum_{j<k} (-1)^k \omega(X_0, \ldots, \nabla_{X_k} X_j, \ldots, \widehat{X_k}, \ldots, X_q).$$

The formula $d = \varepsilon \circ \nabla$ follows because Definition 2.3.2 of the outer product with a 1-form contains a sum over the cyclic permutations $\mathfrak{S}_{q+1}/\mathfrak{S}_1 \times \mathfrak{S}_q$. $\qquad \square$

In comparison with the Koszul formula, the induced connection on the 1-forms takes a very simple form:

Lemma 3.3.5 *The Levi-Civita connection is uniquely determined by*

$$\nabla \alpha = \frac{1}{2} d\alpha + \frac{1}{2} L_{\alpha^\#} g \qquad \text{for all } \alpha \in \Gamma(M, T^* M)$$

as a decomposition into the skew-symmetric and the symmetric part.

Proof. Applying Theorem 3.3.4 one gets $d\alpha(X, Y) = (\nabla_X \alpha)(Y) - (\nabla_Y \alpha)(X)$. By Theorem 3.3.3 it follows for $Z := \alpha^\#$ that

$$(L_Z g)(X, Y) = (\nabla_Z g)(X, Y) + g(X, \nabla_Y Z) + g(\nabla_X Z, Y)$$
$$= (\nabla \alpha)(X, Y) + (\nabla \alpha)(Y, X). \qquad \square$$

By Theorem 3.3.4, the relation between the various extensions of ∇^{TM} to vector-valued forms mentioned in the last section also becomes quite natural:

Corollary 3.3.6 *The connection $\nabla^{TM \otimes \Lambda T^* M}$ induced by a torsion free connection ∇^{TM} as before Lemma 3.2.10, and the extension of ∇^{TM} to $\mathfrak{A}^\bullet(M, TM)$ (before Theorem 3.2.8) satisfy for any $s \in \Gamma(M, TM \otimes \Lambda^\bullet T^* M)$*

$$\nabla^{TM} s = \varepsilon(\nabla^{TM \otimes \Lambda T^* M} s).$$

Proof. The tensor s is a sum of terms of the form $W \otimes \omega$, where W is a vector field and $\omega \in \mathfrak{A}(M)$. By Theorem 3.3.4 it follows that

$$\nabla^{TM}(W \otimes \omega) = W \otimes d\omega + \nabla W \wedge \omega$$
$$= W \otimes \varepsilon(\nabla^{\Lambda T^* M} \omega) + \varepsilon(\nabla W \otimes \omega) = \varepsilon(\nabla^{TM \otimes \Lambda T^* M}(W \otimes \omega)). \qquad \square$$

Exercises

Exercise 3.3.7 Let ∇ be the Levi-Civita connection on the Euclidean plane $\mathbb{R}^2$. Compute $\nabla_X Y$ and $L_Y X$ for the vector fields

$$X_{(x,y)} := (-y, x) \qquad \text{and} \qquad Y_{(x,y)} := \frac{1}{\sqrt{x^2 + y^2}} (x, y).$$

Exercise 3.3.8 Suppose M is the ruled surface associated to (α, w) as in Exercise 3.1.27, where α is the striction curve. Determine at the regular points the volume form and compute $\nabla_{\partial_s} \partial_s, \nabla_{\partial_t} \partial_s, \nabla_{\partial_s} \partial_t$ for the Levi-Civita connection (in terms of λ and $\|w'\|$).

Exercise 3.3.9 Let $G = SO(k)$ be equipped with the Riemannian metric given in Exercise 3.1.23, and let X be a left-invariant vector field. Show that the Levi-Civita connection on G satisfies $\nabla_X X = 0$ (e.g. by applying the Koszul formula).

Exercise 3.3.10 A **Killing vector field** X [7] on a Riemannian manifold (M, g) is a vector field satisfying $L_X g = 0$. If ∇ is the Levi-Civita connection, then show that X is Killing iff

1) ∇X is pointwise a skew-symmetric endomorphism of TM,
2) the flow Φ of X consists of isometries.

Exercise 3.3.11 Let X be a vector field on M, let $p \in M$ be a zero of X, and let ∇ be any connection on TM. Deduce that $(L.X)_{|p} = (\nabla X)_{|p}$ are maps from $T_p M$ to $T_p M$. Therefore both are elements of $\mathrm{End}(T_p M)$.

Exercise 3.3.12 Let (M, g) be a Riemannian manifold and let ∇ the Levi-Civita connection. The **Hessian form** of a function $f \in C^\infty(M, \mathbb{R})$ at $p \in M$ is defined as

$$\mathrm{Hesse}_p(f) := \nabla^{T^*M}(df) \in T_p^* M \otimes T_p^* M.$$

1) Let $\mathrm{grad}\, f := (df)^\# \in \Gamma(M, TM)$ be the gradient. Show that

$$\mathrm{Hesse}_p(f) = \frac{1}{2} L_{\mathrm{grad} f} g.$$

   In particular, $\mathrm{Hesse}_p(f)$ is symmetric.
2) Suppose p is a **critical point** of f, i.e. $df_{|p} = 0$. Without using local coordinates, show that $\mathrm{Hesse}_p(f)$ is independent of the metric g.

Exercise 3.3.13 A critical point $p \in M$ of $f \in C^\infty(M)$ is called **non-degenerate**, if $\mathrm{Hesse}_p(f)$ is non-degenerate. Let the **index** $\mathrm{ind}_p f$ there be the maximal dimension of a subspace of $T_p M$ on which $\mathrm{Hesse}_p(f)$ is negative definite. Show for $X := \mathrm{grad}\, f$ that $\mathrm{sign}\det(L.X)_{|p} = (-1)^{\mathrm{ind}_p f}$.

Exercise* 3.3.14 Let (M, g) be a Riemannian manifold, let ∇ be its Levi-Civita connection and let ∇' be another metric connection on TM with torsion T (cf. Exercise 3.2.22). Consider $S \in T^*M \otimes \mathrm{End}_{\mathrm{schief}} TM$ defined by

$$g(S_X Y, Z) := \frac{1}{2}\big(g(T(X, Y), Z) + g(T(Z, X), Y) + g(T(Z, Y), X)\big).$$

Prove that $\nabla'_X Y - \nabla_X Y = S_X Y$. Use this to conclude that, conversely, for any tensor $T \in \Lambda^2 T^* M \otimes TM$ there is a uniquely determined metric connection with torsion T. Show that S uniquely determines the torsion T by $T(X, Y) = S_X Y - S_Y X$.

Exercise 3.3.15 Determine the parallel transport of tangent vectors along a circle of latitude on S^2 with respect to the Levi-Civita connection.

---

[7] Wilhelm Karl Joseph Killing, 1847–1923.

3.4 Curvature of a Riemannian Manifold

first Bianchi identiy	Gaußian curvature
Riemann curvature tensor	Ricci curvature
sectional curvature	scalar curvature

The curvature Ω of the Levi-Civita connection is the most important pointwise invariant of the metric. At any point $p \in M$, the metric g_p itself has no invariants by Sylvester's Theorem, since it can always be represented by the identity matrix by means of an orthonormal basis. The Levi-Civita connection is a differential operator, but its curvature is a tensor containing information about the first two derivatives of g in a chart. The curvature of the Levi-Civita connection has a lot of symmetries. The first two in the following theorem apply to any metric connection and are listed again here only for the sake of completeness.

Theorem 3.4.1 *Suppose (M, g) is a Riemannian manifold and ∇ is a connection on TM. If X, Y, Z, W are vector fields and $R(X, Y, Z, W) := -g(\Omega(X,Y)Z, W)$, then*

1) $R(X, Y, Z, W) = -R(Y, X, Z, W)$,
2) $R(X, Y, Z, W) = -R(X, Y, W, Z)$, *if ∇ is metric,*
3) $R(X, Y, Z, W) + R(Y, Z, X, W) + R(Z, X, Y, W) = 0$, *if ∇ is torsion free (**1st Bianchi identity**[8]),*
4) $R(X, Y, Z, W) = R(Z, W, X, Y)$, *if ∇ is the Levi-Civita connection.*

The tensor $R \in \Gamma(M, \Lambda^4 T^* M)$ is called the **Riemann curvature tensor**.

Proof. 1) holds because $\Omega \in \Lambda^2 T^* M \otimes \operatorname{End}(TM)$.
  2) is a consequence of Lemma 3.2.16.
  3) follows from the Jacobi identity and Corollary 3.2.9:

$$0 = \big[[X,Y], Z\big] + \big[[Y,Z], X\big] + \big[[Z,X], Y\big] = \nabla_{[X,Y]}Z - \nabla_Z(\nabla_X Y - \nabla_Y X)$$
$$+\nabla_{[Y,Z]}X - \nabla_X(\nabla_Y Z - \nabla_Z Y) + \nabla_{[Z,X]}Y - \nabla_Y(\nabla_Z X - \nabla_X Z)$$
$$= -\Omega(X,Y)Z - \Omega(Y,Z)X - \Omega(Z,X)Y.$$

4) Label the edges of a tetrahedron (or of the complete graph K_4) as in figure 3.6. Apply the combination $R(X, Y, Z, W) = R(Y, X, W, Z)$ of (1),(2) to three of the vertices twice each to see that the sum of the edge labels for each of the vertices equals 0 according to (3). Add the corresponding formulas for the corners marked by "•" and subtract those for the two corners marked by "○". The resulting formula is $0 = 2R(X, Y, Z, W) - 2R(Z, W, X, Y)$.   □

Remark That (1)–(3) imply the 4th identity is ultimately the following fact about the group ring $\mathbb{Q}[\mathfrak{A}_4]$ of the alternating group of four elements (the isometry group of the tetrahedron): Each $\omega \in \mathfrak{A}_4$ acts on $T^* M^{\otimes 4}$ from the left by

$$(\omega R)(X_1, X_2, X_3, X_4) := R(X_{\omega^{-1}(1)}, X_{\omega^{-1}(2)}, X_{\omega^{-1}(3)}, X_{\omega^{-1}(4)}).$$

---

[8] 1880, Ricci-Curbastro.

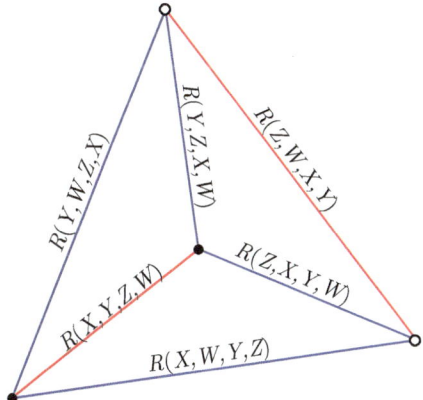

Fig. 3.6 Proof of the 4th symmetry of the curvature tensor.

Consider (in cycle notation) $\widetilde{\tau} := (1\,2)(3\,4)$, $\tau := (1\,3)(2\,4)$, $\sigma := (1\,2\,3) \in \mathfrak{A}_4$ and let e be the neutral element. Then $\widetilde{\tau} - e$ and $\Sigma := e + \sigma + \sigma^2$ are in the annihilator of R because of (1), (2) and (3), respectively. That the 4th identity also holds means that there exist $\alpha, \beta \in \mathbb{Q}[\mathfrak{A}_4]$ such that $\tau - e = (\widetilde{\tau} - e)\alpha + \Sigma\beta$ holds. And indeed, the above proof, when retraced in these terms, shows

$$2(e - \tau) = \Sigma + (\Sigma + (\widetilde{\tau} - e)(\sigma + e))\widetilde{\tau}(e - \sigma - \sigma^2).$$

Alternatively, one can argue that for the symmetric group $\mathfrak{S}_4$ the right ideal in $\mathbb{C}[\mathfrak{S}_4]$ generated by $(1\,2) - e$, $(3\,4) - e$ and Σ has codimension 2 and it is maximal. Since there are nontrivial curvatures with annihilator $\tau - e$, this element must lie in the ideal. This also shows that R cannot have other symmetries of this kind even in more special situations.

According to (1),(2),(4) R defines a symmetric bilinear form $(X \wedge Y, Z \wedge W) \mapsto R(X,Y,Z,W)$ on $\Lambda^2 TM$. In other words if $\omega \in \mathfrak{D}_4 \subset \mathfrak{S}_4$, then $(\omega R)(X,Y,Z,W) = \operatorname{sign} \omega \cdot R(X,Y,Z,W)$ holds, when we consider the action of the dihedral group $\mathfrak{D}_4$ on the square $_W^Y \square_X^Z$. Using the scalar product induced by g on $\Lambda^2 TM$ given by

$$\langle X \wedge Y, \widetilde{X} \wedge \widetilde{Y} \rangle := g(X, \widetilde{X})g(Y, \widetilde{Y}) - g(X, \widetilde{Y})g(Y, \widetilde{X}),$$
$$\text{hence } \|X \wedge Y\|^2 = \|X\|^2 \cdot \|Y\|^2 - g(X,Y)^2,$$

define

$$K_p(X \wedge Y) := \frac{R_p(X,Y,X,Y)}{\|X \wedge Y\|^2} \in \mathbb{R}$$

for non-collinear $X, Y \in T_p M$. If (X,Y) and $(\widetilde{X}, \widetilde{Y})$ span the same plane, i.e. if $X \wedge Y = c \cdot \widetilde{X} \wedge \widetilde{Y}$ holds, it follows that $K_p(X \wedge Y) = K_p(\widetilde{X} \wedge \widetilde{Y})$. Therefore $K_p(X \wedge Y)$ is called the **sectional curvature (in the direction) of the plane** $\mathbb{R} \cdot X + \mathbb{R} \cdot Y \subset T_p M$. In the case of a Riemann surface, $T_p M$ contains only one plane, and the value of

K_p assigns a real number to each point, called the **Gauß curvature**. In general, K_p has several values at each point, depending on the plane. The quadratic form $N := \| \cdot \|_{g_p}^2 \cdot K_p$ on $\Lambda^2 T_p M$ uniquely determines R_p by the polarization identity. In fact, R is even uniquely determined by the values of N on monomials in $\Lambda^2 T^* M$ alone:

Theorem 3.4.2 *At each point $p \in M$, R_p is uniquely determined by the values of $\|X \wedge Y\|^2 \cdot K_p(X \wedge Y)$ for $X, Y \in T_p M$ such that $X \wedge Y \neq 0$.*

Proof. First, because of the symmetry (4) one gets

$$R(X, Y + W, X, Y + W)$$
$$= R(X, Y, X, Y) + R(X, W, X, W) + 2R(X, Y, X, W).$$

Therefore, $R(X, Y, X, W)$ can be represented as a linear combination of three values of N. On the other hand, $R(X + Z, Y, X + Z, W)$ equals

$$R(X, Y, Z, W) + R(Z, Y, X, W) \stackrel{\text{1st Bianchi}}{=} 2R(X, Y, Z, W) - R(X, Z, Y, W)$$

up to such terms. Thus the following linear combination of values of N

$$2R(X + Z, Y, X + Z, W) + R(X + Y, Z, X + Y, W)$$

is equal to $3R(X, Y, Z, W)$ plus a linear combination of values of N.   □

Corollary 3.4.3 *Given an orthonormal basis $(e_j)_j$ of $T_p M$, R_p is uniquely determined by the values $K(e_j \wedge e_\ell)$, $K((e_j + e_k) \wedge e_\ell)$, $K((e_j + e_k) \wedge (e_\ell + e_m))$ $(1 \leq j, k, \ell, m \leq n)$.*

Proof. The statement follows from the construction of the values $R(e_j, e_k, e_\ell, e_m)$ in the last proof.   □

From dimension 4 on, examples can be found of tensors $r \in \text{Sym}^2 \Lambda^2 T^* M$ which satisfy the symmetries (1), (2) and (4), but for which the above theorem does not hold. Thus, the 1st Bianchi identity is essential for the proof. The sectional curvature is a comparatively intuitive way to represent the four-tensor R by a concise map $TM^2 \to \mathbb{R}$. In the next section we will explain in more detail how K is related to intuitive notions of curvature of surfaces. On the other hand, the quotient in the definition of K also makes this notion more difficult to handle; unlike R, it is not a linear map. Algebraically, simpler curvatures can be constructed from R by successively taking traces. However, unlike the sectional curvature, these do not uniquely determine R:

Definition 3.4.4 *The **Ricci curvature**[9] $\text{Ric}(X, Y)$ in the directions $X, Y \in T_p M$ is defined as the trace of the $T_p M$-endomorphism $Z \mapsto -\Omega_p(X, Z)Y$.*

---

[9] Gregorio Ricci-Curbastro, 1853–1925.

Thus with respect to an orthonormal basis $(e_j)_j$ of TM, one gets

$$\text{Ric}(X,Y) = \sum_j R(X, e_j, Y, e_j).$$

Because of the symmetries (1),(2),(4), all other traces over two of the four variables of R are either 0 or equal to $\pm\text{Ric}$.

Corollary 3.4.5 Ric *is a symmetric bilinear form on* T_pM.

Proof. According to Theorem 3.4.1(4) one finds

$$\text{Ric}(Y, X) = \sum_j R(Y, e_j, X, e_j) = \text{Ric}(X, Y). \qquad \square$$

Remark On every manifold of dimension ≥ 3 there exists a metric with Ric $<$ 0 (Lohkamp [Loh]). This shows how little the Ricci curvature reveals about the topology of a manifold.

Given any symmetric bilinear form γ on TM, set $\text{Tr}_g\gamma := \text{Tr}\,(g_{jk})^{-1}(\gamma_{k\ell})$ using the Gram matrix associated to an arbitrary basis. If A is a change of basis, then one finds

$$\text{Tr}\,(\widetilde{g}_{jk})^{-1}(\widetilde{\gamma}_{k\ell}) = \text{Tr}\,(A^t)^{-1}(g_{jk})^{-1}A^{-1}A(\gamma_{k\ell})A^t = \text{Tr}\,(g_{jk})^{-1}(\gamma_{k\ell}).$$

In particular, $\text{Tr}_g\gamma = \sum_j \gamma(e_j, e_j)$ holds if $(e_j)_j$ is an orthonormal basis.

Definition 3.4.6 *The **scalar curvature** $s : M \to \mathbb{R}$ at $p \in M$ is defined as* $s_p :=$ $\text{Tr}_g\text{Ric} = \sum_k \text{Ric}(e_k, e_k) = \sum_{j,k} R(e_k, e_j, e_k, e_j)$ *for any orthonormal basis* $(e_j)_j$. *Thus it equals the trace of* $2R$ *on* $\Lambda^2 T_pM$.

Remark The existence of metrics with $s > 0$ on a given manifold is an important and partially open problem.

Lemma 3.4.7 *Suppose* $(e_j)_j$ *is an orthonormal basis of* T_pM *and* $X \in T_pM$, *then one gets*

1) $\text{Ric}(X, X) = \sum_{\substack{j=1 \\ e_j \neq X}}^{n} K(e_j \wedge X)(\|X\|^2 - g(e_j, X)^2),$

2) $s_p = \sum_{j \neq k} K(e_j \wedge e_k).$

Proof. 1) The right-hand side becomes

$$\sum K(e_j \wedge X)(\|X\|^2 - g(e_j, X)^2) = \sum K(e_j \wedge X)\|X \wedge e_j\|^2$$
$$= \sum R(e_j, X, e_j, X) = \text{Ric}(X, X),$$

2) $\sum \text{Ric}(e_k, e_k) \overset{(1)}{=} \sum_{j \neq k} K(e_j \wedge e_k) = s.$ $\qquad\square$

Example Given a surface M^2, the 1-dimensional space $\Lambda^2 TM$ is spanned by one monomial for which $R(X, Y, X, Y) = K \cdot \|X \wedge Y\|^2$ is given by the Gauß curvature. Thus, by the polarization identity the symmetric form R on $\Lambda^2 TM$ is uniquely determined as $R(X, Y, Z, W) = K \cdot g(X \wedge Y, Z \wedge W)$ and consequently $\mathrm{Ric}(X, Y) = K \cdot g(X, Y)$, $s = 2K$.

Exercises

Exercise 3.4.8 Determine the Levi-Civita connection of the upper half-plane equipped with the metric of Exercise 3.1.22 and compute its curvature tensor Ω.

Exercise* 3.4.9 Assume that the sectional curvature satisfies $K \geq K_0$ (or $> K_0, \leq K_0, < K_0$). Show that then $\mathrm{Ric} - (n-1)K_0 g \geq 0$ (or $>, \leq, <$, respectively) and $s \geq n(n-1)K_0$ (or $>, \leq, <$, respectively) hold.

Exercise 3.4.10 Suppose $G \subset SO(n)$ is a Lie subgroup equipped with the induced metric.

1) Show that the Levi-Civita connection satisfies $\nabla_X X = 0$ for any left-invariant vector field X applying Exercise 3.3.9.
2) Deduce from (a) that for any left-invariant vector fields X, Y one finds

$$\nabla_X Y = \frac{1}{2}[X, Y].$$

3) Compute the curvature as

$$\Omega(X, Y)Z = \frac{1}{4}[Z, [X, Y]] \qquad (X, Y, Z \in \mathfrak{g}).$$

4) If $X, Y, Z, W \in \mathfrak{g}$, prove that

$$R(X, Y, Z, W) = \frac{1}{4}g([X, Y], [Z, W]).$$

Exercise 3.4.11 Determine the sectional curvatures of $SO(3)$.

Chapter 4
The Poincaré–Hopf Theorem and the Chern–Gauß–Bonnet Theorem

The topological observations about the de Rham cohomology are continued in this chapter. They lead to an elegant and far-reaching formula about zeros of sections in vector bundles. Their number (weighted by a sign) is thereby identified with an integral over a certain polynomial in terms of the curvature of the Levi-Civita connection. Following an approach by Mathai and Quillen[1], this formula is more precisely a combination of the classical Theorems of Poincaré–Hopf and Chern–Gauß–Bonnet.

The bridge thus created between differential topological and differential geometrical quantities can be used for numerous unexpected applications. At the end of the chapter, quite a few consequences for vector fields, curvature bounds, extrema of real-valued functions, Lorentzian metrics and circle operations are explained. We largely follow [MaQ] and [BGV, Sect. 1.6].

The contents of this chapter are not a prerequisite for the subsequent chapters. It can therefore be skipped or read at a later date.

4.1 The Mathai–Quillen–Thom Form

Thom form	tautological section
Berezin integral	Mathai–Quillen–Thom form

In this section, as an intermediate goal, we construct a closed form U on the total space of an oriented vector bundle $E \to M$. For compact M, U yields a class in the cohomology $H_c^k(E)$ with compact support such that its integrals $\int_{E_p} U$ over the fibers are equal to 1 for all $p \in M$. A representative of such a class is called a **Thom form** of π. First, U is constructed on a fiber, i.e., a vector space V.

---

[1] 1986, Mathai Varghese, Daniel G. Quillen.

© The Editor(s) (if applicable) and The Author(s), under exclusive license to Springer-Verlag GmbH, DE, part of Springer Nature 2024
K. Köhler, *Differential Geometry and Homogeneous Spaces*, Universitext,
https://doi.org/10.1007/978-3-662-69721-4_4

Definition 4.1.1 *Suppose (V, ω) is a k-dimensional $\mathbb{R}$-vector space equipped with a volume form $\omega \in \Lambda^k V^*$. The **Berezin integral** is defined as the map $\mathcal{T} : \Lambda^\bullet V \to \mathbb{R}$, $\alpha \mapsto \omega(\alpha)$.*

Consider for example an oriented Euclidean vector space $(V, \langle \cdot, \cdot \rangle)$ with orthonormal basis $(e_1, \ldots, e_k)$ and volume form $\omega := \frac{1}{n!} e^1 \wedge \cdots \wedge e^k$. Then one finds for $1 \le j_1 < \cdots < j_m \le k$ that

$$\mathcal{T}(e_{j_1} \wedge \cdots \wedge e_{j_m}) = \begin{cases} 1 & \text{if } k = m \\ 0 & \text{else.} \end{cases}$$

Reversing the orientation changes the sign of $\mathcal{T}$.

Lemma 4.1.2 *Let $(F, h) \to N$ be an oriented Euclidean vector bundle of rank k with a metric connection ∇. Define a product on $\mathfrak{A}^\bullet(N, \Lambda F)$ as*

$$(\rho \otimes \eta) \cdot (\rho' \otimes \eta') := (-1)^{\ell j} (\rho \wedge \rho' \otimes \eta \wedge \eta')$$

with $\rho \in \Lambda^\bullet T^* N, \rho' \in \Lambda^j T^* N, \eta \in \Lambda^\ell F, \eta' \in \Lambda^\bullet F$. Suppose $\alpha \in \mathfrak{A}^\bullet(N, \Lambda F)$ and $s \in \Gamma(N, F^*)$. If $\mathcal{T} : \mathfrak{A}^\bullet(N, \Lambda F) \to \mathfrak{A}^\bullet(N)$ is the Berezin integral induced by the metric on F, then one gets

$$d\mathcal{T}(\alpha) = \mathcal{T}(\nabla \alpha + \iota_s \alpha).$$

Proof. Since $\iota_s \alpha$ has no component in $\mathfrak{A}^\bullet(N, \Lambda^k F)$, $\mathcal{T}(\iota_s \alpha) = 0$ holds. Let $\omega \in \Lambda^k F^*$ be the canonical volume form on F. Then $h(\omega, \omega) \equiv$ const., so $0 = 2h(\nabla \omega, \omega)$ and therefore $\nabla \omega = 0$ because $\Lambda^k F^*$ has dimension 1. It follows that

$$d\mathcal{T}(\alpha) = d(\omega(\alpha)) = (\nabla \omega)(\alpha) + \omega(\nabla \alpha) = \omega(\nabla \alpha) = \mathcal{T}(\nabla(\alpha)). \qquad \square$$

Now let $\pi : (E, h) \to M$ be an oriented Euclidean vector bundle of rank k and let ∇^E be a metric connection on E. Let $\mathfrak{A}^{\ell, m} := \mathfrak{A}^\ell(E, \pi^* \Lambda^m E) \otimes \mathbb{C}$ and choose $\mathcal{T}$ and the product rule as in Lemma 4.1.2 with $N := E$, $F := \pi^* E$. Let $\mathbf{x} \in \Gamma(E, \pi^* E) = \mathfrak{A}^{0,1}$ be the **tautological section** given by $\mathbf{x}_{|s} := s$.

First, for the next theorem, let M be a point. Because $\pi^* E = TE$, $\pi^* E$ has canonical connection d which satisfies $d\mathbf{x} \in \mathfrak{A}^1(E, \pi^* E) = \mathfrak{A}^{1,1}$. The representative of U in the following theorem has no compact support, but it is falling fast enough for $|s| \to \infty$ to yield a corresponding class later. As a compensation U is given by a very easy to handle formula.

Theorem 4.1.3 *The form $U_{|s} := \frac{1}{\sqrt{2\pi}^k} e^{-\|s\|^2/2} dx_1 \wedge \cdots \wedge dx_k \in \mathfrak{A}^k(E)$ satisfies*

$$U = \mathcal{T}(e^{-\|\mathbf{x}\|^2/2 - i \cdot d\mathbf{x}}) \cdot \underbrace{\frac{1}{\sqrt{2\pi}^k} \begin{cases} 1 & \text{if } k \text{ is } \quad even \\ i & \quad\quad\quad\quad odd \end{cases}}_{=:c_k}$$

as well as $\int_{E_p} U \equiv 1$.

Proof. One finds

$$\mathcal{T}(e^{-i\cdot d\mathbf{x}}) = \mathcal{T}\left(\prod_j e^{-i\cdot dx_j \otimes \frac{\partial}{\partial x_j}}\right) = \mathcal{T}\left(\prod_j (1 - i\cdot dx_j \otimes \frac{\partial}{\partial x_j})\right)$$

$$= (-i)^k \cdot (-1)^{\frac{k(k-1)}{2}} dx_1 \wedge \cdots \wedge dx_k.$$

The second part follows from $\int_{\mathbb{R}} e^{-x^2/2}\, dx = \sqrt{2\pi}$. $\qquad\square$

Now let M be general again and set $\nabla := \pi^*\nabla^E : \mathfrak{A}^{\ell,m} \to \mathfrak{A}^{\ell+1,m}$. Let

$$\Omega := \pi^*\langle\Omega^E\cdot,\cdot\rangle := \sum_{j<m\leq k} h(\pi^*\Omega^E e_j, e_m)e_j \wedge e_m \in \mathfrak{A}^2(E, \pi^*\Lambda^2 E) = \mathfrak{A}^{2,2},$$

where Ω^E is the curvature of ∇^E, which is skew-symmetric by Lemma 3.2.16. One could also map Ω to $\mathfrak{A}^2(E, \pi^*\Lambda^2 E^*) = \Gamma(E, \pi^*(\Lambda^2(T^*M \oplus E^*) \otimes \Lambda^2 E^*))$ via h. But then it would be less clear in the calculations which one of the $\Lambda^2 E^*$-factors is used for which operation.

Theorem 4.1.4 *The form* $\widetilde{\Omega} := \frac{\|\mathbf{x}\|^2}{2} + i\nabla\mathbf{x} + \Omega \in \mathfrak{A}^{\bullet,\bullet}$ *satisfies* $(\nabla - i\iota_{\mathbf{x}^*})\widetilde{\Omega} = 0$.

Proof. One finds $\nabla\frac{\|\mathbf{x}\|^2}{2} = 2\langle\nabla\mathbf{x}, \mathbf{x}\rangle = -\iota_{\mathbf{x}^*}\nabla\mathbf{x}$ and $\nabla(\nabla\mathbf{x}) = \iota_{\mathbf{x}^*}\pi^*\langle(\nabla^E)^2\cdot,\cdot\rangle = \iota_{\mathbf{x}^*}\Omega$. Moreover, $\iota_{\mathbf{x}^*}\frac{\|\mathbf{x}\|^2}{2} = 0$ holds, and $\nabla\Omega = 0$ because of the 2nd Bianchi identity. $\qquad\square$

Corollary 4.1.5 *If* $f \in C^\infty(\mathbb{R})$, *then let* $f(\widetilde{\Omega}) := \sum_{j=0}^k \frac{f^{(j)}(\frac{\|\mathbf{x}\|^2}{2})}{j!}(i\nabla\mathbf{x} + \Omega)^{\wedge j}$. *Then one gets* $(\nabla - i\iota_{\mathbf{x}^*})f(\widetilde{\Omega}) = 0$ *and* $f(\widetilde{\Omega}) \in \sum_j \mathfrak{A}^{j,j}$.

Proof. Because $\nabla\mathbf{x} \in \mathfrak{A}^{1,1}$, $\Omega \in \mathfrak{A}^{2,2}$, one gets $f(\widetilde{\Omega}) \in \bigoplus_j \mathfrak{A}^{j,j}$. The term $f(\widetilde{\Omega})$ is obtained by substituting $\widetilde{\Omega}$ into the formal power series of f at $\frac{\|\mathbf{x}\|^2}{2}$, and $(\nabla - i\iota_{\mathbf{x}^*})\widetilde{\Omega}^{\wedge j} = 0$. $\qquad\square$

Corollary 4.1.6 *If* $f \in C^\infty(\mathbb{R})$, *then one gets* $\mathcal{T}(f(\widetilde{\Omega})) \in \mathfrak{A}^k(E)$ *and* $d\mathcal{T}(f(\widetilde{\Omega})) = 0$, *and so* $[\mathcal{T}(f(\widetilde{\Omega}))] \in H^k(E)$ *holds.*

Proof. Because $f(\widetilde{\Omega}) \in \sum_j \mathfrak{A}^{j,j}$, one finds $\mathcal{T}(f(\widetilde{\Omega})) \in \mathfrak{A}^k(E)$. According to Lemma 4.1.2 one gets $d\mathcal{T}(f(\widetilde{\Omega})) = \mathcal{T}((\nabla - i\iota_{\mathbf{x}^*})f(\widetilde{\Omega}))$. $\qquad\square$

Definition 4.1.7 *The **Mathai–Quillen–Thom form**[2] on E is defined as*

$$U := c_k\mathcal{T}(e^{-\widetilde{\Omega}}) \in \mathfrak{A}^k(E).$$

Lemma 4.1.8 *For any* $p \in M$, $\int_{E_p} U = 1$ *holds.*

---

[2] 1952, René Thom, 1986 Mathai–Quillen.

Proof. One finds $\Omega_{|E_p} = 0$, as $\Omega^E \in \mathfrak{A}^2(M, \mathrm{End}(E))$ holds, and $(\nabla \mathbf{x})_{|E_p} = (d\mathbf{x})_{|E_p}$. Hence, by Theorem 4.1.3, $U_{|E_p}$ is the normal distribution density on E_p, for which $\int_{E_p} U = 1$ was shown there. □

Next, we examine the behavior of U when the connection is changed.

Theorem 4.1.9 *The cohomology class of U is independent of ∇.*

Proof. The space of metric connections ∇ on (E, h) is affine. Hence any two connections ∇_0, ∇_1 can be joined by a real parameterized family $u \mapsto \nabla_u^E$ of metric connections on (E, h) (this curve can even be chosen as a straight line). According to Lemma 3.2.15, the derivative $\frac{d\nabla_u^E}{du}$ takes values in the skew-symmetric endomorphisms. Concerning the curvature it follows that

$$\frac{d}{du}(\nabla_u^E)^2 s = \nabla_u^E \left(\frac{d\nabla_u^E}{du} s \right) + \frac{d\nabla_u^E}{du}(\nabla_u^E s) \overset{\text{Leibniz}}{=} \left(\nabla_u^E \frac{d\nabla_u^E}{du} \right) s.$$

If $\vartheta_u := \pi^* \langle \frac{d\nabla_u^E}{du} \cdot, \cdot \rangle \in \mathfrak{A}^{1,2}$, then the derivative of $\widetilde{\Omega}_u$ with respect to u is given by

$$\frac{d\widetilde{\Omega}_u}{du} = i\frac{d\nabla_u}{du}\mathbf{x} + \pi^* \langle ((\nabla_u^E \frac{d\nabla_u^E}{du}) \cdot, \cdot \rangle = -i\iota_{\mathbf{x}^*}\vartheta_u + \nabla_u \vartheta_u = (\nabla_u - i\iota_{\mathbf{x}^*})\vartheta_u.$$

Therefore

$$\frac{d}{du}e^{-\widetilde{\Omega}_u} \overset{\text{Theorem 4.1.4}}{=} -(\nabla_u - i\iota_{\mathbf{x}^*})(\vartheta_u e^{-\widetilde{\Omega}_u})$$

and $\frac{d}{du}U_u = -d\mathcal{T}(\vartheta_u e^{-\widetilde{\Omega}_u}) \cdot c_k$ for the associated Thom forms U_u. Thus one gets

$$U_1 - U_0 = d \int_0^1 \mathcal{T}(-c_k \vartheta_u e^{-\widetilde{\Omega}_u})\, du. \qquad \square$$

Remark Clearly, U has no compact support. However, the pullback of U via $\varphi : E_p \to E_p$, $s \mapsto \frac{s}{\sqrt{1-\|s\|^2}}$ yields a cohomology class $[\varphi^* U] \in H_c^k(E)$ having support in $\{s \in E_p \mid \|s\| \leq 1\}$. The forms used to prove Theorem 4.1.9, whose de Rham derivative is the difference $U_1 - U_0$, behave similarly. It will not be crucial, however, to establish exactly the relationship to $H_c^\bullet$ for the application in the next section, since the proof there does not directly use $H_c^\bullet$ algebraically. The fast falling behavior of these forms is good enough to obtain a result that can be proved topologically by a result about $H_c^\bullet$.

Exercise* 4.1.10 Suppose E_p is a vector space of dimension k. Applying $\varphi^* U \in H_c^k(E_p)$, deduce that $\int_{E_p} : H_c^k(E_p) \to \mathbb{R}$ is well-defined and a vector space epimorphism.

4.2 The Euler Class

Pfaffian	non-degenerate vector field
Euler form	Euler characteristic
Euler class	source
transversal section	sink
Poincaré–Hopf Theorem	non-degenerate critical point
Chern–Gauß–Bonnet Theorem	

In this section, we first construct a canonical cohomology class $\chi(E)$ associated to an oriented real vector bundle E, proving the Poincaré–Hopf Theorem. Then a couple of applications are given.

Definition 4.2.1 *Consider an oriented Euclidian k-dimensional $\mathbb{R}$-vector space V. The **Pfaffian**[3] is defined as the map*

$$\mathrm{Pf}_\Lambda : \Lambda^2 V^* \to \mathbb{R}$$

$$\alpha \mapsto \mathcal{T}(e^\alpha) = \mathcal{T}\left(\sum_{j=0}^\infty \frac{\alpha^{\wedge j}}{j!}\right) = \begin{cases} \mathcal{T}\left(\frac{\alpha^{\wedge k/2}}{(k/2)!}\right) & \text{if } k \text{ is } \begin{matrix} even \\ odd. \end{matrix} \\ 0 \end{cases}$$

If $A \in \mathrm{End}(V)$ is skew-symmetric, then set $\mathrm{Pf}(A) := \mathrm{Pf}_\Lambda(\langle A\cdot, \cdot\rangle)$.

Thus Pf is a homogeneous polynomial of degree $k/2$ in the components of A.

Example Given a matrix representation $\begin{pmatrix} 0 & -\lambda \\ \lambda & 0 \end{pmatrix}$ of A with respect to an orthonormal basis, one gets $\langle A\cdot, \cdot\rangle = \lambda(e^1 \otimes e^2 - e^2 \otimes e^1) = \lambda e^1 \wedge e^2$ and $\mathrm{Pf}\begin{pmatrix} 0 & -\lambda \\ \lambda & 0 \end{pmatrix} = \lambda$.

Theorem 4.2.2 *If $A \in \mathrm{End}(V)$ is skew-symmetric, then $\mathrm{Pf}(A)^2 = \det A$ holds.*

Proof. For k odd both sides vanish. Now let k be even. Using the principal axis theorem, choose an orthonormal basis $\mathcal{B}$ such that

$$\mathrm{Mat}_\mathcal{B}(A) = \begin{pmatrix} 0 & -\lambda_1 & & & \\ \lambda_1 & 0 & & 0 & \\ & & \ddots & & \\ & 0 & & 0 & -\lambda_{k/2} \\ & & & \lambda_{k/2} & 0 \end{pmatrix}.$$

The one finds $\mathrm{Pf}(A) = \lambda_1 \cdots \lambda_{k/2}$ and $\det A = \lambda_1^2 \cdots \lambda_{k/2}^2$. $\qquad\square$

Lemma 4.2.3 *If $A, B \in \mathbb{R}^{k \times k}$ are matrices and A is skew-symmetric, then $\mathrm{Pf}(BAB^t) = \det B \cdot \mathrm{Pf}(A)$ holds.*

---

[3] Johann Friedrich Pfaff, 1765–1825.

Proof. Because $\det(BAB^t) = (\det B)^2 \cdot \det(A)$, both sides of the equation to be shown are equal up to sign. In particular, the identity is true when $\det B = 0$ or $\mathrm{Pf}(A) = 0$. Now assume that $\mathrm{Pf}(A) \neq 0$ and $\det B > 0$. Because $\mathrm{GL}^+(\mathbb{R}^k)$ is connected and by the continuity of both sides, the statement follows as a consequence of the case $B = \mathrm{id}_{\mathbb{R}^k}$. In the case $B \in \mathrm{GL}^-(\mathbb{R}^k)$ choose a basis as in the proof of Theorem 4.2.2. Because of the continuity, it is sufficient to verify the claim for

$$B = \begin{pmatrix} -1 & 0 \\ 0 & \mathrm{id}_{\mathbb{R}^{k-1}} \end{pmatrix}.$$ $\qquad \square$

Remark 4.2.4 If $s \in \Gamma(M, E)$, then because $\pi \circ s = \mathrm{id}_M$, one finds

$$s^* \widetilde{\Omega} = \frac{\|s\|^2}{2} + s^* \nabla^{\pi^* E} \mathbf{x} + s^* \Omega$$

$$= \frac{\|s\|^2}{2} + \nabla^{s^* \pi^* E} s^* \mathbf{x} + s^* \pi^* \langle \Omega^E \cdot, \cdot \rangle = \frac{\|s\|}{2} + \nabla^E s + \langle \Omega^E \cdot, \cdot \rangle.$$

Definition 4.2.5 *Using the zero section* $s_0 : M \to E, p \mapsto 0$, *the **Euler form** is defined as* $\chi(\nabla^E) := s_0^* U$, *i.e.*

$$\chi(\nabla^E) = c_k \mathcal{T}(e^{-\Omega}) = c_k \mathrm{Pf}(-(\nabla^E)^2) = \begin{cases} \mathrm{Pf}\left(\frac{-1}{2\pi}\Omega^E\right) & \text{if } k \text{ is} \quad \text{even} \\ 0 & \quad \text{odd}. \end{cases}$$

The Euler form represents a cohomology class that is canonically assigned to E:

Theorem 4.2.6 *The Euler form has the following properties:*

1) $\chi(\nabla^E) \in \mathfrak{A}^k(M)$ *and* $d\chi(\nabla^E) = 0$ *hold.*
2) *The **Euler class** $[\chi(\nabla^E)] \in H^k(M)$ depends only on the orientation of E and not on the choice of h or ∇^E.*
3) *If $s \in \Gamma(M, E)$, then $[s^* U] = [\chi(\nabla^E)]$ in $H^k(M)$.*

Proof. 1) holds because $U \in \mathfrak{A}^k(E)$ with $dU = 0$.

2) The independence of ∇^E follows from Theorem 4.1.9. Next, let h_0, h_1 be two metrics on E. Then there exists a (with respect to h_0) positive definite endomorphism $\sqrt{H} \in \mathrm{End}(E)$ satisfying $h_1 = h_0(\sqrt{H} \cdot, \sqrt{H} \cdot)$. Let $\mathrm{Pf}_\wedge^j$ be the Pfaffian associated to the metric h_j. Then by the definition of the Berezin integral one finds $\mathrm{Pf}_\wedge^0 = \det H \cdot \mathrm{Pf}_\wedge^1$. Given an h_0-compatible connection ∇_0, $\nabla_1 := \sqrt{H}^{-1} \nabla_0 \sqrt{H}$ is an h_1-compatible connection and

$$\chi(\nabla_1) = \mathrm{Pf}_\wedge^1\left(h_1\left(-\frac{\nabla_1^2}{2\pi}\cdot, \cdot\right)\right) = \mathrm{Pf}_\wedge^1\left(h_0\left(\sqrt{H}\sqrt{H}^{-1}\left(-\frac{\nabla_0^2}{2\pi}\right)\sqrt{H}\cdot, \sqrt{H}\cdot\right)\right)$$

$$\overset{4.2.3}{=} \det H \cdot \mathrm{Pf}_\wedge^1\left(h_0\left(-\frac{\nabla_0^2}{2\pi}\cdot, \cdot\right)\right) = \chi(\nabla_0).$$

3) follows from Corollary 2.4.7 applied to the homotopy $t \mapsto ts$, $t \in \mathbb{R}$, from s_0 (at $t = 0$) to s (at $t = 1$). $\qquad \square$

Given an oriented Riemannian manifold M, let $\chi(M, g) := \chi(\nabla^{\text{Levi}-\text{Civita}})$.

Example If M is a surface, then $\chi(M, g) = \frac{1}{2\pi} K \cdot d\text{vol}$ and thus this value in $H^2(M)$ is independent of the choice of metric.

Definition 4.2.7 *A section $s \in \Gamma(M, E)$ is called **transversal (to the zero section)** if for a connection ∇^E on E and all zeros p of s the homomorphism $(\nabla^E s)_{|p} \in \text{Hom}(T_p M, E_p)$ is invertible. For E oriented set $\text{sign}(s, p) = \pm 1$ depending on whether $(\nabla^E s)_{|p}$ is orientation-preserving or -reversing.*

Remark According to the Thom Transversality Theorem ([BtD2, (14.6)]), arbitrarily close to any given section there is a transversal section.

Since two connections on E differ by an element of $\mathfrak{A}^1(M, \text{End } E)$, $(\nabla^E s)_{|p}$ and hence also $\text{sign}(s, p)$ are independent of the choice of ∇^E. Therefore, the following relation now holds between an integer invariant of a non-degenerate vector field and the cohomology class canonically associated to the oriented manifold, which itself is a polynomial in the coefficients of the curvature tensor:

Poincaré–Hopf Theorem 4.2.8 ([Hopf][4]) *Suppose M is an oriented compact n-dimensional manifold, $E \to M$ is oriented with $\text{rank } E = n$, $s \in \Gamma(M, E)$ is a transversal section, and ∇^E is a metric connection associated to a metric on E. Then the identity*

$$\sum_{p \text{ zero of } s} \text{sign}(s, p) = \int \chi(\nabla^E)$$

holds.

Remark Formulated in this way, the theorem can be seen more accurately as a combination of the Poincaré–Hopf Theorem with the **Chern–Gauß–Bonnet Theorem** ([Chern][5]). One can show that $\int \chi(M, g) = \sum_{q=0}^{n}(-1)^q \dim H^q(M)$, and a theorem of de Rham implies that this invariant does not depend on the C^∞-structure ([BoTu]). Theorem 4.2.8 can also be generalized to more general zeros and non-isolated zeros.

Proof. Around each zero p, choose a trivialization $h : \pi^{-1}(\widetilde{U}_p) \to \widetilde{U}_p \times \mathbb{R}^n$ of E. For the flat connection d induced by h on E, $d(h \circ s)$ is invertible. Hence there is a neighborhood U_p of p such that $\overline{U_p} \subset \widetilde{U}_p$, on which $\psi := \pi_2 \circ h \circ s : U_p \to U_p \times \mathbb{R}^n \overset{\text{can.}}{\to} \mathbb{R}^n$ is a diffeomorphism onto its image $V \subset \mathbb{R}^n$. Let the U_p be chosen such that $\varphi_p^{-1}(\overline{U_p}) \cap \varphi_q^{-1}(\overline{U_q}) = \emptyset$ holds for different zeros p, q. Since $E_{|U_p} \to U_p$ is a trivial bundle, $(\psi^{-1})^* E \to V$ is also trivial and canonically isomorphic to TV. The map ψ^* changes orientation by $\text{sign}(s, p)$. Then $(\psi^{-1})^* s \in \Gamma(V, TV)$ is given by $((\psi^{-1})^* s)_{|\mathbf{x}} = s_{|(h \circ s)^{-1}(\mathbf{x})} = \mathbf{x}$. Using the Euclidean metric on TV and the associated Levi-Civita connection $d =: \nabla^{TV}$, it follows that $d((\psi^{-1})^* s)_{|v} = \text{id}_{TV}$.

---

[4] 1881 in the case of surfaces by H. Poincaré; 1926, Heinz Hopf, 1894–1971.

[5] 1944, Shiing-Shen Chern, 1911–2004; 1848, Pierre Ossian Bonnet, 1819–1892 in the case of surfaces.

Pull back the metric on TV via ψ^* and extend it to all of $E \to M$ outside of $\bigcup_p U_p$ by a partition of unity. Extend $\psi^* d$ to a metric connection ∇^E. Then $(\nabla^E s)_{|U_p} = \nabla^{\psi^* TV}(\psi^* \mathbf{x}) = \psi^* \mathrm{id}_{TV}$ holds. Here ψ^* is the pullback of 1-forms on T^*V and the pullback of vector bundles on TV.

Since M is compact, there is a $\delta > 0$ with $\|s\|^2 > \delta$ on $M \setminus \bigcup_p U_p$. Let $\tau : \mathbb{R}^+ \to [0,1]$ be a test function satisfying $\tau(x) = \begin{cases} 0 & \text{if } x > \delta \\ 1 & x < \delta/2 \end{cases}$. If $t \in \mathbb{R}^+$ is positive, then

$$\int_M \chi(\nabla^E) = \int_M (ts)^* U = \int_M \underbrace{(1 - \tau(\|s\|^2))(tX)^* U}_{\text{support in } \|s\|^2 > \delta/2} + \int_M \underbrace{\tau(\|s\|^2)(ts)^* U}_{\text{support in } U_p}.$$

Now $(ts)^* U = e^{-t^2 \|s\|^2/2} \cdot \text{polynomial}(t)$ holds, and so it follows that

$$\int_M (1 - \tau(\|s\|^2))(ts)^* U = \int_{\|s\|^2 > \delta/2} (1 - \tau(\|s\|^2))(ts)^* U$$
$$= O(e^{-t^2 \delta/4}) \overset{t \to \infty}{\to} 0.$$

The second summand equals

$$\int_M \tau(\|s\|^2)(ts)^* U$$
$$= \sum_p \mathrm{sign}(s, p) \cdot \int_V (\psi_p^{-1})^* (\tau(\|s\|^2)(ts)^* U)$$
$$= \sum_p \mathrm{sign}(s, p) \cdot \int_{\mathbb{R}^n} \tau(\|x\|^2) c_n e^{-t^2 \|x\|^2/2} t^n dx_1 \wedge \cdots \wedge dx_n$$
$$\overset{u := xt}{=} \sum_p \mathrm{sign}(s, p) \cdot \int_{\mathbb{R}^n} \tau\left(\frac{\|u\|^2}{t^2}\right) c_n e^{-\|u\|^2/2} du_1 \wedge \cdots \wedge du_n$$
$$\overset{t \to \infty}{\to} \sum_p \mathrm{sign}(s, p). \qquad \square$$

Remark A strong generalization of this theorem is the Atiyah–Singer Index Theorem, which identifies integrals over more general polynomials of curvature terms with integer invariants.

The remainder of the section is used to list, by way of example, some first consequences of the Poincaré–Hopf Theorem.

Corollary 4.2.9 *One finds* $\int_M \mathrm{Pf}(\frac{-\Omega^E}{2\pi}) \in \mathbb{Z}$.

Suppose $X \in \Gamma(M, TM)$ is a vector field and $p \in M$ such that $X_p = 0$. Then for a Riemannian metric on M and the associated Levi-Civita connection $(L.X)_{|p} = (\nabla.X)_{|p} \in \mathrm{End}(T_p M)$ holds. Hence, the transversality condition from Definition 4.2.7 becomes the following:

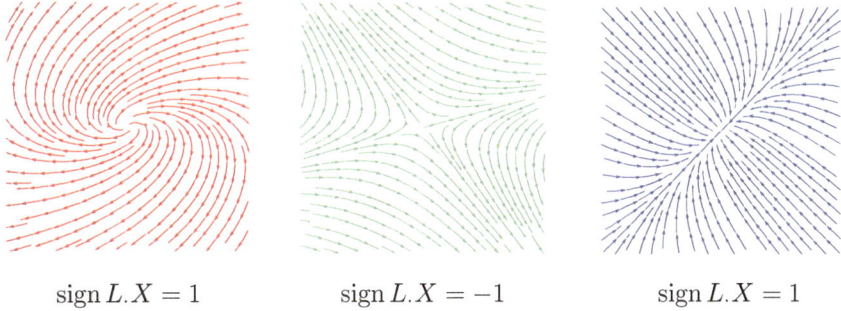

$$\text{sign } L.X = 1 \qquad\qquad \text{sign } L.X = -1 \qquad\qquad \text{sign } L.X = 1$$

Fig. 4.1 Examples of sources and sinks.

Definition 4.2.10 *A vector field X on M is called **non-degenerate** if for all zeros p of X the endomorphism $L.X_{|p} \in \text{End}(T_p M)$ is invertible. The **Euler characteristic** of M is defined as $\chi(M) = \int \chi(M, g)$.*

Given any two vector bundles $E, F \to M$, $\chi(\nabla^{E \oplus F}) = \chi(\nabla^E) \wedge \chi(\nabla^F)$ holds, hence it follows in particular that $\chi(M \times N) = \chi(M) \cdot \chi(N)$ for any two oriented compact manifolds M, N (Exercise 4.2.21). A zero p of X satisfying sign $\det(L.X_{|p}) = 1$ is called a **source**, one with sign $\det(L.X_{|p}) = -1$ is called a **sink** (fig.4.1). Thus $\int \chi(M, g) = \#\text{Quellen} - \#\text{Senken}$ holds.

Example On every torus there is a metric such that $\Omega \equiv 0$, hence #sources = #sinks holds there.

Example The vector field $X_{|x} = x$ on $M := \mathbb{R}^n$ satisfies $dX_{|0} = $ id, hence 0 is a source of X. The vector field $X_{|x} = -x$ satisfies sign $\det(L.X_{|0}) = (-1)^n$. The vector field $X_{\binom{x}{y}} = \binom{x}{-y}$ on $\mathbb{R}^2$ has a sink.

Corollary 4.2.11 *The number $\sum_{p \text{ zero of } s} \text{sign}(s, p)$ is independent of the choice of the transversal section $s \in \Gamma(E, M)$. In particular, the number #sources – #sinks is the same for all non-degenerate vector fields X on a compact oriented manifold.*

This is because the right-hand side in the Poincaré–Hopf Theorem does not depend on s.

Corollary 4.2.12 *Suppose $\dim M$ is odd. Then $\sum_{p \text{ zero of } s} \text{sign}(s, p) = 0$ for every transverse section $s \in \Gamma(E, M)$. In particular,*

$$\#\text{sources} = \#\text{sinks}$$

for every non-degenerate vector field.

This is because the Pfaffian vanishes in odd dimensions.

Corollary 4.2.13 *The following implications hold for any compact oriented manifold M:*

There exists a vector field without zeros.
$\Rightarrow$ *For all non-degenerate vector fields, #sources = #sinks holds.*
$\Leftrightarrow \chi(M) = 0.$

Remark "$\Leftarrow$" is also valid in the 2nd line ([Hopf]). This can be shown (sketched) as follows: Suppose X is a vector field with non-degenerate (hence isolated) zeros and let $U \subset M$ be a contractible open subset containing the zeros of X. Then $\frac{X}{\|X\|} : \partial U \to S^{n-1}$ has mapping degree 0. Then, by the Hopf Theorem (or in more modern terms, because $\pi_n(S^n) \cong \mathbb{Z}$ via the mapping degree), there is a homotopy H_t from a constant map $\partial U \to Y_0 \in S^{n-1}$ to $\frac{X}{\|X\|}$ such that $t \in [0,1]$. Given a diffeomorphism $g : U \to B_1^n(0)$ let

$$Y_q := H_{\|g(q)\|}\left(g^{-1}\left(\frac{g(q)}{\|g(q)\|}\right)\right) \cdot \left\|X_{g^{-1}\left(\frac{g(q)}{\|g(q)\|}\right)}\right\|$$

if $g(q) \neq 0$ and $Y_{g^{-1}(0)} := Y_0$. Then Y is continuous, $Y_{|\partial U} = X_{|\partial U}$ and $\|Y\| \geq \min_{q \in \partial U} \|X_q\|$ hold. Apply the Weierstraß Approximation Theorem to smooth Y to a C^∞ vector field.

Corollary 4.2.14 *The number of zeros of any non-degenerate vector field X is at least as large as $|\chi(M)|$. If $\chi(M) \neq 0$ holds, then TM is not a trivial bundle.*

Example Suppose $M := S^n$ is a sphere with north pole $N = (1, 0, \ldots, 0)^t$. Let $X_{|p} := N - \langle N, p \rangle p \in T_p S^n$. X has zeros exactly at the poles $\mathbb{R} \cdot N \cap S^n$; at the north pole X points toward it, hence sign $\det(L.X_{|N}) = (-1)^n$, and at the south pole X points outward, hence sign $\det(L.X_{|N}) = 1$. More precisely, $\frac{\partial}{\partial p} X = -\langle N, p \rangle \mathrm{id} - \langle N, \cdot \rangle p$, so $(\frac{\partial}{\partial p} X_{|N})_{|TS^n} = \mathrm{id}$ holds. Therefore one gets

$$\chi(S^n) = 1 + (-1)^n = \begin{cases} 2 & \text{even} \\ 0 & \end{cases} \text{if } n \text{ is } \begin{matrix} \text{even} \\ \text{odd.} \end{matrix}$$

In particular, the nontriviality of TS^n for n even follows once again. Similarly, for complex projective spaces $\chi(\mathbb{P}^n\mathbb{C}) = n+1$ follows. Thus $\mathbb{P}^n\mathbb{C} \neq S^{2n}$ holds for $n > 1$.

Example Suppose M is an oriented compact surface with g handles. By orthogonal projection of a constant vector field it follows that

$$\chi(M) = \sum_p \text{sign} \det(L.X)_{|p} = 2 - 2g.$$

If $g > 2$, then the surface must have points of negative curvature.

Remark The oriented compact surfaces can be uniquely characterized by $g \in \mathbb{N}_0$ up to homeomorphism (see [Hi, ch. 9.3]). Because $\chi(M) = \frac{1}{2\pi} \int_M K \cdot d\text{vol}$, it follows from $K > 0$ everywhere that $M = S^2$, and from $K = 0$ everywhere it follows that $M = T^2$.

Corollary 4.2.15 *Consider a function $f \in C^\infty(M)$ on a compact orientable manifold whose **critical points are non-degenerate**, i.e., at points p satisfying $df_p = 0$ the Hessian form $\mathrm{Hesse}_p(f) := \nabla df$ is non-degenerate (Exercise 3.3.12). Then using the index $\mathrm{ind}_p f$ of f at the critical points (Exercise 3.3.13), one obtains*

$$\chi(M) = \sum_{p \text{ critical point}} (-1)^{\mathrm{ind}_p f}$$

$$= \#\{p \mid \mathrm{ind}_p f \text{ even}\} - \#\{p \mid \mathrm{ind}_p f \text{ odd}\}.$$

The following application generalizes Remark 3.1.5.

Corollary 4.2.16 *If M carries a Lorentzian metric g_L, then $\chi(M) = 0$ holds.*

Proof. Choose any Riemannian metric g on M. Then there is a symmetric endomorphism $g^{-1} g_L \in \Gamma(M, \mathrm{End}(TM))$ such that $g_L = g(\cdot, (g^{-1} g_L)\cdot)$. Let $\mathcal{L}$ be the line bundle associated to the negative eigenvalue $g^{-1} g_L$ and let $E \subset TM$ be the g-orthogonal complement to $\mathcal{L}$. Then $\Omega^{\mathcal{L}} = 0$ holds for the metric connection $\nabla^{\mathcal{L}}$ on $\mathcal{L}$ induced by the Levi-Civita connection to g. Therefore one obtains

$$\chi(M, g) = \mathrm{Pf}\left(\frac{-1}{2\pi} \begin{pmatrix} \Omega^E & 0 \\ 0 & \Omega^{\mathcal{L}} \end{pmatrix} \right) = 0. \qquad \square$$

Remark The converse follows from the Hopf Theorem as in Corollary 4.2.13 .

Exercises

Exercise* 4.2.17 If $A \in \mathrm{GL}(2n)$ is skew-symmetric, show that

$$\mathrm{Pf}(A)\mathrm{Pf}(-A^{-1}) = 1$$

holds.

Exercise* 4.2.18 Determine $\mathrm{Pf}(A)$ and $\det A$ for the matrix

$$A = \begin{pmatrix} 0 & -a & -b & -c \\ a & 0 & -e & -f \\ b & e & 0 & -g \\ c & f & g & 0 \end{pmatrix} \in \mathbb{R}^{4 \times 4}$$

with respect to the standard scalar product.

Exercise 4.2.19 If $A \in \mathbb{R}^{2k \times 2k}$ is skew-symmetric and $n \in \mathbb{N}_0$, prove that

$$\mathrm{Pf}(A^{2n+1}) = (-1)^{kn}\mathrm{Pf}(A)^{2n+1}.$$

Exercise* 4.2.20 In the case of rank $E = 2k$, show the independence of $[\chi(\nabla^E)] \in H^{2k}(M)$ from the choice of connection in a more direct way.

Exercise 4.2.21 Suppose $E, F \to M$ are oriented vector bundles with metrics h^E, h^F and metric connections ∇^E, ∇^F.

1) Show that $\chi(\nabla^{E \oplus F}) = \chi(\nabla^E) \wedge \chi(\nabla^F)$.
2) Given any two oriented compact manifolds M, N, deduce that $\chi(M \times N) = \chi(M) \cdot \chi(N)$.

Exercise 4.2.22 The curvature of the sphere S^n is given by

$$g(\Omega(X, Y)Z, U) = g(X, U)g(Y, Z) - g(X, Z)g(Y, U).$$

Applying this formula in the case when n is even, calculate $\mathrm{Pf}(\frac{-\Omega}{2\pi})$ as a multiple of the volume form, and use $\chi(S^n)$ to obtain the formula for the volume of even-dimensional spheres.

Exercise 4.2.23 Let $\Phi : S^1 \times M \to M$ be a C^∞ action of the circle on a manifold M (i.e. the flow Φ_t has period 2π in t). Let X be the vector field associated to the flow $t \mapsto \Phi_t$.

1) Choose any Riemannian metric g' on M. Prove that

$$g_p(Y, Z) := \int_0^{2\pi} (\Phi_t^* g')_p(Y, Z) \, dt \qquad (p \in M, Y, Z \in T_p M)$$

   is an S^1-invariant metric on M.
2) Show that $\dim M$ is even if Φ has at least one isolated fixed point (e.g., by interpreting X as a Killing vector field, see Exercise 3.3.10). To do so you may assume at each fixed point p that $\det(L.X)_{|p} \neq 0$ (or you can prove this).
3) Assume that Φ finitely many fixed points. Prove the identity

$$\#\{\text{fixed points of the } S^1\text{-operation}\} = \sum_{p \text{ zeros of } X} \operatorname{sign} \det(L.X)_{|p}.$$

4) Conclude: If a circle acts with finitely many fixed points on an oriented compact manifold M, then the number of fixed points equals $\chi(M)$. In particular, $\chi(M) \geq 0$ holds, and the number of fixed points does not depend on the action.

Exercise 4.2.24 Compute $\chi(\mathbb{P}^n\mathbb{C})$ by using a circle action

$$e^{i\varphi} \cdot (z_0 : \cdots : z_n) := (e^{ik_0\varphi} z_0 : \cdots : e^{ik_n\varphi} z_n)$$

for suitable $k_0, \ldots, k_n \in \mathbb{Z}$.

Chapter 5
Geodesics

In this chapter more objects used to describe Riemannian manifolds are introduced. In the first section the differential geometry of submanifolds is treated by means of the 2nd fundamental form, and of these again the hypersurfaces are studied in more detail. Since our visual intuition is mainly restricted to surfaces in three-dimensional Euclidean space, one can understand the geometric meaning of the notion of curvature a little better in this particular case. In the second section, embedded curves with vanishing 2nd fundamental form are considered, the geodesics. These enable us to construct a canonical chart around a chosen point p on a Riemannian manifold associated to an orthonormal basis on $T_p M$, the normal coordinates. The description of metrics, connection and curvatures in this canonical chart leads to further interpretations of the sectional curvature, as it dominates the volume of small balls and spheres. In the last section, we develop a simple criterion for when any two points on M can be connected by shortest paths.

5.1 Immersions

second fundamental form	Hopf Theorem
first fundamental form	Weingarten equation
Gauß equation	Mainardi–Codazzi equation
hypersurface	Codazzi–Mainardi equation
Weingarten map	catenoid
principal curvatures	helical surface
principal directions	Ricci equation
Gauß–Kronecker curvature	

By Theorem 3.3.2, for any Riemannian submanifold $M \subset \widetilde{M}$ there is a close relation between the Levi-Civita connections on M and $\widetilde{M}$ which makes a corresponding comparison of curvatures plausible. More generally, the same is true for immersions, since locally on the immersed manifold these are embeddings. This is the content

K. Köhler, *Differential Geometry and Homogeneous Spaces*, Universitext, https://doi.org/10.1007/978-3-662-69721-4_5

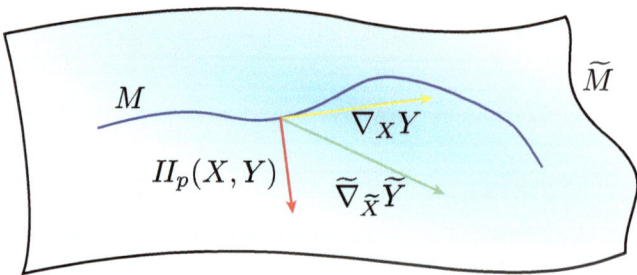

Fig. 5.1 Definition of the second fundamental form.

of this chapter, which in particular opens a further approach to the curvature of submanifolds of the Euclidean $\mathbb{R}^n$. Thus some intuitive interpretations of curvature become possible. There is a special calculus for the case $\dim \widetilde{M} = \dim M + 1$, which is explained in the middle of the section before the techniques for arbitrary codimension are treated.

Definition 5.1.1 *Let $\iota : M \to \widetilde{M}$ be a Riemannian immersion, let $\nabla, \widetilde{\nabla}$ be the Levi-Civita connections and let locally $\widetilde{X}, \widetilde{Y}$ be vector fields on $\widetilde{M}$ such that $X := \widetilde{X}_{|M}$, $Y := \widetilde{Y}_{|M}$ are vector fields on M. The **second fundamental form** at $p \in M$ is defined as*

$$II_p(\widetilde{X}, \widetilde{Y}) := \widetilde{\nabla}_{\widetilde{X}}\widetilde{Y} - \nabla_X Y \in T_p\widetilde{M}.$$

First fundamental form is another name for the Riemannian metric. The immersion induces a vector bundle monomorphism $TM \to \iota^* T\widetilde{M}$. Given the normal bundle $N := (TM)^\perp \subset \iota^* T\widetilde{M}$ and the orthogonal projections $\iota^* T\widetilde{M} \to TM, X \mapsto X^{TM}$ and $\iota^* T\widetilde{M} \to N, X \mapsto X^N =: X^\perp$ onto the components $\iota^* T\widetilde{M} = TM \oplus N$, one finds by Theorem 3.3.2 that

$$II(\widetilde{X}, \widetilde{Y}) = (\nabla_X^{\iota^* T\widetilde{M}}\widetilde{Y})^N$$

is the part of $\widetilde{\nabla}_{\widetilde{X}}\widetilde{Y}$ that is orthogonal to $T_p M$ (Fig. 5.1). Thus it has values in the normal bundle N. Although the definition appears to involve first-order derivatives, II is actually a much more manageable object:

Theorem 5.1.2 *The map II is symmetric and a tensor $II_p : T_p M \times T_p M \to N_p$.*

Proof. One obtains

$$II(\widetilde{X}, \widetilde{Y}) \quad = \quad (\widetilde{\nabla}_{\widetilde{X}}\widetilde{Y})^\perp = (\widetilde{\nabla}_{\widetilde{Y}}\widetilde{X} - [\widetilde{X}, \widetilde{Y}])^\perp$$

$$\overset{\text{Lemma } 1.4.7}{=} (\widetilde{\nabla}_{\widetilde{Y}}\widetilde{X} - \underbrace{[X, Y]}_{\in TM})^\perp = II(\widetilde{Y}, \widetilde{X}).$$

By definition II is tensorial in the 1st variable and because of the symmetry it is therefore also tensorial in the 2nd variable. We have $\widetilde{X}_p, \widetilde{Y}_p \in T_p M$, and in Theorem 3.3.2 it had been shown that every vector $X \in T_p M$ locally has an extension $\widetilde{X}$ on $\widetilde{M}$ (globally for embeddings). $\qquad\qquad\square$

Using the 2nd fundamental form, the curvature of M can be obtained from that of $\widetilde{M}$:

Theorem 5.1.3 (Gauß equation) *If $\iota : M \to \widetilde{M}$ is a Riemannian immersion and $X, Y, Z, W \in T_p M$, then the curvatures $R, \widetilde{R}$ of $M, \widetilde{M}$ satisfy*

$$R(X, Y, Z, W) = \widetilde{R}(X, Y, Z, W) + \widetilde{g}(II(X, Z), II(Y, W))$$
$$- \widetilde{g}(II(Y, Z), II(X, W)).$$

Proof. On M one finds

$$g(\nabla_{[Y,X]} Z, W) - \widetilde{g}(\widetilde{\nabla}_{[\widetilde{Y},\widetilde{X}]} \widetilde{Z}, \widetilde{W}) \overset{\text{Lemma 1.4.7}}{=} g(\nabla_{[Y,X]} Z, W) - \widetilde{g}(\widetilde{\nabla}_{[Y,X]} \widetilde{Z}, \widetilde{W})$$
$$\overset{\text{Th. 3.3.2}}{=} 0$$

and

$$g(\nabla_Y \nabla_X Z, W) - \widetilde{g}(\widetilde{\nabla}_{\widetilde{Y}} \widetilde{\nabla}_{\widetilde{X}} \widetilde{Z}, \widetilde{W})$$
$$= Y.g(\nabla_X Z, W) - \widetilde{Y}.\widetilde{g}(\widetilde{\nabla}_{\widetilde{X}} \widetilde{Z}, \widetilde{W}) - g(\nabla_X Z, \nabla_Y W) + \widetilde{g}(\widetilde{\nabla}_{\widetilde{X}} \widetilde{Z}, \widetilde{\nabla}_{\widetilde{Y}} \widetilde{W})$$
$$\overset{\text{Th. 3.3.2}}{=} \widetilde{g}(II(X, Z), II(Y, W)). \qquad \square$$

Definition 5.1.4 *Consider a **hypersurface** $M^n \subset \widetilde{M}$, i.e. $\dim \widetilde{M} = \dim M + 1$. On a neighborhood $U \subset M$ of $p \in M$ let $\mathfrak{n}$ be a normal vector (hence $\mathfrak{n}$ is uniquely determined up to sign). The **shape operator**[1] (or **Weingarten map**) $\mathcal{W} \in \Gamma(U, \mathrm{End}(TU))$ is defined by $\widetilde{g}(II(X, Y), \mathfrak{n}) = g(X, \mathcal{W}Y)$. According to Theorem 5.1.2, $\mathcal{W}$ is symmetric (with respect to g). The real eigenvalues $\lambda_1, \ldots, \lambda_n$ are called **principal curvatures** and the vectors of an orthonormal basis $(e_1, \ldots, e_n)$ of eigenvectors are called **principal directions**.*

Thus, the λ_j change sign with $\mathfrak{n}$, the e_j are independent of the choice of $\mathfrak{n}$, and $II(e_j, e_k) = \lambda_j \delta_{jk} \mathfrak{n}$.

Corollary 5.1.5 *For any distinct indices j, k, the sectional curvatures of $M, \widetilde{M}$ satisfy $K(e_j \wedge e_k) = \widetilde{K}(e_j \wedge e_k) + \lambda_j \lambda_k$. In particular if $\widetilde{M} = (\mathbb{R}^{n+1}, g_{\mathrm{Eucl}})$, then*

$$R(e_j, e_k, e_\ell, e_m) = \lambda_j \lambda_k (\delta_{j\ell} \delta_{km} - \delta_{jm} \delta_{k\ell})$$

and according to Corollary 3.4.7 one finds

$$K(e_j \wedge e_k) \overset{j \neq k}{=} \lambda_j \lambda_k, \quad \mathrm{Ric}(e_j, e_j) = \sum_{\substack{k=1 \\ j \neq k}}^n \lambda_j \lambda_k = \lambda_j \cdot (\mathrm{Tr}\, \mathcal{W} - \lambda_j),$$

---

[1] Julius Weingarten, 1836–1910.

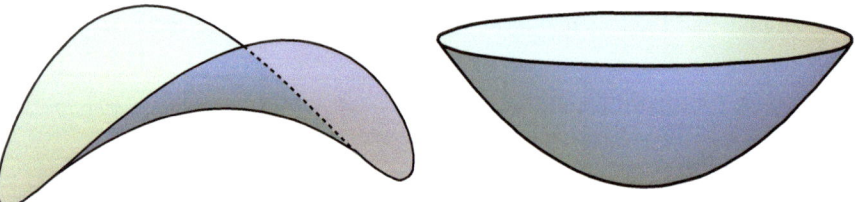

Fig. 5.2 Surfaces of negative and positive Gauß curvature.

$$\text{Ric}(e_j, e_\ell) \overset{j \neq \ell}{=} 0, \quad s = \sum_{\substack{j,k=1 \\ j \neq k}}^{n} \lambda_j \lambda_k = (\text{Tr}\,\mathcal{W})^2 - \text{Tr}\,\mathcal{W}^2 = 2\text{Tr}\,(\mathcal{W} \wedge \mathcal{W}).$$

As a nonlinear function, K is not necessarily uniquely determined by the values $K(e_j \wedge e_k)$. If $X = \sum_j \alpha_j e_j, Y = \sum_k \beta_k e_k$, then K is more generally given by

$$K(X \wedge Y) = \frac{\sum_{j<k}(\alpha_j \beta_k - \alpha_k \beta_j)\lambda_j \lambda_k}{\sum_{j<k}(\alpha_j \beta_k - \alpha_k \beta_j)}.$$

Lemma 5.1.6 *Consider a Riemannian submanifold $M^n \subset \mathbb{R}^{n+k}$ and let $\gamma : V \to U$ be a parametrization. Then for all $x \in V, W, Z \in T_x\mathbb{R}^n$ we have*

$$\gamma''_{|x}(W, Z)^\perp = II_{\gamma(x)}(T_x\gamma(W), T_x\gamma(Z)).$$

Proof. Let $c : I \to M$ be a curve and let Y be a local extension of the vector field $\dot{c}$. Because $\dot{c}(t) = Y_{|c(t)}$, it follows by differentiating that

$$\ddot{c}(t) = dY_{c(t)}(\dot{c}(t)) = \nabla_Y^{\mathbb{R}^{n+k}} Y_{|c(t)}$$

and therefore

$$\ddot{c}^\perp = II(\dot{c}, \dot{c}).$$

If $c = \gamma \circ \widetilde{c}$ with $x = \widetilde{c}(0), W = \widetilde{c}'(0)$, then $\gamma''_{|x}(W, W) = II(T_x\gamma(W), T_x\gamma(W))$ and the polarization identity proves the assertion. $\qquad\square$

In particular, given a hypersurface $M^n \subset \mathbb{R}^{n+1}$ and $\dot{\gamma}(0) = e_j$ one obtains $\ddot{\gamma}(0)^\perp = \lambda_j \mathfrak{n}$. If $K(e_j \wedge e_k) > 0$, then M curves in both directions e_j, e_k either toward $\mathfrak{n}$ or away from $\mathfrak{n}$. If $K(e_j \wedge e_k) < 0$, then M curves away from $\mathfrak{n}$ in one direction and toward $\mathfrak{n}$ in the other (Fig. 5.2).

Corollary 5.1.7 *1) Let $M^n \subset \mathbb{R}^{n+k}$ be locally given as $f^{-1}(0)$ with a map $f : \mathbb{R}^{n+k} \to \mathbb{R}^k$ such that $\text{rg}\,Tf = k$. Then $II_p(X, Y) \in (\ker T_p f)^\perp = \text{im}\,(T_p f)^t$ is uniquely determined by $T_p f(II_p(X, Y)) = -f''_{|p}(X, Y)$ as*

$$II_p(X, Y) = -(T_p f)^t \cdot (T_p f \cdot (T_p f)^t)^{-1} f''_{|p}(X, Y).$$

2) *If $k = 1$, then using* $\operatorname{grad} f = df^{\#}$ *and the normal field* $\mathfrak{n} := \frac{\operatorname{grad} f}{\|df\|}$ *one obtains*

$$g(X, WY) = \frac{-1}{\|df\|} f''(X, Y) \text{ and } II(X, Y) = -f''(X, Y) \frac{\operatorname{grad} f}{\|df\|^2}.$$

Proof. 1) As $f \circ \gamma = 0$, differentiating shows that $\operatorname{im} T\gamma = \ker Tf = (\operatorname{im}(Tf)^t)^{\perp}$. And if $X = T_x\gamma(W)$, $Y = T_x\gamma(Z)$, then

$$\begin{aligned} 0 &= f''_{|\gamma(x)}(T\gamma(W), T\gamma(Z)) + T_{\gamma(x)}f(\gamma''(W, Z)) \\ &\overset{5.1.6}{=} f''_{|\gamma(x)}(X, Y) + T_{\gamma(x)}f(II(X, Y)). \end{aligned}$$

Because $T_p f \cdot (T_p f)^t \in \mathbb{R}^{k \times k}$ has maximal rank, the equation follows using the pseudoinverse via the approach $II(X, Y) = T_p f^t \cdot v$.
  2) is a special case of (1). $\qquad\qquad\square$

Example 5.1.8 If $f : \mathbb{R}^{n+1} \to \mathbb{R}, \mathbf{x} \to \|\mathbf{x}\|^2 - 1$, then $M = S^n$, $df(Y) = 2\langle \mathbf{x}, Y \rangle$ and $f''(X, Y) = 2\langle X, Y \rangle$. Hence $W = -\mathrm{id}$ and $K \equiv 1$.

Corollary 5.1.9 *Suppose $M^n \subset \mathbb{R}^{n+1}$ is determined by $x_0 = f(x_1, \ldots, x_n)$ satisfying $f(0) = 0$, $df_{|0} = 0$, and thus $T_0 M = \{0\} \times \mathbb{R}^n$. If $X, Y \in T_0 M$, then $g(X, WY) = f''_{|0}(X, Y)$, i.e.. II is the Hessian form of f.*

Proof. Set $\gamma(x) = \begin{pmatrix} f(x) \\ x \end{pmatrix}$. The result follows from Lemma 5.1.6 because $\gamma''_{|0} = \begin{pmatrix} f''(0) \\ 0 \end{pmatrix} \perp T_0 M$. $\qquad\qquad\square$

Theorem 5.1.10 *Let $M^n \subset \mathbb{R}^{n+1}$ be a hypersurface and $p \in M$. Then the following are equivalent:*

1) *$K_p > 0$ for any plane $\subset T_p M$,*
2) *M is strictly convex at p, i.e. $M \setminus \{p\}$ lies locally on one side of $T_p M$.*

Proof. $K > 0 \Rightarrow$ for all distinct indices j, k, the inequality $\lambda_j \lambda_k > 0$ holds for all principal curvatures,
  $\Leftrightarrow$ either $\lambda_j > 0$ holds for all j or $\lambda_j < 0$ holds for all j,
  $\Leftrightarrow$ $g(\cdot, W\cdot)$ is positive or negative definite. By Corollary 5.1.9, M around p (up to a translation of p to 0 and a rotation of $\mathbb{R}^{n+1}$) is given by $\mathbb{R}^n \to \mathbb{R}^{n+1}, x \mapsto \begin{pmatrix} \frac{1}{2}g(x, Wx) + O(|x|^3) \\ x \end{pmatrix}$. Therefore sufficiently close to 0 it lies either above or below $\{0\} \times \mathbb{R}^n$.
  If $g(\cdot, W\cdot)$ is positive or negative definite and X, Y are linearly independent, the Cauchy–Schwarz inequality states

$$K(X \wedge Y) \overset{\substack{\text{Gauß} \\ \text{equation}}}{=} \frac{g(X, WX)g(Y, WY) - g(X, Wy)^2}{\|X \wedge Y\|^2} > 0. \qquad\square$$

Theorem 5.1.3 states for a hypersurface that in terms of the shape operator

$$R(X, Y, Z, W) = \widetilde{R}(X, Y, Z, W) + g(\mathcal{W}X, Z)g(\mathcal{W}Y, W) - g(\mathcal{W}Y, Z)g(\mathcal{W}X, W)$$

or

$$\Omega(X, Y)Z = \widetilde{\Omega}(X, Y)Z - g(\mathcal{W}X, Z)\mathcal{W}Y + g(\mathcal{W}Y, Z)\mathcal{W}X.$$

This can be written in a shorter way by dualizing in the 3rd variable as $\Omega^{\#} = \widetilde{\Omega}^{\#}_{|TM^{\otimes 4}} + \frac{1}{2}\mathcal{W} \wedge \mathcal{W}$, where $\Omega^{\#} \in \Lambda^2 T^*M \otimes \Lambda^2 TM \subset \Lambda T^*M \otimes \mathrm{End}(TM)^{\#}$ because

$$\sum_{j,k,\ell,m} a_{jk}a_{\ell m}e^j \otimes e^\ell \otimes e_k \wedge e_m = \frac{1}{2} \sum_{j,k,\ell,m} a_{jk}a_{\ell m}e^j \wedge e^\ell \otimes e_k \wedge e_m.$$

Here $\mathcal{W}$ is understood as a vector-valued 1-form. Analogous to Lemma 4.1.2, we calculate independently in the outer algebras spanned by the forms and the vectors. Using the convention $\mathcal{W}, \Omega^{\#}, \widetilde{\Omega}^{\#} \in \Lambda T^*M \wedge \Lambda TM$, in the same way one obtains $\Omega^{\#} = \widetilde{\Omega}^{\#} - \frac{1}{2}\mathcal{W} \wedge \mathcal{W}$.

Theorem 5.1.11 *Suppose $M^{2n} \subset (\mathbb{R}^{2n+1}, g_{\mathrm{Eucl}})$ is an oriented compact submanifold. Then*

$$\mathrm{Pf}(-\Omega) = \frac{(2n)!}{2^n n!} \det \mathcal{W} \, d\mathrm{vol}.$$

In particular, one gets

$$\chi(M) = \frac{(2n-1)!!}{(2\pi)^n} \int_M \lambda_1 \cdots \lambda_{2n} \, d\mathrm{vol}.$$

The function $\det \mathcal{W} \in C^\infty(M)$ *is called the* **Gauß–Kronecker curvature**.

Proof. One finds

$$\frac{1}{n!}\mathcal{T}\left(\left(-\frac{1}{2}\mathcal{W} \wedge \mathcal{W}\right)^{\wedge n}\right) = \frac{1}{(-2)^n n!}\mathcal{T}(\mathcal{W}^{\wedge 2n})$$

$$= \frac{(2n)!}{(-2)^n n!} \det \mathcal{W} \cdot T(e^1 \wedge e_1 \wedge \cdots \wedge e^{2n} \wedge e_{2n})$$

$$= (2n-1)!!(-1)^n \det \mathcal{W} \cdot (-1)^{n(2n-1)} e^1 \wedge \cdots \wedge e^{2n}.$$

$\square$

Hopf Theorem 5.1.12 *Let $M^{2n} \subset (\mathbb{R}^{2n+1}, \langle \cdot, \cdot \rangle_{\mathrm{Eucl}})$ be an oriented compact submanifold and let $\mathfrak{n} : M^{2n} \to S^{2n}$ be a covering of degree m. Define $\varepsilon := 1$ or -1 if $\mathfrak{n}$ is orientation-preserving or -reversing, respectively. Then $\mathfrak{n}^* d\mathrm{vol}_{S^{2n}} = \varepsilon \det \mathcal{W} \, d\mathrm{vol}$ and $\chi(M) = 2\varepsilon m$.*

Proof. According to Lemma 5.1.14, one has $\langle \mathcal{W}X, Y \rangle_{\mathrm{Eucl}} = -\langle T\mathfrak{n}(X), Y \rangle_{\mathrm{Eucl}}$, and so

$$\mathfrak{n}^* d\mathrm{vol}_{S^{2n}} = \varepsilon \det \mathcal{W} \, d\mathrm{vol}_g.$$

Therefore one finds

$$
\begin{aligned}
\chi(M) &= \frac{(2n-1)!!}{(2\pi)^n} \int_M \det \mathcal{W} \, d\mathrm{vol} = \frac{(2n-1)!!}{(2\pi)^n} \varepsilon \int_M \mathfrak{n}^* d\mathrm{vol}_{S^{2n}} \\
&= \frac{(2n-1)!!}{(2\pi)^n} \varepsilon m \int_{S^{2n}} d\mathrm{vol}_{S^{2n}} = \frac{(2n-1)!!}{(2\pi)^n} \varepsilon m \cdot \mathrm{vol}\, S^{2n}. \qquad \square
\end{aligned}
$$

Remark These results can also be used the other way around to show the Poincaré–Hopf Theorem for hypersurfaces.

The following definition by O'Neill [ON1] subsumes the 2nd fundamental form and a generalization of the shape operator.

Definition 5.1.13 *If $\iota : M \to \widetilde{M}$ is a Riemannian immersion of arbitrary codimension, let $T \in \Gamma(M, (\iota^* T^* \widetilde{M})^{\otimes 2} \otimes \iota^* T\widetilde{M})$ denote the tensor given by*

$$
T_X Y := (\nabla^{\iota^* T\widetilde{M}}_{X^{TM}} Y^{TM})^N + (\nabla^{\iota^* T\widetilde{M}}_{X^{TM}} Y^N)^{TM}.
$$

If $X, Y \in \Gamma(M, TM)$, then $II(X, Y) = T_X Y$. Because $T_{\mathfrak{n}} = 0$, T does not have the symmetry of the 2nd fundamental form. The orthogonal projection of $\widetilde{\nabla}$ onto N induces a metric connection ∇^N satisfying $\nabla^N_X \mathfrak{n} := (\nabla^{\iota^* T\widetilde{M}}_X \mathfrak{n})^N$ for any $X \in TM, \mathfrak{n} \in \Gamma(M, N)$. Thus, analogous to the definition of the 2nd fundamental form, the following lemma holds:

Lemma 5.1.14 (Weingarten equation) *If $X \in \Gamma(M, TM)$, then the tensor T_X is skew-symmetric. If $\mathfrak{n} \in \Gamma(M, N)$, then $T_X \mathfrak{n} = \nabla^{\iota^* T\widetilde{M}}_X \mathfrak{n} - \nabla^N_X \mathfrak{n}$.*

The right-hand summand in the definition of T is therefore the negative of the adjoint of the left-hand summand. Given a hypersurface $M \subset \widetilde{M}$ and a normal vector $\mathfrak{n}$ of norm 1, one thus obtains

$$
\widetilde{g}(Y, -T_X\mathfrak{n}) = \widetilde{g}(T_X Y, \mathfrak{n}) = \widetilde{g}(II(X, Y), \mathfrak{n}) = \widetilde{g}(Y, \mathcal{W}X),
$$

in other words $T_X \mathfrak{n} = -\mathcal{W}X$.

Proof. ∇^{TM} and ∇^N induce a metric connection $\nabla^{TM \oplus N}$ on $TM \oplus N \cong \iota^* T\widetilde{M}$, and T equals the difference $\nabla^{\iota^* T\widetilde{M}} - \nabla^{TM \oplus N}$. According to Lemma 3.2.15, T is skew-symmetric. $\qquad \square$

Analogous to the proof of the Gauß equation one obtains:

Theorem 5.1.15 (Mainardi–Codazzi equation[2]) *Consider a Riemannian immersion $M \to \widetilde{M}$. If $Z \in T_p M, \mathfrak{n} \in N_p$, then*

$$
\widetilde{R}(X, Y, Z, \mathfrak{n}) = \widetilde{g}((\nabla_Y T)_X Z, \mathfrak{n}) - \widetilde{g}((\nabla_X T)_Y Z, \mathfrak{n}).
$$

---

[2] 1853, Karl M. Peterson (thesis), 1828–1881; 1856, G. Mainardi; 1860, D. Codazzi.

In this equation T is differentiated using the connection ∇ induced by ∇^{TM}, ∇^N on $T^*M^{\otimes 2} \otimes N$. However, when applying $\nabla^{\iota^*\widetilde{TM}}$ the equation is also correct. The skew-symmetry of $\nabla_Y T$ yields the transformation

$$\widetilde{R}(X,Y,Z,\mathfrak{n}) = -\widetilde{g}(Z,(\nabla_Y T)_X \mathfrak{n}) + \widetilde{g}(Z,(\nabla_X T)_Y \mathfrak{n}). \tag{5.1}$$

If the connection is extended to $\mathfrak{A}^1(M, TM \otimes N)$, this can also be written as

$$g(\nabla^{TM\otimes N} II^\#, Z) = (\widetilde{\Omega}_{|TM^{\otimes 2}} Z)^N.$$

Given a hypersurface $M \subset \widetilde{M}$ and $\|\mathfrak{n}\| = 1$, $\widetilde{g}(\widetilde{\nabla}\mathfrak{n}, \mathfrak{n}) = 0$, so $\nabla^N \mathfrak{n} = 0$. If, in addition, $\widetilde{M}$ is the Euclidean $\mathbb{R}^{n+1}$, equation (5.1) implies $\nabla^{TM}(T\mathfrak{n}) \equiv 0 \in \mathfrak{A}^2(M, TM)$, where $T\mathfrak{n}$ is interpreted as an element of $\mathfrak{A}^1(M, TM)$.

Proof. Consider $X, Y, Z \in \Gamma(M, TM)$, $\mathfrak{n} \in \Gamma(M, N)$ with extensions $\widetilde{X}, \widetilde{Y}, \widetilde{Z}, \widetilde{\mathfrak{n}}$ to $\widetilde{M}$. Then

$$\begin{aligned}
\widetilde{g}(\widetilde{\nabla}_X \widetilde{\nabla}_{\widetilde{Y}} \widetilde{Z}, \mathfrak{n}) &= X.\widetilde{g}(\widetilde{\nabla}_Y \widetilde{Z}, \mathfrak{n}) - \widetilde{g}(\widetilde{\nabla}_Y \widetilde{Z}, \widetilde{\nabla}_X \widetilde{\mathfrak{n}}) \\
&= X.\widetilde{g}(II(Y,Z), \mathfrak{n}) - \widetilde{g}(II(Y,Z), \nabla_X^N \mathfrak{n}) - g(\nabla_Y Z, T_X \mathfrak{n}) \\
&= \widetilde{g}(\nabla_X^N(II(Y,Z)), \mathfrak{n}) + \widetilde{g}(II(X, \nabla_Y Z), \mathfrak{n}) \\
&= \widetilde{g}((\nabla_X II)(Y,Z), \mathfrak{n}) + \widetilde{g}(II(\nabla_X Y, Z), \mathfrak{n}) \\
&\quad + \widetilde{g}(II(Y, \nabla_X Z), \mathfrak{n}) + \widetilde{g}(II(X, \nabla_Y Z), \mathfrak{n}).
\end{aligned}$$

Thus one obtains because of the torsion freeness

$$\begin{aligned}
\widetilde{g}(\widetilde{\nabla}_X \widetilde{\nabla}_{\widetilde{Y}} \widetilde{Z} - \widetilde{\nabla}_Y \widetilde{\nabla}_{\widetilde{X}} \widetilde{Z} &- \widetilde{\nabla}_{[X,Y]} \widetilde{Z}, \mathfrak{n}) \\
&= \widetilde{g}((\nabla_X II)(Y,Z), \mathfrak{n}) - \widetilde{g}((\nabla_Y II)(X,Z), \mathfrak{n}). \qquad \square
\end{aligned}$$

Remark There is another curvature identity for $R(X, Y, \mathfrak{n}, \mathfrak{n}')$, the Ricci equation (Exercise 5.1.20).

Exercises

Exercise 5.1.16 Let $\gamma : U \to \mathbb{R}^{n+1}$, $U \subset \mathbb{R}^n$, be a parametrization of a hyper surface with normal vector $\mathfrak{n} : U \to S^n$. Show that $II(X,Y) = \langle \gamma''(X,Y), \mathfrak{n}\rangle \mathfrak{n}$ and $\mathcal{W}X = -d\gamma^{-1}(d\mathfrak{n}(X)))$ for all $X, Y \in TU$.

Exercise 5.1.17 On a surface $M \subset \mathbb{R}^3$, choose an orientation on $U \subset M$. Given a local oriented basis (e_1, e_2), let $\mathfrak{n} : U \to S^2$ be the normal vector such that the basis $(e_1, e_2, \mathfrak{n})$ of $\mathbb{R}^3$ is standard oriented.

1) Applying Exercise 5.1.16, show that $\mathfrak{n}^* d\mathrm{vol}_{S^2} = K\, d\mathrm{vol}_M$, where $K \in C^\infty(M)$ is the sectional curvature.
2) Deduce that $\int_M K\, d\mathrm{vol}_M = 4\pi$ holds, if $\mathfrak{n} : M \to S^2$ is a diffeomorphism.

Exercise 5.1.18 Consider the **catenoid**

$$f_1 : \mathbb{R}^2 \to \mathbb{R}^3, (x, y) \mapsto (\cosh x \cos y, \cosh x \sin y, x)$$

and the **helical surface**

$$f_2 : \mathbb{R}^2 \to \mathbb{R}^3, (x, y) \mapsto (\sinh x \cos y, \sinh x \sin y, y).$$

Find a local isometry between these surfaces. Sketch both surfaces and determine the 2nd fundamental forms and the shape operators on $T\mathbb{R}^2$ as well as the principal curvatures.

Exercise* 5.1.19 Let M a ruled surface as in Exercise 3.1.27 with a parametrization u associated to (α, w), where α is the striction curve. Show that the curvature is given by

$$K_{u(t,s)} = -\frac{\lambda(t)^2}{(\lambda(t)^2 + s^2)^2} \leq 0.$$

Exercise* 5.1.20 Let $\iota : M \hookrightarrow \widetilde{M}$ be an immersion with normal bundle N, let $\widetilde{g}$ be a Riemannian metric on $\widetilde{M}$ and let g, g^N be the induced metrics on TM, N. Analogously to the Gauß and Mainardi–Codazzi equations, deduce the **Ricci equation**

$$\widetilde{R}(X, Y, \mathfrak{n}, \mathfrak{n}') = -g^N(\Omega^N(X, Y)\mathfrak{n}, \mathfrak{n}') + g(T_Y\mathfrak{n}, T_X\mathfrak{n}') - g(T_X\mathfrak{n}, T_Y\mathfrak{n}').$$

5.2 Geodesics

geodesic	radial vector field
exponential map	Gauß's Lemma
normal coordinate system	energy
normal coordinates	totally geodesic
normal neighborhood	torsion
geodesic polar coordinates	

In this section, shortest paths on manifolds are studied. For a chosen orthonormal basis $(e_j)_j$ of T_pM, these provide for a point $p \in M$ a canonical chart around this point, the normal coordinates.

Definition 5.2.1 *A **geodesic** on a Riemannian manifold (M, g) is a curve c with parallel velocity field, i.e. $\nabla^{c^*TM} \dot{c} \equiv 0$, where ∇^{TM} is the Levi-Civita connection.*

Remark Since the tangent space of $\mathbb{R}$ is spanned by the vector $\frac{\partial}{\partial t}$, the defining condition is equivalent to the vanishing of the "acceleration" $\nabla^{c^*TM}_{\partial/\partial t} \dot{c}$. Lemma 3.2.19 shows that $\|\dot{c}\| \equiv$ const., hence geodesics have constant velocity.

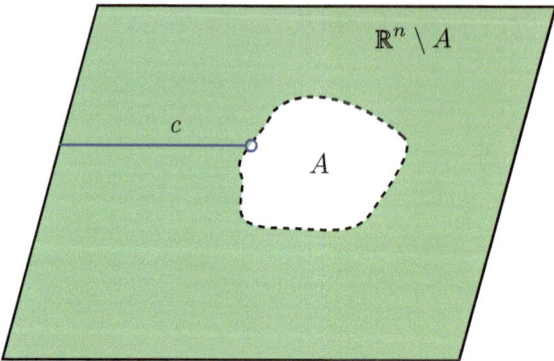

Fig. 5.3 A geodesic not definable on all of $\mathbb{R}$.

Example If $M = (\mathbb{R}^n, g_{\text{Eucl}})$, then this equation becomes $\ddot{c} \equiv 0$ and thus $c(t) = at + b$. Because of the local isometry of a cylinder $S^1 \times \mathbb{R} \subset \mathbb{R}^3$ to $\mathbb{R}^2$, this also determines the geodesics on the cylinder.

As the simple example $M = \mathbb{R}^n \setminus A$ for a closed subset A already shows, geodesics need not be definable on the whole of $\mathbb{R}$ (Fig. 5.3).

Lemma 5.2.2 *Consider* $p \in M$ *and* $X \in T_p M \setminus \{0\}$. *Then there exist an* $\varepsilon > 0$ *and a neighborhood* $\widetilde{U} \subset TM$ *of* X *such that for any* $Y \in \widetilde{U}$ *there exists exactly one geodesic* $c :] - \varepsilon, \varepsilon[\to M$ *satisfying* $\dot{c}(0) = Y$. *This geodesic* c *depends* C^∞ *on* Y.

Example 5.2.3 Suppose $M = S^n$. Rather than solving the differential equation in local coordinates, it is much easier to determine the geodesics with the above lemma by a symmetry argument. Let $p \in S^n$, $X \in T_p M$ with $\|X\| = 1$ and let $H \subset \mathbb{R}^{n+1}$ be the 2-dimensional plane spanned by p, X. Then S^n, p, X are invariant under reflection at H. Thus, because of its uniqueness, the geodesic c with initial vector X is also invariant under reflection. Hence it must lie on the great circle $H \cap S^N$. Since geodesics have constant velocity, it follows that $c(t) = p \cdot \cos t + X \cdot \sin t$.

Proof. Let $\varphi : U \to V$ be a chart around p and in this chart let $\nabla^{TM} = d + \Gamma$. Then the defining equation for geodesics reads

$$0 = \frac{\partial^2 \varphi(c(t))}{\partial t^2} + \Gamma_{|\varphi(c(t))} \left(\frac{\partial \varphi(c(t))}{\partial t}, \frac{\partial \varphi(c(t))}{\partial t} \right).$$

This is a second-order ordinary differential equation. Hence, around $\varphi(p) \in V$ and $T\varphi(X)$ there exist neighborhoods $\widetilde{V} \subset \mathbb{R}^n$ and $W \subset \mathbb{R}^n$, respectively, and an $\varepsilon > 0$, such that this differential equation has a uniquely determined solution $\varphi \circ c :$ $] - \varepsilon, \varepsilon[\to V$ for initial values $\varphi(c(0)) \in \widetilde{V}$, $T\varphi(\dot{c}(0)) \in W$. Set $\widetilde{U} := T\varphi^{-1}(\widetilde{V} \times W)$. The solution depends smoothly on the initial values since the coefficients of the equation are smooth. $\qquad \square$

Corollary 5.2.4 *Given any $p \in M$, there exist a neighborhood U and a $\delta > 0$ such that for all $Y \in TU$ with $\|Y\| < \delta$ there exists a uniquely determined geodesic $c :] - 2, 2[\rightarrow M$ satisfying $\dot{c}(0) = Y$.*

Proof. Let $c :] - \varepsilon, \varepsilon[\times \widetilde{U} \rightarrow M$ be a solution as obtained in Lemma 5.2.2 with $\widetilde{V} \times W \subset T\varphi(\widetilde{U})$ for a neighborhood W of $T\varphi(X) = 0$. Rescale c as $\widetilde{c}(t) := c(\frac{\varepsilon t}{2})$. Then $\widetilde{c}$ is a geodesic with initial velocity $\dot{\widetilde{c}}(0) = \frac{\varepsilon}{2}\dot{c}(0)$, since the defining differential equation is invariant under rescaling of t with constants. Thus, solutions exist on $] - 2, 2[$ for any initial values in $T\varphi^{-1}(\widetilde{V} \times \frac{\varepsilon}{2}W)$. □

Definition 5.2.5 *Let $W \subset TM$ be an open neighborhood of $(q, 0) \in TM$ such that $c(1)$ exists for all geodesics c with initial values in W. The **exponential map** is defined as the map $\exp_p : W \rightarrow M, (p, X) \mapsto \exp_p X := c_X(1)$, where c_X is the geodesic with $c(0) = p, \dot{c}(0) = X$.*

Example If $M = (\mathbb{R}^n, g_{\mathrm{Eucl}})$ and $p, X \in \mathbb{R}^n$, then $\exp_p X - p + X$.

The relation between this exponential map and the one for a Lie group G depends, of course, on the choice of a metric on G. On many Lie groups there is no metric that makes these two maps equal.

Lemma 5.2.6 *If $p \in M$, then $T_0 \exp_p = \mathrm{id}_{T_p M}$ (where $T_0 \exp_p : T_0 T_p M \rightarrow T_p M$ is understood as an endomorphism of $T_p M$).*

Proof.

$$\frac{\partial}{\partial \varepsilon}\Big|_{\varepsilon=0} \exp_p(\varepsilon X) = \frac{\partial}{\partial \varepsilon}\Big|_{\varepsilon=0} c_{\varepsilon X}(1) = \frac{\partial}{\partial \varepsilon}\Big|_{\varepsilon=0} c_X(\varepsilon) = \dot{c}_X(0) = X. \qquad \square$$

Corollary 5.2.7 *The map $\exp_p$ is a local diffeomorphism of a neighborhood $V \subset T_p M$ of $0_{T_p M}$ onto a neighborhood U of p in M, i.e., a local parametrization of M.*

(This follows from the Implicit Function Theorem.) Thus, any coordinate system on $T_p M$ induces local coordinates on M. Choose an orthonormal basis $(e_j)_j$ of $T_p M$, i.e. an isometry from $(T_p M, g_p)$ to the Euclidian $\mathbb{R}^n$, to which charts map according to the definition of manifolds. The parameterization by $\underbrace{\widetilde{V}}_{\subset \mathbb{R}^n} \overset{(e_j)}{\rightarrow} \underbrace{V}_{\subset T_p M} \overset{\exp}{\rightarrow} \underbrace{U}_{\subset M}$

is called the **normal coordinate system** (or **normal coordinates**) on the **normal neighborhood** U. Associated to polar coordinates on $\mathbb{R}^n$ there are corresponding

geodesic polar coordinates $\underbrace{\widehat{V}}_{\subset \mathbb{R}^+ \times S^{n-1}} \overset{(e_j)}{\rightarrow} \underbrace{V}_{\subset T_p M} \overset{\exp}{\rightarrow} \underbrace{U}_{\subset M}$. The images of radial

straight lines are therefore geodesics.

Definition 5.2.8 *The **radial vector field** is defined as*

$$\mathfrak{R} := \exp_{p*} \frac{\partial}{\partial r} = \exp_{p*} \sum x_\ell e_\ell.$$

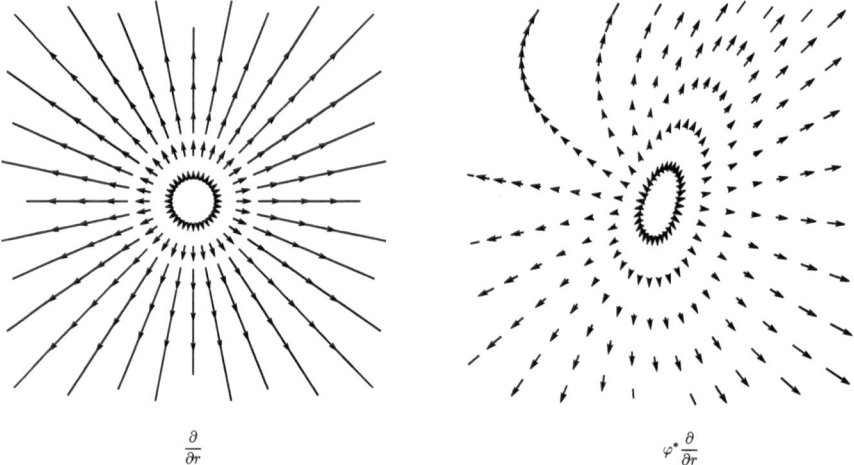

$$\frac{\partial}{\partial r}$$ $$\varphi^* \frac{\partial}{\partial r}$$

Fig. 5.4 Radial vector field.

Thus $\mathfrak{R}_{|\exp X} = (T_X \exp_p)(X)$ (via $T_X(T_pM) \overset{\text{can.}}{\cong} T_pM$) or $\mathfrak{R}_{c(t)} = t\dot{c}(t)$ if c is a geodesic starting at p (Fig. 5.4). Thus, this vector field is more or less the velocity field of the radial geodesic, but unlike the latter, it is also well-defined at p. Given $e_{j|(p,X)} \in T_X(T_pM) \overset{\text{can.}}{\cong} T_pM$, let $b_{j|\exp X} := T_X \exp(e_{j|p}) = (\exp_p)_* e_{j|(p,X)}$ be the vector fields associated to Cartesian coordinates on U. Since $\exp$ is a diffeomorphism on U, $(b_j)_j$ is a C^∞-basis on TU, though in general not an orthonormal basis.

Given $X \in T_pM$, let $e_{j|\exp tX}$ be the parallel transport of $e_{j|p}$ along $\exp_p tX$, i.e. $\nabla_\mathfrak{R} e_j \equiv 0$. Since the parallel transport depends smoothly on the initial value and is an isometry, $(e_j)_j$ is a C^∞ orthonormal basis on TU. Because $b_{j|p} = T_0 \exp e_j \overset{5.2.6}{=} e_{j|p}$, it follows that $b_{j|\exp X} = e_j + O(\|X\|)$.

Lemma 5.2.9 *The radial vector field satisfies*

1) $\nabla_\mathfrak{R} \mathfrak{R} = \mathfrak{R}$,

2) $\mathfrak{R}_{|\exp_p \sum x_j e_j} = \sum x_j b_j$,

3) $\mathfrak{R}_{|\exp_p \sum x_j e_j} = \sum x_j e_j$, *in particular* $\|\mathfrak{R}\|_{|\exp X}^2 = \|X\|^2$.

Proof. 1) Given the geodesic $c(t) := \exp_p tX$, one obtains in c^*TM

$$c^* \nabla_\mathfrak{R} \mathfrak{R} = \nabla_{t\partial/\partial t}^{c^*TM}(t\dot{c}) = t\frac{\partial t}{\partial t} \cdot \dot{c} = \mathfrak{R}.$$

2) $$\mathfrak{R}_{|\exp_p \sum x_j e_j} = \exp_{p*} \sum x_j e_j = \sum x_j \exp_{p*} e_j = \sum x_j b_j.$$

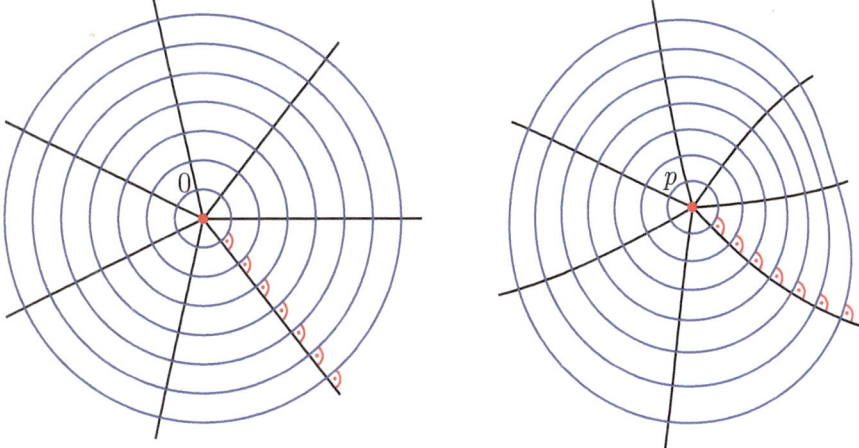

Fig. 5.5 Polar coordinates on $T_p M$ and geodesic polar coordinates on M.

3) Choose fixed $x_1, \ldots, x_n \in \mathbb{R}$. Then $\frac{1}{t}\mathfrak{R}$ is parallel along $\exp(t\sum x_j e_j)$ as the velocity field of the geodesic, $\sum x_j e_j$ is parallel along the same curve by the construction of e_j, and

$$\lim_{t\searrow 0} \frac{1}{t}\mathfrak{R}_{\exp(t\sum x_j e_j)} \overset{(2)}{=} \lim_{t\searrow 0} \frac{1}{t}\sum tx_j b_{j|\exp(t\sum x_j e_j)} = \sum x_j e_{j|p}.$$

The uniqueness of the parallel transport with the same initial values implies that $\mathfrak{R}_{|\exp_p \sum x_j e_j} = \sum x_j e_j$. $\qquad\square$

Gauß's Lemma 5.2.10 *If $X, Y \in V \subset T_p M$, then*

$$g_{\exp X}(\overbrace{\exp_* X, \exp_* Y}^{=\mathfrak{R}}) = g_p(X,Y).$$

If Y is proportional to X, then this is again the equation $\|\mathfrak{R}\|^2_{|\exp X} = \|X\|^2$. In the case $Y \perp X$ the new statement is that the image vectors are also perpendicular to each other. Hence the images of spheres around $0_{T_p M}$ intersect the radial geodesics perpendicularly (Fig. 5.5).

Proof. The equation is linear in Y, thus it suffices to consider the case $Y = e_j$. Then the left-hand side equals $g_{\exp \sum x_k e_k}(\mathfrak{R}, b_j)$ and the right-hand side equals

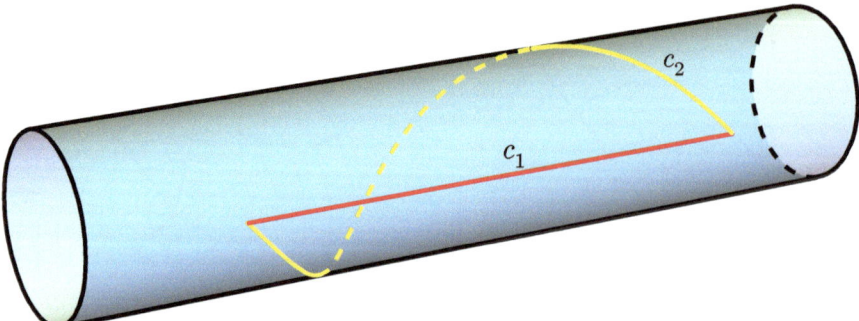

Fig. 5.6 The geodesic c_1 is shorter than the geodesic c_2.

$g_p(\underbrace{\sum x_k e_k}_{=:X}, e_j) = x_j.$ Furthermore one obtains

$$
\begin{aligned}
\mathfrak{R}.(g(\mathfrak{R}, b_j))_{|\exp \sum x_k e_k} \\
&= g(\nabla_{\mathfrak{R}}\mathfrak{R}, b_j) + g(\mathfrak{R}, \nabla_{\mathfrak{R}} b_j) \\
&\overset{5.2.9(1)}{=} g(\mathfrak{R}, b_j) + g(\mathfrak{R}, [\mathfrak{R}, b_j]) + g(\mathfrak{R}, \nabla_{b_j}\mathfrak{R}) \\
&= g(\mathfrak{R}, b_j) + g(\mathfrak{R}, \exp_* \underbrace{[\sum x_k e_k, e_j]}_{=-e_j \text{ as vector field on } T_p M}) + \frac{1}{2} b_j.g(\mathfrak{R}, \mathfrak{R}) \\
&\overset{5.2.9(3)}{=} \frac{1}{2}\frac{\partial}{\partial x_j}\sum x_k^2 = x_j.
\end{aligned}
$$

On the other hand, $\mathfrak{R}.x_j = \exp_*(\sum_k x_k e_k.x_j) = x_j$ also holds. Thus there is a constant $y \in \mathbb{R}$ such that $g(\mathfrak{R}, b_j) = x_j + y$ along any radial geodesic. Because $g(\mathfrak{R}, b_j)_p = 0$, it follows that $y = 0$ and therefore

$$
g_{\exp \sum x_k e_k}(\mathfrak{R}, b_j) = x_j. \qquad \square
$$

Theorem 5.2.11 *Let $V \subset T_p M$ be ball-shaped and be the preimage of a normal neighborhood. Then*

$$
\operatorname{dist}(p, \exp_p X) = \|X\| \qquad \text{for all } X \in V.
$$

Geodesics are locally the shortest connecting paths. Conversely, shortest paths (even globally) are always geodesics.

Globally, a geodesic does not have to be the shortest path. See for example Fig. 5.6, which shows two different geodesics. Using the example of Fig. 5.3 one can also see that shortest paths between two points do not always exist.

Proof. Suppose c is a curve from p to $\exp_p X$.

1st case: c stays in the normal neighborhood. Then $r \cdot u := \exp_p^{-1} \circ c$ is a curve in V from 0 to X with $r : I \to \mathbb{R}_0^+, u : I \to S^{n-1}$ (hence $\|u\| \equiv 1$). Thus

$$
\begin{aligned}
\|\dot{c}\|^2 \quad &= \quad \|T_{r(t)u(t)} \exp_p (\dot{r}u + r\dot{u})\|^2 \\
&= \quad \dot{r}^2 \cdot \|T_{ru} \exp u\|^2 + r^2 \|T_{ru} \exp \dot{u}\|^2 \\
&\quad + 2\dot{r} \cdot g_{ru}(T \exp(ru), T \exp(\dot{u})) \\
&\overset{\text{Gauß's Lemma}}{=} \dot{r}^2 \|u\|^2 + r^2 \|T_{ru} \exp \dot{u}\|^2 + 2\dot{r}r \cdot g_p(u, \dot{u}) \\
&= \quad \dot{r}^2 + r^2 \|T_{ru} \exp \dot{u}\|^2 \geq \dot{r}^2.
\end{aligned}
$$

Thus length$(c) = \int_a^b \|\dot{c}\| \, dt \overset{\text{=, if } \dot{u} \equiv 0}{\geq} \int_a^b |\dot{r}| \, dt \overset{\text{=, if } r \text{ is monotonic}}{\geq} |r(b) - r(a)| = \|X\|.$
Hence the minimum is taken exactly at the radial geodesic.
2nd case: c leaves the normal neighborhood. Then according to the 1st case c is at least as long as the radius of ∂V and thus longer than $\|X\|$.
The converse follows because in the normal neighborhood of any point $c(t_0)$ the curve c must be a geodesic. Hence it satisfies $\nabla^{c^* TM} \dot{c} \equiv 0$ everywhere. $\qquad \square$

Remark According to this proof, geodesics minimize (in the above sense) **any** functional of the form $\int f(\|\dot{c}\|) \, dt$ with f monotonically increasing, in particular the **energy** $\int \|\dot{c}\|^2 dt$.

Corollary 5.2.12 *Let $c \neq \widetilde{c}$ be geodesics of the same length between $c(0)$ and $c(t_0)$. Then for all $t_1 > t_0$, c is not a shortest path from $c(0)$ to $c(t_1)$.*

Proof. Without restriction, suppose that $c(t_0) = \widetilde{c}(t_0)$, i.e., $\|\dot{c}\| = \|\dot{\widetilde{c}}\|$. Consider $\gamma : t \mapsto \begin{cases} \widetilde{c}(t) & \text{if } \; t \leq t_0 \\ c(t) & t_0 < t < t_1 \end{cases}$ Then length γ =length $c_{|[0,t_1]}$. If c were a shortest path, then γ would also be a shortest path, i.e., a geodesic. But because of the uniqueness of geodesics with initial vector $-\dot{c}(t_1)$, $\gamma = c$ follows. $\quad \notni \qquad \square$

Exercises

Exercise 5.2.13 Let $G \subset SO(k)$ be a Lie subgroup and let $\mathfrak{so}(k)$ be the vector space of skew-symmetric $k \times k$-matrices. For any $A, B \in \mathfrak{g} \subset \mathfrak{so}(k)$ set $g_{\mathrm{id}}(A, B) = -\mathrm{Tr}\, AB$. Show that for $A \subset \mathfrak{g}$, $\exp_{\mathrm{id}}(A)$ is given by

$$
e^A = \sum_{j=0}^{\infty} \frac{A^j}{j!}
$$

(for example by applying Exercise 3.4.10).

Exercise 5.2.14 An immersion $\iota : M \hookrightarrow \widetilde{M}$ is called **totally geodesic** if $II \equiv 0$. Show that this is equivalent to stating that every geodesic on M is also a geodesic on $\widetilde{M}$.

Exercise 5.2.15 Show that the common fixed point set M^Γ of a set Γ of isometries is a totally geodesic submanifold. In particular, any one-dimensional connected component of the fixed point set is the path of a geodesic.

Exercise 5.2.16 Consider an isometry γ of a compact Riemannian manifold M.

1) Show that a fixed point $x \in M$ of γ is isolated if and only if none of the eigenvalues of $T_x\gamma$ is equal to 1.
2) Prove that the fixed point set M^γ is finite if all fixed points are isolated.

Exercise 5.2.17 Using Exercise 5.2.15, prove that the common fixed point set M^Γ of a finite group of diffeomorphisms is a submanifold.

Exercise 5.2.18 Let γ_t be a one-parameter group of isometries and let $X := \frac{\partial}{\partial t}\big|_{t=0}\gamma_t$. Let M^X be the zero set of X. Show that

1) X is a Killing vector field.
2) On a neighborhood $U \subset M$ such that $\overline{U}$ is compact and for t sufficiently small, M^X is the fixed point set of γ_t.
3) $\ker \nabla X_{|M^X} = T(M^X)$.
4) ∇X is parallel along M^X.
5) The normal bundle N to M^X can be decomposed as an orthogonal sum of the eigenspaces of $(\nabla X)^2$.
6) The vector bundle N has even rank and it can be given a complex structure.
7) If M is orientable, then every connected component of M^X is orientable

Exercise* 5.2.19 Suppose X is a Killing field of constant norm. Show that the flow trajectories are geodesics.

Exercise 5.2.20 Consider a submanifold N of a Riemannian manifold M.

1) Let c be a geodesic in M whose path lies in N. Deduce that c is also a geodesic in N (in particular, straight lines on submanifolds of Euclidean space are geodesics).
2) Let c be a curve in N. Show that c is a geodesic in N if and only if $\nabla^{c^*TM}_{\partial/\partial t}\dot{c} \perp c^*TN$.

Exercise 5.2.21 Let $M = \mathbb{C}/(\mathbb{Z} + \tau\mathbb{Z})$ be a flat 2-dimensional torus associated to $\tau \in \mathbb{C}\backslash\mathbb{R}$. Determine the points reached by at least two different geodesics of minimal length starting at 0. Draw a sketch of this set of points in a fundamental domain. To do so you can assume that $|\tau| \geq 1$ and $0 \leq \operatorname{Re}\tau \leq 1/2$ (up to scaling, this can always be achieved by rotations, reflections and the choice of grid basis).

Exercise 5.2.22 Let G be a Lie group.

1) Show that the condition "if Y is any left-invariant vector field, then $\nabla^L Y = 0$" defines a connection ∇^L on TG.
2) Compute the curvature Ω^L.
3) Compute the **torsion**

$$T^L : \mathfrak{g} \times \mathfrak{g} \to \mathfrak{g}$$

$$(X, Y) \mapsto \nabla^L_X Y - \nabla^L_Y X - [X, Y].$$

4) Show that for ∇^L, the Lie group exponential map and the geodesic exponential map $\exp^L$ agree (the latter is defined analogously to the case of the Levi-Civita connection).

Exercise* 5.2.23 Find a simpler proof of Theorem 3.1.17(3),(4) applying Theorem 5.2.11.

5.3 Jacobi Fields

Jacobi field	geodesic disk
variation of a geodesic	rolling over the plane
geodesic sphere	osculating developable surface
geodesic ball	Minding's Theorem
geodesic circle	

Jacobi fields are infinitesimal variations of geodesics. In this section they are used to determine the first terms of the Taylor expansion of the Riemannian metric in normal coordinates. Thus, at the end of the section, the behavior of small geometric objects around a point $p \in M$ can be related to the curvature at p.

Definition 5.3.1 *The **Jacobi fields** $Y \in \Gamma(I, c^*TM)$ along a geodesic c are defined as the solutions of the differential equation*

$$(\nabla^{c^*TM}_{\partial/\partial t})^2 Y = \Omega_{c(t)}(\dot{c}, Y)\dot{c}.$$

Example 5.3.2 Because $\nabla^{c^*TM}_{\partial/\partial t}\dot{c} = 0$, both $\dot{c}$ and $Y_t = t \cdot \dot{c}(t)$ are Jacobi fields.

Lemma 5.3.3 *The vector space of all Jacobi fields is isomorphic to $(T_{c(0)}M)^2$ via the map $Y \mapsto (Y_0, \nabla_{\partial/\partial t}Y_t|_{t=0})$.*

Proof. Uniqueness and global existence of solutions of ordinary linear differential equations (of 2nd order). $\qquad\square$

A **variation of a geodesic** $c : I \to M$ is defined as a C^∞-map $H : U \to M$, $U \subset \mathbb{R}^2$ with $U \cap 0 \times \mathbb{R} = I$, $H(0, \cdot) = c$, such that $H(s, \cdot)$ is a geodesic for each s.

Theorem 5.3.4 *The Jacobi fields are exactly the vector fields Y such that there exists such a variation H that satisfies $Y = \frac{\partial}{\partial s}|_{s=0}H$.*

Intuitively Y can be regarded as the infinitesimal distance to neighboring geodesics in H.

Example The Examples 5.3.2 can be obtained when considering the two geodesic variations $H(s, t) = c(t+s)$ and $H(s, t) = c((s+1)t)$. Since geodesics have constant velocity, all reparametrizations of c yielding geodesics again are affine.

Proof. 1) Let H be a variation of c. Locally around a point $H(s,t) \in M$ let $\widetilde{X}, \widetilde{Y}$ be extensions of $\frac{\partial H}{\partial t}, \frac{\partial H}{\partial s}$, respectively. According to Lemma 1.4.7 one thus obtains $[\widetilde{Y}, \widetilde{X}]_{|H(s,t)} = [\frac{\partial}{\partial s}, \frac{\partial}{\partial t}].H = 0$. Considering $Y := \frac{\partial}{\partial s}_{|s=0} H$ and $H^* : TM \to H^*TM$, $H^*\Omega = \Omega^{H^*TM} \in \mathfrak{A}^2(U, \mathrm{End}(H^*TM))$ at $s = 0$ is therefore given by

$$\Omega(\frac{\partial H}{\partial t}, \frac{\partial H}{\partial s})\frac{\partial H}{\partial t}_{|s=0}$$

$$= \Omega^{H^*TM}(\frac{\partial}{\partial t}, \frac{\partial}{\partial s})\dot{c}$$

$$= \nabla^{H^*TM}_{\partial/\partial t} \nabla^{H^*TM}_{\partial/\partial s} \frac{\partial H}{\partial t} - \nabla^{H^*TM}_{\partial/\partial s} \underbrace{\nabla^{H^*TM}_{\partial/\partial t} \frac{\partial H}{\partial t}}_{=0} - \underbrace{\nabla^{H^*TM}_{[\partial/\partial t, \partial/\partial s]} \frac{\partial H}{\partial t}}_{=0}$$

$$= \nabla^{H^*TM}_{\partial/\partial t} H^* \nabla^{TM}_{\widetilde{Y}} \widetilde{X}$$

$$= \nabla^{H^*TM}_{\partial/\partial t} \nabla^{H^*TM}_{\partial/\partial t} \frac{\partial H}{\partial s} + \nabla^{H^*TM}_{\partial/\partial t} H^* \underbrace{[\widetilde{Y}, \widetilde{X}]}_{=0} \stackrel{s=0}{=} (\nabla^{c^*TM}_{\partial/\partial t})^2 Y.$$

The pullback $H^* : TM \to H^*TM$ of vector bundles should not be mistaken here for the pullback of vector fields by diffeomorphisms.

2) Since a geodesic is uniquely determined by the initial vector, geodesic variations $H : I \times J \to M$ uniquely correspond to maps $\widetilde{H} : J \to TM, s \mapsto \frac{\partial H}{\partial t}(s, 0)$. Hence, the vector fields $\frac{\partial H}{\partial s}_{|s=0}$ are mapped bijectively and linearly to $\frac{\partial H}{\partial s}_{|s=0}$ in the vector space $T_{\frac{\partial H}{\partial t}(0,0)} TM$ of dimension $\dim TM = 2n$. On the other hand, by Lemma 5.3.3 this is the dimension of the space of Jacobi fields, so the map in part (1) is surjective. □

Corollary 5.3.5 *Consider a geodesic $c : I \to M, t \mapsto \exp_p tX_0, 0 \in I$. Then the Jacobi fields along c having a zero at p are the sections $Y \in \Gamma(I, c^*TM)$ of the form $Y_t = T_{tX_0} \exp_p(tV)$ with $V \in T_pM$ (Fig. 5.7).*

Therefore these Jacobi fields provide the derivative of $\exp_p$ at a general point in a general direction.

Proof. Differentiating the variation $H(s, t) := \exp_p(t(X_0 + sV))$ of radial geodesics yields these Jacobi fields. Then $Y_0 = 0$ and because $T_0 \exp_p =\mathrm{id}$, one obtains

$$\nabla_{\partial/\partial t} Y_{t|t=0} = \left((T_{tX_0} \exp_p)(V) + t \cdot \nabla_{\partial/\partial t}(T_{tX_0} \exp_p)(V) \right)_{|t=0} = V.$$

Thanks to the uniqueness to given initial data, these are all of the Jacobi fields having a zero at p. □

In terms of the basis $(b_j)_j$ in a normal neighborhood, these Jacobi fields have the form $Y_t = t \cdot \sum v_j b_j$ with $v_j \in \mathbb{R}, j = 1, \dots, n$.

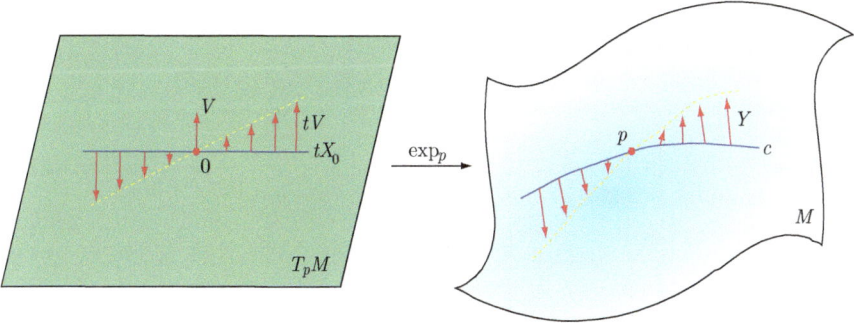

Fig. 5.7 Jacobi field having a zero at p.

Theorem 5.3.6 *Locally, g has the Taylor expansion*

$$(\exp^* g)_{|X} = g_p + \frac{1}{3} g_p(\overbrace{\Omega_p(X,\cdot)X,\cdot}^{R(X,\cdot,\cdot,X)}) + O(\|X\|^3)$$

$$= g_p + \frac{1}{3} \sum_{k,\ell=1}^{n} g_p(\Omega_p(e_k,\cdot)e_\ell,\cdot) x_k x_\ell + O(\|X\|^3)$$

for $X \in T_pM$.

In a ball-shaped neighborhood of p, $\|X\|$ is the metric distance to p and thus this can be taken as a Taylor expansion by distance. This formula is in fact Riemann's original definition of the curvature tensor.

Proof. Both sides are symmetric bilinear forms; for the 2nd order term this follows from $R(X,Y,X,Z) = R(X,Z,X,Y)$. Thus, according to the polarization identity, it suffices to show the equation for the associated quadratic form. For this we check for the Jacobi field $Y_t = (\exp_p)_*(tV)$ along a geodesic $c(t) = \exp t X_0$, that

$$t^2(\exp^* g)(V,V) = g_{c(t)}(Y,Y) = t^2 g_p(V,V) + \frac{t^4}{3} g_p(\Omega_p(X_0,V)X_0,V) + O(t^5).$$

The form of the left side shows that $\|Y\|_{|p}^2 = 0$, $\frac{d}{dt}_{|t=0} \|Y\|^2 = 0$ and

$$\frac{d^2}{dt^2}_{|t=0} \|Y\|^2 = 2\|T_0 \exp V\|_{|p}^2 = 2\|V\|^2.$$

Because $Y_0 = 0$, one also finds that $(\nabla_{\partial/\partial t})^2 Y_0 = 0$. Therefore it follows that

$$\frac{d^3}{dt^3}\Big|_{t=0} \|Y\|^2 = 2g((\nabla_{\partial/\partial t})^3 Y, \underbrace{Y}_{=0\,\text{at}\,t=0}) + 6g(\underbrace{(\nabla_{\partial/\partial t})^2 Y, \nabla_{\partial/\partial t} Y}_{=0\,\text{at}\,t=0}) = 0,$$

$$\frac{d^4}{dt^4}\Big|_{t=0} \|Y\|^2 = 8g((\nabla_{\partial/\partial t})^3 Y, \nabla_{\partial/\partial t} Y) = 8g(\nabla_{\partial/\partial t}(\Omega(\dot c, Y)\dot c), V)$$

$$\overset{Y_0=0}{=} 8g(\Omega(\dot c, \underbrace{\nabla_{\partial/\partial t} Y}_{=V\,\text{at}\,t=0})\dot c, V). \qquad \square$$

Corollary 5.3.7 *The Riemannian volume form in normal coordinates has the Taylor expansion*

$$(\exp_p^* d\mathrm{vol}_g)_{|X} = \left(1 - \frac{1}{6}\mathrm{Ric}(X, X) + O(\|X\|^3)\right) d\mathrm{vol}_{g_p}.$$

Proof. If $A \in \mathbb{R}^{n \times n}$, then

$$\sqrt{\det(\mathrm{id} + \varepsilon A + O(\varepsilon^2))} = \sqrt{\mathrm{id} + \varepsilon\, \mathrm{Tr}\, A + O(\varepsilon^2)}$$

$$= \mathrm{id} + \frac{\varepsilon}{2}\mathrm{Tr}\, A + O(\varepsilon^2).$$

Thus, given an orthonormal basis $(e_j)_j$ associated to g_p and

$$A_{jk} := -\frac{1}{3}R(e_j, X, e_k, X),$$

one finds

$$\exp_p^* d\mathrm{vol}_{g|X} = \sqrt{\det((\exp_p^* g)(e_j, e_k))_{jk}} \cdot d\mathrm{vol}_{g_p}$$

$$= (1 - \frac{1}{6}\sum_k R(e_k, X, e_k, X) + O(\|X\|^3)) \cdot d\mathrm{vol}_{g_p}$$

$$= \left(1 - \frac{1}{6}\mathrm{Ric}(X, X) + O(\|X\|^3)\right) e^1 \wedge \cdots \wedge e^n. \qquad \square$$

For any symmetric bilinear form γ on TM set $\mathrm{Tr}_g \gamma := \mathrm{Tr}\,(g_{jk})^{-1}(\gamma_{k\ell})$ as before definition 3.4.6 using the Gram matrices associated to any basis.

Proposition 5.3.8 *Let $P(X) = \alpha + \beta(X) + \gamma(X, X) + \delta(X, X, X)$ be any polynomial of degree ≤ 3 on an n-dimensional Euclidean vector space (V, g), where β is a linear form and γ, δ are symmetric multilinear forms. Then the ball $B_r := B_r(0)$ of radius $r > 0$ around 0 satisfies*

$$\int_{B_r} P(X)\, d\mathrm{vol}_g = \mathrm{vol}(B_r) \cdot (\alpha + \frac{r^2}{n+2}\mathrm{Tr}_g \gamma)$$

and the sphere $S_r^{n-1} = \partial B_r(0)$ *satisfies*

$$\int_{S_r^{n-1}} P(X)\, d\mathrm{vol}_{S_r^{n-1}} = \mathrm{vol}(S_r^{n-1}) \cdot \left(\alpha + \frac{r^2}{n}\mathrm{Tr}_g\gamma\right).$$

Proof. Because of the point symmetry of the spheres and balls around 0, integrals over the terms of odd order vanish: Using $\sigma : V \to V$, $X \mapsto -X$ one obtains

$$\int_{B_r} P(X)d\mathrm{vol}_g = \mathrm{sign}\det T\sigma \int_{B_r} P(-X)\sigma^* d\mathrm{vol}_g = \int_{B_r} P(-X)d\mathrm{vol}_g.$$

Considering the sphere S_r^{n-1} of radius r and coordinates $\sum_j x_j e_j$ and $1 \le k \le n$, one finds

$$r^2 \mathrm{vol}(S_r^{n-1}) = \int_{S_r^{n-1}} \sum_j x_j^2\, d\mathrm{vol}_{S_r^{n-1}} = n \int_{S_r^{n-1}} x_k^2\, d\mathrm{vol}_{S_r^{n-1}}.$$

If $(e_j)_j$ is a γ diagonalizing g-orthonormal basis, one gets

$$\int_{S_r^{n-1}} \gamma(X,X)\, d\mathrm{vol}_{S_r^{n-1}} = \sum_j \int_{S_r^{n-1}} x_j^2\gamma(e_j,e_j)\, d\mathrm{vol}_{S_r^{n-1}} = \frac{r^2}{n}\mathrm{Tr}_g\gamma \cdot \mathrm{vol}\, S_r^{n-1}.$$

Therefore because $\mathrm{vol}\, B_{r_0} = \int_0^{r_0} r^{n-1}\mathrm{vol}\, S_1^{n-1}\, dr = \frac{r_0^n}{n}\mathrm{vol}\, S_1^{n-1}$ and $dr = \frac{1}{r}\sum x_j\, dx_j$, $d\mathrm{vol}_{S_r^{n-1}} = \iota_{X/r} d\mathrm{vol}_g$, hence $dr \wedge d\mathrm{vol}_{S_r^{n-1}} = d\mathrm{vol}_g$, one obtains

$$\int_{B_{r_0}} \gamma(X,X)\, d\mathrm{vol}_g = \int_0^{r_0} \frac{r^2\mathrm{Tr}_g\gamma}{n}\mathrm{vol}(S_r^{n-1})\, dr$$

$$= \int_0^{r_0} r^{n+1}\, dr \cdot \frac{\mathrm{Tr}_g\gamma}{r_0^n}\mathrm{vol}\, B_{r_0} = \frac{r_0^2}{n+2}\mathrm{Tr}_g\gamma \cdot \mathrm{vol}\, B_{r_0}. \qquad \square$$

Corollary 5.3.9 *If $p \in M$ and $r > 0$, then the **geodesic sphere** $S_r(p) := \{q \in M \mid \mathrm{dist}(p,q) = r\}$ and the **geodesic ball** $B_r(p) := \{q \in M \mid \mathrm{dist}(p,q) < r\}$ satisfy*

$$\mathrm{vol}\, S_r(p) = \mathrm{vol}\, S_{\mathrm{Eucl}}^{n-1} \cdot r^{n-1}\left(1 - \frac{r^2 s_p}{6n} + O(r^4)\right),$$

$$\mathrm{vol}\, B_r(p) = \mathrm{vol}\, B_{\mathrm{Eucl}}^n \cdot r^n\left(1 - \frac{r^2 s_p}{6(n+2)} + O(r^4)\right).$$

Remark For r large, $S_r(p)$ is not necessarily a submanifold, compare Fig. 5.8.

Proof. According to Theorem 5.2.11, $S_r(p), B_r(p)$ are images of spheres and balls in T_pM under $\exp$ for small r. As $\mathrm{Tr}_g(-\frac{1}{6}\mathrm{Ric}) = -\frac{s_p}{6}$, the corollary follows from Proposition 5.3.8. According to the proposition, the 3rd order term of the Taylor

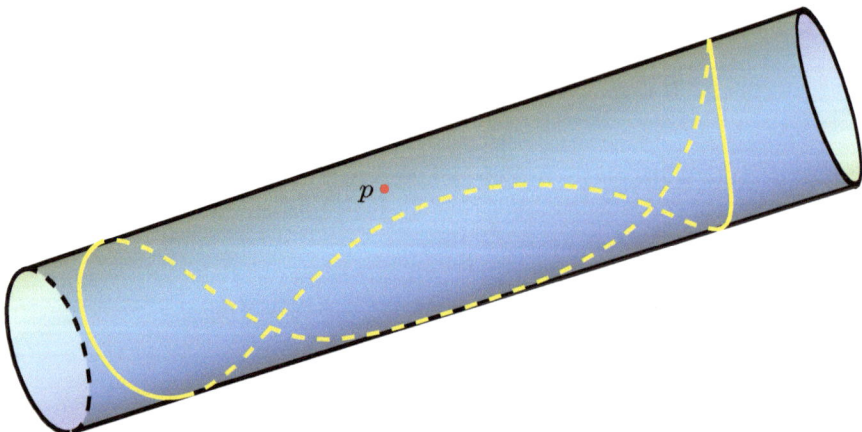

Fig. 5.8 $\exp_p(S_r(0))$ on a cylinder.

expansion of g does not contribute, therefore the equation holds with $O(r^4)$ and not only $O(r^3)$. □

Corollary 5.3.10 *Consider $X, Y \in T_p M$, $X \perp Y$, $\|X\| = \|Y\| = 1$ and let D be the intersection of $\mathbb{R} \cdot X + \mathbb{R} \cdot Y$ with a ball-shaped normal neighborhood of p. Let c_r be the **geodesic circle** of radius $r > 0$ around p in in the directions X, Y, i.e.*

$$c_r([0, 2\pi[) = S_r(p) \cap \exp_p(D).$$

*Let $D_r := B_r(p) \cap \exp_p(D)$ be the **geodesic disk**. Then for $r \to 0$, one gets*

$$\text{length}(c_r) = 2\pi r - \frac{\pi r^3}{3} K_p(X \wedge Y) + O(r^5),$$

$$\text{surface area}(D_r) = \pi r^2 - \frac{\pi r^4}{12} K_p(X \wedge Y) + O(r^6).$$

Proof. As shortest paths, the geodesics $t \mapsto \exp_p(t \cdot (aX + bY))$ are also geodesics in the submanifold $N := \exp_p(D)$. Hence $\exp_p$ and $\exp_p^N$ coincide on D. In particular, $\exp_p^* g_{|D} = \exp_p^{N*} g^N$ and by Theorem 5.3.6 $s_p^N = 2K_p^N(X \wedge Y) = 2K_p(X \wedge Y)$. Therefore the formulas follow from the ones in Lemma 5.3.9 for the two-dimensional case. □

On a sphere, for instance, c_r is shorter than a circle of the same radius in Euclidean space (Fig. 5.9). Of course, the Taylor expansion of the metric also yields an expansion of the Levi-Civita connection:

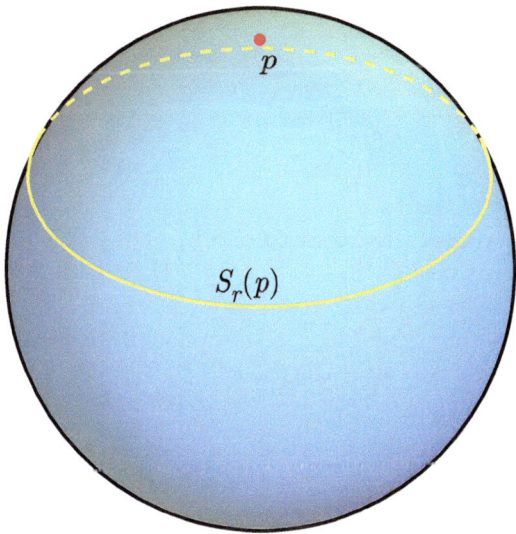

Fig. 5.9 $S_r(p)$ on a sphere.

Corollary 5.3.11 *Consider $X, Y, Z \in T_p M$ and pull back Y, Z onto $TT_p M$ as constant vector fields. Then*

$$\left(\nabla_Y^{\exp_p^* TM} Z\right)_{|X} = \Gamma_X(Y, Z) = \frac{1}{3}\Omega_p(X, Y)Z + \frac{1}{3}\Omega_p(X, Z)Y + O(\|X\|^2).$$

Proof. If $W \in TT_p M$ is constant, one gets

$$W.((\exp_p^* g)_{|X}(Y, Z)) \doteq \frac{1}{3}W.(g_p(\Omega_p(Y, X)Z, X) + O(\|X\|^3))$$

$$= \frac{1}{3}g_p(\Omega_p(Y, W)Z, X) + \frac{1}{3}g_p(\Omega_p(Y, X)Z, W) + O(\|X\|^2)$$

$$= \frac{1}{3}g_p(\Omega_p(Z, X)Y, W) + \frac{1}{3}g_p(\Omega_p(Y, X)Z, W) + O(\|X\|^2)$$

and thus the Koszul formula shows that

$$2(\exp^* g)_{|X}(\nabla_Y Z, W)$$

$$= Y.(\exp^* g(Z, W)) + Z.(\exp^* g(Y, W)) - W.(\exp^* g(Y, Z))$$

$$= \frac{2}{3}g_p(\Omega(X, Y)Z, W) + \frac{2}{3}g_p(\Omega(X, Z)Y, W) + O(\|X\|^2). \qquad \square$$

In the two-dimensional case the metric in normal coordinates takes a particularly concise form:

Lemma 5.3.12 (Jacobi's equation) *Suppose that M is 2-dimensional. Then in geodesic polar coordinates* $\exp^* g = dr^2 + G^2 d\vartheta^2$ *holds, where*

$$-\frac{1}{G}\frac{\partial^2 G}{\partial r^2}\Big|_{(r,\vartheta)} = K_{|\exp(r,\vartheta)}, \qquad G = r + O(r^2)$$

for $r \searrow 0$.

Proof. The radial geodesics have constant velocity, and by Gauß's Lemma $\frac{\partial}{\partial\vartheta} \perp \frac{\partial}{\partial r}$ holds. Thus there is a map $G :]0, r_{\max}[\times \mathbb{R}/2\pi\mathbb{Z} \to \mathbb{R}^+$ such that $\exp^* g$ takes the form $dr^2 + G^2 d\vartheta^2$. Then $G^2 = \exp^* g\left(\frac{\partial}{\partial\vartheta}, \frac{\partial}{\partial\vartheta}\right) = g_p\left(\frac{\partial}{\partial\vartheta}, \frac{\partial}{\partial\vartheta}\right) + O(r^3) = r^2 + O(r^3)$ and hence $G = r + O(r^2)$. Now let Y be the Jacobi fields $\exp_*\frac{\partial}{\partial\vartheta}$ along a radial geodesic c parameterized by arc length and let $V := Y/G$. Then $V \perp \dot{c}$ and $\|V\| \equiv 1$, therefore V is parallel along c. Thus it follows by the Jacobi ODE that

$$\Omega(\dot{c}, Y)\dot{c} = (\nabla_{\partial/\partial t})^2 Y = (\nabla_{\partial/\partial t})^2(GV) = \frac{\partial^2 G}{\partial r^2} \cdot V$$

holds and hence

$$K = -g(\Omega(\dot{c}, V)\dot{c}, V) = -\frac{1}{G}\frac{\partial^2 G}{\partial r^2}. \qquad \square$$

Remark 5.3.13 If $M \subset \mathbb{R}^3$ is a surface, then the parallel transport of tangent vectors can be interpreted in a more intuitive way. One **rolls** the surface M along the curve c **over** the Euclidean plane $\mathbb{R}^2$:

Given intervals I with $II_{c(t)}$ non-degenerate for all $t \in I$ choose smoothly a vector field W along c such that $II(W_t, c'(t)) \neq 0$. Then the ruled surface (see Exercise 3.1.27) $\widetilde{M}$ with $\widetilde{f}(t, s) = c'(t) + s \cdot W_t$ is the unique ruled surface such that $K_{\widetilde{M}} \equiv 0$ and for all $t \in I$ we have $T_{c(t)}\widetilde{M} = T_{c(t)}M$. The surface M is called the **osculating developable surface**.For by Exercise 5.1.19, $K_{\widetilde{M}} = 0$ holds if and only if $\nabla^{c^*T\mathbb{R}^3}_{\partial_t} = W'_t \in \text{span}(c'(t), W) = T_{c(t)}\widetilde{M}$, i.e. if and only if $\widetilde{II}(c', W_t) = 0$. And because $T_{c(t)}\widetilde{M} = T_{c(t)}M$, one gets $\widetilde{II}(c', W_t) = II(c', W_t)$.

Exercise 5.3.20 shows that $\widetilde{M}$ is locally isometric to a plane. Conversely, it can be shown that any surface in $\mathbb{R}^3$ with $K \equiv 0$ and II non-degenerate is a ruled surface ([Kl1, Th. 3.7.9]). Thus there is a unique way to roll M along the curve c over a Euclidean plane, such that $T_{c(t)}M$ is always identified with the plane.

The parallel transport along c is the same in $\widetilde{M}$ and M, since the tangent spaces coincide. And the parallel transport in the plane is trivial, which thus yields the parallel transport on M.

Exercises

Exercise 5.3.14 Show that a Killing vector field X (see Exercise 3.3.10) on a Riemannian manifold M is a Jacobi field along any geodesic c.

Exercise 5.3.15 Let M be connected, $p \in M$ be fixed, and let X, Y be Killing vector fields with $X_p = Y_p$, $(\nabla X)_p = (\nabla Y)_p$. Conclude that $X = Y$ (e.g. with Exercise 5.3.14).

Exercise* 5.3.16 Explicitly determine a three-dimensional subspace of Jacobi fields along the generating curve $c : I \to \mathbb{R}^3$ parametrized by arc length of a surface of revolution M.

Exercise* 5.3.17 Show that on the space of Jacobi fields along a geodesic c

$$\omega(Y, \widetilde{Y}) := g(Y_t, \nabla_{\partial/\partial t}^{c^*TM} \widetilde{Y}_t) - g(\nabla_{\partial/\partial t}^{c^*TM} Y_t, \widetilde{Y}_t)$$

yields a symplectic form independent of t.

Exercise* 5.3.18 Verify the differential equation of Jacobi fields for the fields in Corollary 5.3.5.

Exercise* 5.3.19 Calculate the 3rd order term of the Taylor expansion in Theorem 5.3.6 as

$$\frac{1}{6} g_p \left((\nabla_X \Omega_p)(X, \cdot)X, \cdot \right).$$

Exercise 5.3.20 (Minding's Theorem) Let M be a surface of constant curvature $K \in \mathbb{R}$. Apply Lemma 5.3.12 to explicitly determine the metric in geodesic polar coordinates and conclude that all surfaces with this constant curvature K are locally isometric.

Exercise* 5.3.21 Applying the Jacobi differential equation, show once again that S^n has constant sectional curvature 1 using the explicit variation of great circles

$$c_s(t) = p \cdot \cos t + \sin t \cdot (X \cos s + V \sin s)$$

where $p \in S^n$, $X, V \in T_p S^n$, $\|X\| = \|V\| = 1$ and $X \perp V$.

5.4 The Hopf–Rinow Theorem

geodesically complete	complete
Hopf–Rinow Theorem	

In this section, we will use a simple criterion for the distance metric to characterize when the exponential map is globally defined.

Proposition 5.4.1 *Given any $p, q \in M$, there is a geodesic $c_0(t) = \exp_p t X_0$ and $p_0 = c_0(r_0) \neq p$ satisfying*

$$\mathrm{dist}(p, q) = \mathrm{dist}(p, p_0) + \mathrm{dist}(p_0, q),$$

such that c_0 is the uniquely determined shortest path from p to p_0 parameterized by arc length.

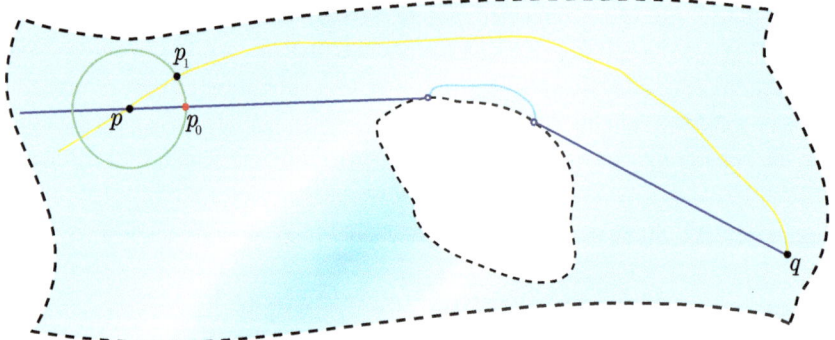

Fig. 5.10 p_0 in the case of a noncomplete manifold.

Although there need not be a shortest path from p to q, according to this proposition there is a direction X_0 at p in which the path length is arbitrarily close to the
infimum (cf. Fig. 5.10 for the case $M \setminus A$ where A is a closed subset). Of course,
there can be several points p_0 of this kind.

Proof. Let U be a ball-shaped normal neighborhood around p. For $q \in U$ set $p_0 := q$.
Otherwise, let $S_{r_0}(p) = \{p_1 \in M \mid \text{dist}(p, p_1) = r_0\} \subset U$ be a geodesic sphere. Since
S_{r_0} is compact, there exists a point $p_0 \in S_{r_0}$ such that $\text{dist}(S_{r_0}, q) = \text{dist}(p_0, q)$.
Consider a curve c from p to q. According to the intermediate value theorem c must
meet the sphere at a point p_1, and

$$\text{length}(c) \geq \underbrace{\text{dist}(p, p_1)}_{=r_0} + \text{dist}(p_1, q) \geq \underbrace{\text{dist}(p, p_0)}_{=r_0} + \text{dist}(p_0, q).$$

Therefore $\text{dist}(p, q) = \inf_c \text{length}(c) \geq \text{dist}(p, p_0) + \text{dist}(p_0, q)$. The converse is
the triangle inequality.   □

Theorem 5.4.2 (Hopf–Rinow) *Suppose that $p \in M$ is such that $\exp_p$ is defined at
least on a ball $B_r(0) \subset T_p M$ of radius r. Then any $q \in M$ satisfying $\text{dist}(p, q) < r$
can be connected to p by a shortest path.*

In contrast to the previous section, we do not require $\exp_p$ to be invertible on the
ball. Accordingly, there can also be several shortest paths from p to q.

Proof. Choose c_0, r_0 as in Proposition 5.4.1. The curve $c_0(t)$ is defined on
$|t| < r$ by assumption. Set $I := \{t \in \mathbb{R} \mid t + \text{dist}(c_0(t), q) = \text{dist}(p, q)\}$.
I is closed and nonempty because $r_0 \in I$. Because of the closedness, there ex-
ists $t_1 := \max I \leq \text{dist}(p, q)$. Assume that $t_1 < \text{dist}(p, q)$. Then by Proposition 5.4.1
there exist c_2, r_2 (Fig. 5.11) such that

$$\text{dist}(p, p_2) \overset{\Delta \text{ inequality}}{\geq} \text{dist}(p, q) - \text{dist}(p_2, q)$$
$$= \text{dist}(p, q) - \text{dist}(c_0(t_1), q) + \text{dist}(c_0(t_1), p_2) = t_1 + r_2.$$

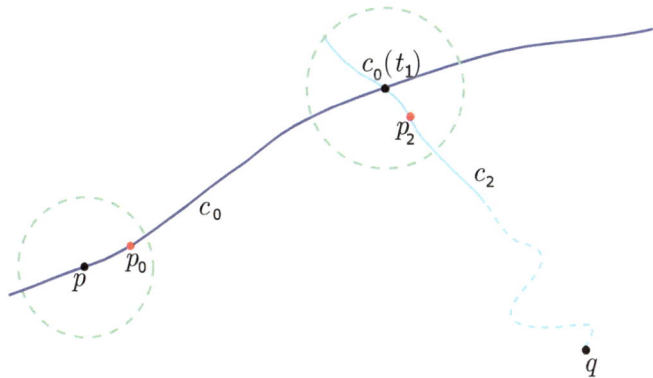

Fig. 5.11 Proof of the Hopf–Rinow Theorem.

On the other hand, the union $\bar{c}$ of $c_{0|[0,t_1]}$ with the minimum geodesic c_2 from $c_0(t_1)$ to p_2 satisfies

$$\text{dist}(p, p_2) \leq \text{length}\,\bar{c} = t_1 + r_2.$$

Hence $\text{dist}(p, p_2) = t_1 + r_2$ and $\bar{c}$ is a shortest path, thus geodesic. Therefore $\bar{c} = c_0$ and

$$\text{dist}(q, \underbrace{p_2}_{=c_0(t_1+r_2)}) + t_1 + r_2 = \text{dist}(p, q)$$

holds, hence $t_1 + r_2 \in I \, \frac{\ell}{\ell}$.   □

Corollary 5.4.3 *If M is **geodesically complete**, i.e. if exp is defined on all of TM, then any two points can be connected by shortest paths.*

The converse does not hold. For example, on a bounded interval in $\mathbb{R}$, any two points can be connected by a minimal geodesic even though the exponential function is not globally defined.

Hopf–Rinow Theorem 5.4.4 ([HoRi][3]) *The following statements about a Riemannian manifold M are equivalent:*

1) any bounded closed subset $K \subset M$ is compact,
2) as a metric space, M is complete, i.e., every Cauchy sequence converges,
3) M is geodesically complete,
4) there exists a point $p \in M$ such that $\exp_p$ is defined on all of $T_p M$.

Accordingly, a Riemannian manifold M subject to these conditions is called **complete**

Proof. $(1) \Rightarrow (2)$: Every Cauchy sequence is bounded, so by (1) it is contained in a compactum, thus it has a convergent subsequence, so it converges.

---

[3] 1931, Heinz Hopf, 1894–1971, Willi Rinow, 1907–1979.

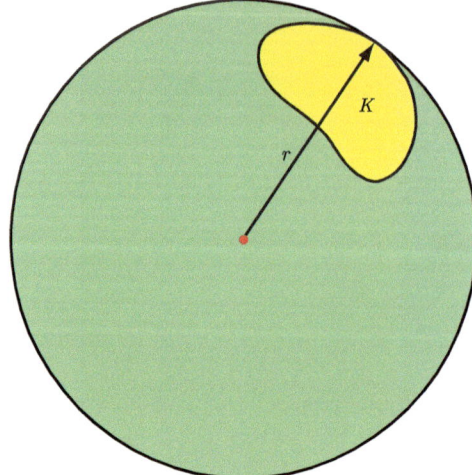

Fig. 5.12 Step $(4) \Rightarrow (1)$ of the proof.

$(2) \Rightarrow (3)$: Suppose c is a geodesic parametrized by arc length that is maximally defined on an open interval. Let $t_0 = \sup I < \infty$ be the limit of a sequence $(t_j)_j$ in I. Because $\text{dist}(c(t_j), c(t_k)) \leq |t_k - t_j|$, the limit $\lim_{t \nearrow t_0} c(t) =: p$ exists according to (2). Thus c can be extended in a normal neighborhood of p ↯. Likewise for $\inf I$.

$(3) \Rightarrow (4)$: Obvious.

$(4) \Rightarrow (1)$: Let $\exp_p$ be defined on $T_p M$. Because of the boundedness of K there is a finite $r > \sup\{\text{dist}(p, q) \mid q \in K\}$. By Theorem 5.4.2, $K \subset \exp_p \overline{B_r(0)}$, and $\exp_p \overline{B_r(0)}$ is compact as an image of a compactum under a continuous map. Thus the closed subset K is also compact (Fig. 5.12).   □

Remark 5.4.5 A closed submanifold $N \subset M$ of a complete manifold M (with respect to its distance metric dist_M) is complete (with respect to dist_N). This follows easily without the Hopf–Rinow Theorem: Given $p, q \in N$, $\text{dist}_M(p, q) \leq \text{dist}_N(p, q)$, hence every Cauchy sequence in N is also one in M. Because of the closedness, the limit p lies in N. And since the topology on N is the subspace topology to that of M, the sequence converges to p with respect to the topology on N as well. For example, every closed Lie group $G \subset \mathbb{R}^{n \times n}$ is complete.

Corollary 5.4.6 (Examples)

1) Let $G \subset \text{SO}(n)$ be a (closed, i.e. compact) Lie subgroup, then $\exp_{\text{id}} A = e^A$ is defined on the whole $\mathfrak{g} = T_{\text{id}} G$. By Hopf–Rinow, $\exp_{\text{id}}$ is surjective, i.e., every $h \in G$ can be written as $h = e^A$ with an $A \in \mathfrak{g}$.

2) Suppose $G = \text{SL}_2(\mathbb{R})$. Exercise 1.6.30 shows that $A \mapsto e^A$ is not surjective, since $\text{Tr}\, e^A \geq -2$. Hence, according to Hopf–Rinow, in general $\exp_{\text{id}} A \neq e^A$.

Corollary 5.4.7 Let M be compact. Then $\exp$ is defined on all TM, and any two points can be connected by shortest geodesics.

Exercises

Exercise 5.4.8 Let M be a complete Riemannian manifold and c be a geodesic. Suppose that there is no shorter geodesic than c from $c(a)$ to $c(b)$. Conclude that c is the shortest path from $c(a)$ to $c(b)$. Find a counterexample to this statement for noncomplete M.

Chapter 6
Homogeneous Spaces

The Lie groups have already given us nontrivial and, on the other hand, relatively easy to study examples of Riemannian manifolds. Nevertheless, as their trivial tangent bundle already shows, they are a very special class of manifolds for which one cannot study many more general effects. Much more interesting examples, and to some extent similarly easy to understand, can be found by dividing Lie groups by subgroups. These homogeneous spaces, which can also be understood as Riemannian manifolds with a transitive isometric group, are studied in this chapter. First, hyperbolic space is discussed in more detail. Together with the Euclidean space and the round sphere, it is one of the most regular kinds of metric space. These three spaces are characterized as spaces of constant curvature. Then, after studying subspaces of manifolds from the beginning, submersions and quotients of Riemannian manifolds are also considered in general and O'Neill's Formulas for curvatures and geodesics are worked out. These are applied to the study of homogeneous spaces.

6.1 Hyperbolic Space

hyperbolic space	Lorentz isometry
Minkowski form	upper half-space model

In direct analogy to corresponding results for the sphere, some basic properties of hyperbolic space are discussed in this section (cf. Example 3.1.3).

Lemma 6.1.1 *As a set, S^n can be identified canonically with the equivalence classes* $SO(n + 1)/SO(n)$ *with* $SO(n) \cong \begin{pmatrix} 1 & 0 \\ 0 & SO(n) \end{pmatrix} \subset SO(n + 1)$.

Proof. $G := SO(n + 1)$ acts transitively on S^n: Given $\mathbf{x}, \mathbf{y} \in S^n$, choose a $\mathbf{z} \in S^n \setminus \{\mathbf{x}, \mathbf{y}\}$. Let $A \in G$ denote the composition of the reflection $\mathbf{w} \mapsto \mathbf{w} - 2 \frac{\mathbf{x}-\mathbf{z}}{\|\mathbf{x}-\mathbf{z}\|^2} \langle \mathbf{x}-\mathbf{z}, \mathbf{w} \rangle$

© The Editor(s) (if applicable) and The Author(s), under exclusive license to Springer-Verlag GmbH, DE, part of Springer Nature 2024
K. Köhler, *Differential Geometry and Homogeneous Spaces*, Universitext,
https://doi.org/10.1007/978-3-662-69721-4_6

at $(\mathbf{x} - \mathbf{z})^{\perp}$ with that at the hyperplane $(\mathbf{y} - \mathbf{z})^{\perp}$. Then $A(\mathbf{y}) = \mathbf{x}$ holds. The stabilizer of $N := (1, 0, \cdots, 0)^t$ equals

$$G_N = \{h \in \mathrm{SO}(n+1) \mid h(1, 0, \cdots, 0)^t = (1, 0, \cdots, 0)^t\} = \begin{pmatrix} 1 & 0 \\ 0 & \mathrm{SO}(n) \end{pmatrix}.$$

Hence $\mathrm{SO}(n+1)/\mathrm{SO}(n) \to S^n$, $[h] \mapsto hN$ is well-defined and bijective. $\square$

In the case of the sphere, the stereographic projection $\varphi_- : S^n \to \mathbb{R}^n$, $\binom{x_0}{x} \mapsto \frac{x}{1+x_0}$ yields a chart with associated parametrization $\varphi_-^{-1} : u \mapsto \frac{1}{1+\|u\|^2} \binom{1-\|u\|^2}{2u}$. On this chart, the Riemannian metric equals $((\varphi_-^{-1})^* g)_u = \frac{4}{(1+\|u\|^2)^2} \langle \cdot, \cdot \rangle_{\mathrm{Eucl}}$ by Exercise 2.2.14. Now instead of the Euclidean metric we take the **(canonical) Minkowski form on** $\mathbb{R}^{n+1}$, i.e. the symmetric bilinear form

$$\langle (x_0, \ldots, x_n)^t, (y_0, \ldots, y_n)^t \rangle_L := x_0 y_0 - \sum_{j>0} x_j y_j.$$

This definition is also frequently found in the literature with opposite signs. As an abbreviation, $\langle v, v \rangle_L$ is written as $\|v\|_L^2$, while $\| \cdot \|^2$ continues to refer to Euclidean norms. A **Lorentzian isometry** is an isometry of $(\mathbb{R}^{n+1}, \langle \cdot, \cdot \rangle_L)$. Set

$$H^n := \{\mathbf{x} = \binom{x_0}{x} \in \mathbb{R}^+ \times \mathbb{R}^n \mid x_0^2 - \|x\|_{\mathrm{Eucl}}^2 = \|\mathbf{x}\|_L^2 = 1\}$$

by analogy with S^n (one "sheet" of a two-sheet hyperboloid). This hyperboloid is contained in the double cone $\{\mathbf{x} \in \mathbb{R}^{n+1} \mid \|\mathbf{x}\|_L^2 > 0\}$. It approaches the cone boundary for large x_0. By differentiating, the tangent space turns out to be $T_{\mathbf{x}} H^n = \{X \in \mathbb{R}^{n+1} \mid \langle \mathbf{x}, X \rangle_L = 0\}$, analogous to the sphere. Just as in the case of S^n, there is a stereographic projection $\varphi : H^n \to B_1^n(0)$, $\binom{x_0}{x} \mapsto \frac{x}{1+x_0}$ (Fig. 6.1). This is a chart of H^n, because the corresponding parametrization is $\varphi^{-1} : B_1^n(0) \to H^n$, $u \mapsto \frac{1}{1-\|u\|^2} \binom{1+\|u\|^2}{2u}$, for $\|\varphi^{-1}(u)\|_L^2 = 1$ and

$$\varphi(\varphi^{-1}(u)) = \frac{\frac{2u}{1-\|u\|^2}}{\frac{1+\|u\|^2}{1-\|u\|^2} + 1} = \frac{2u}{2} = u.$$

Theorem 6.1.2 $(B_1^n(0), -(\varphi^{-1})^* \langle \cdot, \cdot \rangle_L)$ *is the hyperbolic space as in Example 3.1.3.*

Proof. We compute $-(\varphi^{-1})^* \langle \cdot, \cdot \rangle_L = -\langle T\varphi^{-1} \cdot, T\varphi^{-1} \cdot \rangle_L$ and show that this provides the hyperbolic metric on $B_1^n(0)$. Set

$$f : \mathbb{R}^{n+1} \setminus \{\|\mathbf{x} - S\|_L^2 = 0\} \to \mathbb{R}^{n+1} \setminus \{\|\mathbf{x} - S\|_L^2 = 0\}$$

$$\mathbf{x} \mapsto S + \frac{2(\mathbf{x} - S)}{\|\mathbf{x} - S\|_L^2},$$

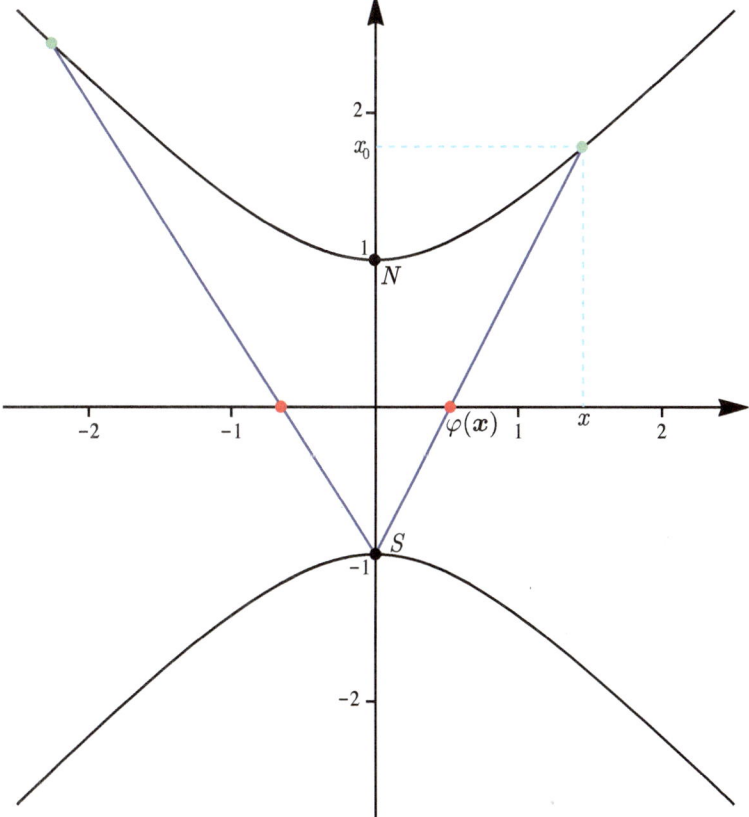

Fig. 6.1 The stereographic projection of the hyperbolic space. In the figure, the axis of rotation with parameter x_0 is chosen to be vertical.

with $S := (-1, 0, \cdots, 0)^t$. Then one finds

$$f\left(\begin{pmatrix} 0 \\ u \end{pmatrix}\right) = \begin{pmatrix} -1 \\ 0_{\mathbb{R}^n} \end{pmatrix} + \frac{2\begin{pmatrix} 1 \\ u \end{pmatrix}}{1 - \|u\|^2} = \varphi^{-1}(u).$$

Furthermore,

$$T_{\mathbf{x}} f(Y) = \frac{2Y}{\|\mathbf{x} - S\|_L^2} - \frac{4(\mathbf{x} - S)\langle \mathbf{x} - S, Y \rangle_L}{\|\mathbf{x} - S\|_L^4}$$

$$= \frac{2}{\|\mathbf{x} - S\|_L^2} \cdot \underbrace{\left[Y - 2\frac{\langle \mathbf{x} - S, Y \rangle}{\|\mathbf{x} - S\|_L^2}(\mathbf{x} - S) \right]}_{\text{reflection of } Y \text{ at } (\mathbf{x} - S)^\perp,\ \text{hence a Lorentzian isometry}}.$$

Therefore $f^*\langle\cdot,\cdot\rangle_L = \frac{4}{\|x-S\|_L^4}\langle\cdot,\cdot\rangle_L$. In particular, on $\{0\}\times B_1^n(0)$ this shows that

$$-(\varphi^{-1})^* g_{|\binom{0}{u}} = -f^*\langle\cdot,\cdot\rangle_{L|\binom{0}{u}} = \frac{-4}{(1-\|u\|^2)^2}\langle\cdot,\cdot\rangle_{L|\{0\}\times\mathbb{R}^n}$$

$$= \frac{4}{(1-\|u\|^2)^2}\langle\cdot,\cdot\rangle_{\text{Eucl}|\{0\}\times\mathbb{R}^n}. \qquad \square$$

In fact, even $f\circ f =$ id is true, and so $f_{|H^n} = \varphi$ also holds.

Remark In Special Relativity, the hyperboloid H^3 represents all possible values of the energy-momentum vector of a particle of rest mass 1. In particular, in quantum field theory, it is necessary to integrate over the space of these possibilities using the measure induced by the Minkowski form, i.e., the volume form of the hyperbolic space.

Analogous to Example 5.1.8, it follows that:

Lemma 6.1.3 *The sectional curvature of H^n equals $K \equiv -1$.*

Proof. This follows by applying the Gauß equation 5.1.3 for the 2nd fundamental form associated to the immersion $(H^n, g) \subset (\mathbb{R}^{n+1}, -\langle\cdot,\cdot\rangle_L)$. This formula had been proved for Riemannian metrics, but not for Lorentzian metrics, where each $T_p M$ carries a Minkowski form. But the proof and that of Theorem 3.3.2 did not use any assumptions about the signature of the non-degenerate bilinear forms on the tangent space. Considering the radial vector field $\mathfrak{R} = \sum_{j=0}^n x_j \frac{\partial}{\partial x_j}$, vector fields X, Y on H^n and an extension $\widetilde{Y}$ of Y to $\mathbb{R}^{n+1}$, it follows with $-\langle\cdot,\cdot\rangle_{L|H^n}$ as Riemannian metric that

$$II(X,Y) = (\nabla_X^{\mathbb{R}^{n+1}}\widetilde{Y})^{\perp TH^n} = -\langle X.\widetilde{Y}, \mathfrak{R}\rangle_L \cdot \mathfrak{R}$$
$$= (\langle Y, X.\mathfrak{R}\rangle_L - X.\langle\widetilde{Y}, \mathfrak{R}\rangle_L)\mathfrak{R} = \langle X, Y\rangle_L \mathfrak{R}.$$

Given orthonormal X, Y, one thus gets

$$K(X\wedge Y) = K^{\mathbb{R}^{n+1}}(\widetilde{X}\wedge\widetilde{Y}) + (-\langle II(X,X), II(Y,Y)\rangle_L) - (-\|II(X,Y)\|_L^2)$$
$$= -\|X\|_L^2\|Y\|_L^2\|\mathfrak{R}_{|H^n}\|_L^2 = -1. \qquad \square$$

Proposition 6.1.4 *If $\mathbf{x}, \mathbf{y} \in H^n$ and $\mathbf{x} \neq \mathbf{y}$, then $\|\mathbf{x}-\mathbf{y}\|_L^2 < 0$ and $\langle\mathbf{x},\mathbf{y}\rangle_L > 1$.*

Proof. If $\mathbf{x} = \binom{x_0}{x}, \mathbf{y} = \binom{y_0}{y}$, then $\|x\|^2 + 1 = x_0^2, \|y\|^2 + 1 = y_0^2$. Cauchy–Schwarz shows that

$$x_0 y_0 = \sqrt{1+\|x\|^2}\sqrt{1+\|y\|^2} = \left\|\binom{1}{x}\right\|_{\text{Eucl}} \cdot \left\|\binom{1}{y}\right\|_{\text{Eucl}}$$

$$\overset{x\neq y}{>} \left\langle\binom{1}{x}, \binom{1}{y}\right\rangle_{\text{Eucl}} = 1 + \langle x, y\rangle_{\text{Eucl}}.$$

Therefore $\langle \mathbf{x}, \mathbf{y} \rangle_L = \sqrt{1 + \|x\|^2}\sqrt{1 + \|y\|^2} - \langle x, y \rangle > 1$ and $\|\mathbf{x} - \mathbf{y}\|_L^2 = 2 - 2\langle \mathbf{x}, \mathbf{y} \rangle_L < 0$.

$\square$

Definition 6.1.5 *Define* $O(1, n) := \{A \in GL_{n+1}(\mathbb{R}) \mid \|A\mathbf{x}\|_L^2 = \|\mathbf{x}\|_L^2\}$ *and* $SO(1, n) := \{A \in O(1, n) \mid \det A = 1\}$ *(as in the case of* $O(n)$,

$$A^t \begin{pmatrix} -1 & & & \\ & 1 & & \\ & & \ddots & \\ & & & 1 \end{pmatrix} A = \begin{pmatrix} -1 & & & \\ & 1 & & \\ & & \ddots & \\ & & & 1 \end{pmatrix}$$

implies that $\det A = \pm 1$). *Clearly, every* $A \in O(1, n)$ *maps the two-sheet hyperboloid* $\{\mathbf{x} \mid \|\mathbf{x}\|_L^2 = 1\}$ *onto itself. Thus because of continuity it maps* H^n *either onto* H^n *or* $-H^n$. *Let* $SO_0(1, n) \subset O_0(1, n)$ *be the subgroups of those* A *which map* H^n *onto itself.*

Theorem 6.1.6 $SO_0(1, n)$ *acts transitively and isometrically on* H^n *(and thus the same holds for* $O_0(1, n)$).

Proof. Consider $\mathbf{x}, \mathbf{y} \neq \mathbf{z} \in H^n$. The reflection $A_1 \in O(1, n)$ at $(\mathbf{x} - \mathbf{z})^\perp$

$$A_1 \mathbf{y} := \mathbf{y} - 2 \frac{\mathbf{x} - \mathbf{z}}{\|\mathbf{x} - \mathbf{z}\|_L^2} \langle \mathbf{x} - \mathbf{z}, \mathbf{y} \rangle_L$$

is well-defined according to Proposition 6.1.4. Because $\langle \mathbf{x} + \mathbf{z}, \mathbf{x} - \mathbf{z} \rangle_L = \|\mathbf{x}\|_L^2 - \|\mathbf{z}\|_L^2 = 0$, it satisfies $A_1 \mathbf{x} = \mathbf{z}$. In particular, $A_1 \in O_0(1, n)$ holds. Now let A_2 be the reflection at $(\mathbf{z} - \mathbf{y})^\perp$. Then $(A_2 A_1)\mathbf{x} = \mathbf{y}$ and $A_2 A_1 \in SO_0(1, n)$, hence $SO_0(1, n)$ acts transitively.

Since $O_0(1, n)$ consists of isometries of the Minkowski form and the metric on H^n is induced by $\langle \cdot, \cdot \rangle_L$, $O_0(1, n)$ also operates isometrically on H^n. $\square$

Thus, in contrast to the case of S^n, one of 4 components of $O(1, n)$ is sufficient here.

Corollary 6.1.7 *Let* $N := (1, 0, \cdots, 0)^t$. *Then* $SO_0(1, n)/SO(n) \to H^n$, $[h] \mapsto hN$ *with* $SO(n) \cong \begin{pmatrix} 1 & 0 \\ 0 & SO(n) \end{pmatrix}$ *is well-defined and bijective.*

In Section 6.4, $SO_0(1, n)/SO(n)$ is given the structure of a Riemannian manifold and in Section 6.7 this manifold is identified with H^n.

Proof. If $h \in SO_0(1, n)$, then $hN = N \Leftrightarrow h \in \begin{pmatrix} 1 & 0 \\ 0 & SO(n) \end{pmatrix}$. $\square$

Theorem 6.1.8 *The (paths of) geodesics on* H^n *are the intersections of* H^n *with 2-dimensional planes* E *through* 0.

Proof. 1) Let $E \subset \mathbb{R}^{n+1}$ be a 2-dimensional subspace intersecting H^n (in more than one point) along a curve. Given $\mathbf{x} \neq \mathbf{y}$ in $H^n \cap E$, Proposition 6.1.4 shows that $\det \begin{pmatrix} \langle \mathbf{x}, \mathbf{x} \rangle_L & \langle \mathbf{x}, \mathbf{y} \rangle_L \\ \langle \mathbf{y}, \mathbf{x} \rangle_L & \langle \mathbf{y}, \mathbf{y} \rangle_L \end{pmatrix} = 1 - \langle \mathbf{x}, \mathbf{y} \rangle_L^2 < 0$. Therefore $\mathbf{x}, \mathbf{y}$ are linearly independent, i.e. $E = \mathbb{R}\mathbf{x} + \mathbb{R}\mathbf{y}$, and $\langle \cdot, \cdot \rangle_{L|E}$ is non-degenerate. Thus the reflection $A = \{ \begin{smallmatrix} 1 \\ -1 \end{smallmatrix} \text{ on } \begin{smallmatrix} E \\ E^\perp \end{smallmatrix}$ at

E is well-defined. Hence $H^n \cap E$ is the fixed point set of the action of the isometry A on H^n. Thus by Exercise 5.2.15 $H^n \cap E$ is the path of a geodesic.

2) Conversely, for any $\mathbf{x} \neq \mathbf{y} \in H^n$ there is exactly one plane $E = \mathbb{R}\mathbf{x} + \mathbb{R}\mathbf{y}$ through $0, \mathbf{x}$ and $\mathbf{y}$. In $H^n \cap E$, the points $\mathbf{x}, \mathbf{y}$ are joined by the connected curve

$$c : [0, 1] \to H^n, \quad t \mapsto \frac{t\mathbf{x} + (1-t)\mathbf{y}}{\|t\mathbf{x} + (1-t)\mathbf{y}\|_L}.$$

c is well-defined, as

$$\|t\mathbf{x} + (1-t)\mathbf{y}\|_L^2 = t^2 + (1-t)^2 + 2t(1-t) \overbrace{\langle \mathbf{x}, \mathbf{y} \rangle_L}^{>1 \text{ by } 6.1.4} > 0.$$

Thus, choosing $\mathbf{y}$ in a normal neighborhood of $\mathbf{x}$ gives all geodesics through $\mathbf{x}$ in all directions up to reparametrization. $\square$

For example, the geodesic through N in the direction $\begin{pmatrix} 0 \\ X \end{pmatrix}$ with $\|X\|^2 = 1$ can be explicitly written as $c(t) = \begin{pmatrix} \cosh t \\ X \sinh t \end{pmatrix}$. For $\|c(t)\|_L^2 = 1$ and $-\|\dot{c}(t)\|_L^2 = 1$. In particular, these geodesics are defined on all of $\mathbb{R}$.

Corollary 6.1.9 *According to Hopf–Rinow, H^n is complete. Any two distinct points on H^n are connected by exactly one geodesic, so all geodesic connections are minimal.*

Remark In Exercise 6.6.9, a left-invariant metric is constructed on a Lie group, which is then isometric to H^n.

Exercises

Exercise 6.1.10 Show that the geodesic through $\mathbf{x} \in H^n$ in the direction $X \in T_\mathbf{x} H^n$ with $\|X\|_L^2 = -1$ has the form

$$c(t) := \mathbf{x} \cosh t + X \sinh t.$$

Exercise 6.1.11 Let $N := (1, 0, \ldots, 0)^t \in \mathbb{R}^n$ and

$$\varphi : B_1^n(0) \to \mathbb{R}^+ \times \mathbb{R}^{n-1}, x \mapsto 2\frac{x + N}{\|x + N\|_{\text{Eucl}}^2} - N.$$

Show that φ is a chart and that the hyperbolic metric g on $B_1^n(0)$ satisfies

$$(\varphi^{-1})^* g = \frac{g_{\text{Eucl}}}{x_1^2}.$$

This is the **upper half-space model** of H^n (cf. Exercise 3.1.22).

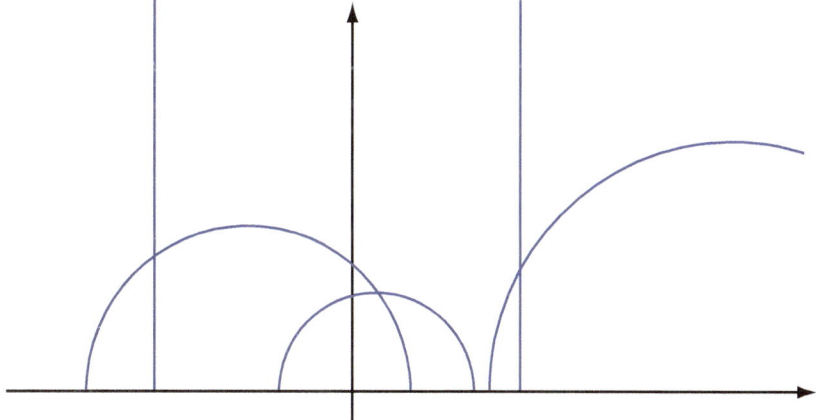

Fig. 6.2 Geodesics in the hyperbolic plane.

Exercise* 6.1.12 Parts (1)–(3) of this exercise can be solved by elementary geometry.

1) Given any sphere in Euclidean space and a straight line through zero intersecting this sphere at two points p, q, show that $\|p\| \cdot \|q\|$ is a constant independent of the straight line.
2) Using (1), prove that the map φ from Exercise 6.1.11 maps circles or straight lines to circles or straight lines (with respect to the Euclidean metric on $B_1^n(0), \mathbb{R}^+ \times \mathbb{R}^{n-1}$). To do this, you can first simplify φ by translations and show that sufficiently general spheres are mapped to spheres.
3) Show that φ preserves angles with respect to the Euclidean metrics on $B_1^n(0)$ and $\mathbb{R}^+ \times \mathbb{R}^{n-1}$.
4) For the ball model as well as the upper half-space model of H^n, show that the geodesics are sections of those circles and straight lines which intersect the boundary perpendicularly (Fig. 6.2).

Exercise* 6.1.13 Consider the ball model of hyperbolic space and calculate the distance between 0 and a point $u \in H^n$ with Euclidean distance $r < 1$ from 0.

Exercise* 6.1.14 Determine the sectional curvature of H^n independently of Lemma 6.1.3, by considering a variation $c_s(t) = \begin{pmatrix} \cosh t \\ X_s \sinh t \end{pmatrix}$, $X_s \in S^n$ of geodesics as in Exercise 5.3.21.

Exercise 6.1.15 Compute the sectional curvature of H^n by applying the Taylor expansion of g_{hyp} at the point N.

Exercise 6.1.16 Show that the identification $SO_0(1, n)/SO(n) \rightarrow H^n$ is a homeomorphism when $SO_0(1, n)/SO(n)$ is given the quotient topology. Deduce that $SO_0(1, n)$ is connected.

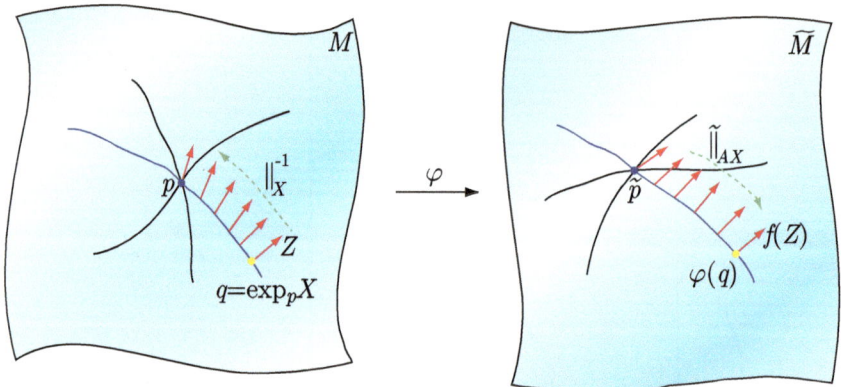

Fig. 6.3 Maps considered in Cartan's Theorem.

6.2 Cartan's Theorem and Spaces of Constant Curvature

Cartan's Theorem

Cartan's Theorem is a general result that shows the local isometry of two manifolds M and $\widetilde{M}$ if the curvature tensors of M and $\widetilde{M}$ satisfy a certain form of "equality". One quickly observes with counterexamples that given a map $\varphi : M \to \widetilde{M}$, the requirement $\varphi^*\widetilde{\Omega} = \Omega$ would not suffice. The identification of the curvature tensors must be made with more precision.

Cartan's Theorem 6.2.1[1] *Let $(M, g), (\widetilde{M}, \widetilde{g})$ be Riemannian manifolds, $p \in M, \widetilde{p} \in \widetilde{M}$ with normal neighborhoods $B_\varepsilon(p), \widetilde{B}_\varepsilon(\widetilde{p})$ and let $A : T_pM \to T_{\widetilde{p}}\widetilde{M}$ be an isometry. Set $\varphi := \exp_{\widetilde{p}} \circ A \circ \exp_p^{-1} : B_\varepsilon(p) \to \widetilde{B}_\varepsilon(\widetilde{p})$. Denote the radial parallel transports by $||_X : T_pM \to T_{\exp_p X}M$, $\widetilde{||}_{AX} : T_{\widetilde{p}}\widetilde{M} \to T_{\exp_{\widetilde{p}} AX}\widetilde{M}$ and set*

$$f := \widetilde{||}_{AX} \circ A \circ ||_X^{-1} : T_{\exp_p X}M \to T_{\exp_{\widetilde{p}} AX}\widetilde{M}$$

(i.e. $f(\sum z_j e_j) = \sum z_j \widetilde{e}_j$ in terms of the bases $(e_j), (\widetilde{e}_j)_p := (Ae_{j|p})$ of 5.2.9). If locally for all $X, Y, Z \in TM$

$$f(\Omega_q(X, Y)Z) = \widetilde{\Omega}_{\varphi(q)}(f(X), f(Y))f(Z)$$

holds, then $\varphi : B_\varepsilon(p) \to \widetilde{B}_\varepsilon(\widetilde{p})$ is an isometry (Fig. 6.3).

---

[1] 1920, Élie Joseph Cartan, 1869–1951.

Proof. Associate to $V \in T_p M$ the Jacobi field $Y_t = T_{tX} \exp tV$ along $\exp_p tX$. As φ maps radial geodesics to radial geodesics, φ_* maps the Jacobi field $Y_t = \frac{\partial}{\partial s} c_s(t)$ to a Jacobi field, namely

$$\varphi_* Y_t = \varphi_* T_{tX} \exp_p(tV) = T_{tAX} \exp_{\tilde{p}}(tAV).$$

According to the Jacobi ODE, $f(Y)$ is also a Jacobi field along $\varphi(\exp tX)$. This is because every vector field $Z(t) = \sum z_j(t) e_j$ along $\exp tX$ satisfies

$$\tilde{\nabla}_{\partial/\partial t} f(Z) = \sum \frac{\partial z_j}{\partial t} \tilde{e}_j = f(\sum \frac{\partial z_j}{\partial t} e_j) = f(\nabla_{\partial/\partial t} Z).$$

Therefore

$$\tilde{\nabla}^2_{\partial/\partial t} f(Y) = f(\nabla^2_{\partial/\partial t} Y) = f(\frac{1}{t^2} \Omega(\mathfrak{R}, Y)\mathfrak{R}) \overset{\text{Vor.}}{=} \frac{1}{t^2} \Omega(\mathfrak{R}, f(Y))\mathfrak{R}).$$

Because $f(Y_0) = 0 = \varphi_* Y_0$ and

$$\tilde{\nabla}_{\partial/\partial t} f(Y_t)|_{t=0} = f(\nabla_{\partial/\partial t} Y_t|_{t=0}) = f(V) = AV = \left(\tilde{\nabla}_{\partial/\partial t} \varphi_* Y_t\right)_{|t=0},$$

one obtains $f(Y_t) = \varphi_* Y_t$. On a normal neighborhood of p every tangent vector can be written as Y_t (except at p), as the bijectivity of $\exp_p$ implies the bijectivity of $T \exp_p$. Thus $T\varphi = f$ is an isometry of Euclidian vector spaces and φ is a local isometry. □

Remark If $\tilde{M}$ is complete, the proof shows that for every ball-shaped normal neighborhood $U \subset M$, $\varphi_{|U}$ is a local isometry onto its image.

The classification of spaces of constant curvature follows as a corollary. The following proposition is a first step toward this classification.

Proposition 6.2.2 *Let M be a Riemannian manifold of dimension ≥ 3 whose sectional curvature $K_p(X \wedge Y) \equiv K_p$ at any point $p \in M$ is independent of the plane $X \wedge Y \in \Lambda^2 TM$. Then $K_p \equiv K$ is independent of p.*

Proof. Define $\Omega^1_p(X, Y)Z := g(Y, Z)X - g(X, Z)Y$ for $X, Y, Z \in T_p M$. The definition of sectional curvature implies

$$R_p(X, Y, X, Y) = -g(\Omega^1_p(X, Y)X, Y)K_p(X \wedge Y).$$

Hence here one finds $\Omega_p = \Omega^1_p \cdot K_p$ according to the polarization identity for symmetric forms on $\Lambda^2 TM$. Because ∇ is metric, one gets $\nabla \Omega^1 = 0$. Thus applying the 2nd Bianchi identity shows that

$$0 = \nabla \Omega = dK \wedge \Omega^1 \in \Lambda^3 T^* M \otimes \text{End}(TM).$$

Therefore for X, Y, Z orthonormal (which exist because dim $M \geq 3$) it follows that

$$
\begin{aligned}
0 &= g((dK \wedge \Omega^1)(X, Y, Z)Y, Z) \\
&= dK(X) \cdot \underbrace{g(\Omega^1(Y, Z)Y, Z)}_{=-1} + dK(Y) \cdot \underbrace{g(\Omega^1(Z, X)Y, Z)}_{=0} \\
&\quad + dK(Z) \cdot \underbrace{g(\Omega^1(X, Y)Y, Z)}_{=0} = -X.K.
\end{aligned}
$$

Hence K is constant. $\qquad\qquad\qquad\qquad\qquad\qquad\qquad\qquad\qquad\qquad\qquad\qquad\qquad$ □

Theorem 6.2.3 *Assume a Riemannian manifold $\widetilde{M}$ of dimension ≥ 3 to have at each point p sectional curvature K_p independent of the choice of plane. Then $\widetilde{M}$ is locally isometric to $M = S^n_{1/\sqrt{K}}$, $\mathbb{R}^n$ or to $(H^n, \frac{g_{\mathrm{hyp}}}{-K})$.*

Theorem 6.5.8 is a more precise result for the case of complete manifolds.

Proof. By Proposition 6.2.2, $(\widetilde{M}, g)$ has constant sectional curvature $K \in \mathbb{R}$ and $\Omega = K \cdot \Omega^1$. Because $f : T_{\exp X}M \to T_{\exp AX}\widetilde{M}$ is an isometry of Euclidean vector spaces in Cartan's Theorem, it follows that

$$
\begin{aligned}
f(\Omega(X, Y)Z) &= K \cdot f(\Omega^1(X, Y)Z) = K \cdot \widetilde{\Omega}^1(f(X), f(Y))f(Z) \\
&= \widetilde{\Omega}(f(X), f(Y))f(Z).
\end{aligned}
$$

Hence $\widetilde{M}$ and M are locally isometric. $\qquad\qquad\qquad\qquad\qquad\qquad\qquad\qquad$ □

In the case dim $\widetilde{M} = 2$, a corresponding theorem holds only under the stricter assumption $K \equiv \mathrm{const.}$ (Minding's Theorem, Exercise 5.3.20).

6.3 Riemannian Submersions

vertical tangent bundle	warped product
horizontal tangent bundle	O'Neill's tensor
horizontal component	O'Neill's Theorem
vertical component	proper
horizontal lift	Ehresmann's Fibration Theorem
Riemannian submersion	Hermann's Fibration Theorem
lift	O'Neill's Formulae

Dual to the considerations on immersions and the second fundamental form, in this section we consider submersions of Riemannian manifolds in general. Via the embeddings of the fibers, the second fundamental form will again play an important role here. The comparison of geodesics on the base and the total space becomes

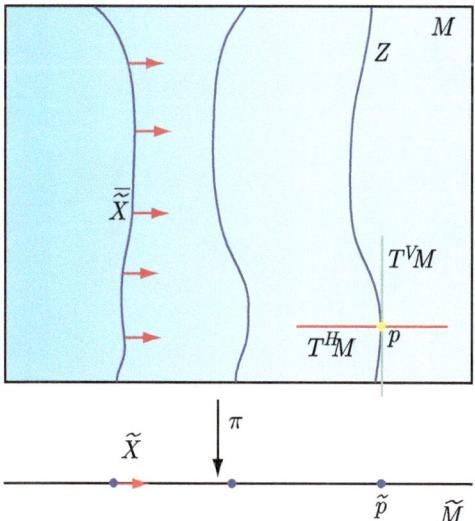

Fig. 6.4 Riemannian submersion

particularly straightforward. As a nice application Hermann's Fibration Theorem 6.3.10 follows.

According to Lemma 1.1.3 the fibers $\pi^{-1}(p)$ of a submersion $\pi : M \to \widetilde{M}$ are submanifolds.

Definition 6.3.1 *Given a submersion* $\pi : M \to \widetilde{M}$, *the vector bundle* $T^V M :=$ $\ker T\pi \subset TM$ *is called the **vertical tangent bundle**. Thus, for any fiber* $Z = \pi^{-1}(\widetilde{p})$ *over* $\widetilde{p} \in \widetilde{M}$, $TZ = T^V M_{|Z}$ *holds. If* $g = \langle \cdot, \cdot \rangle$ *is a Riemannian metric on* M, *the bundle* $T^H M := (T^V M)^\perp$ *is called the **horizontal tangent bundle** on* M. *In particular,* $T_p \pi : T_p^H M \to T_{\pi(p)} \widetilde{M}$ *is a vector space isomorphism. Let* $X^H \in T_p^H M$ *and* $X^V \in T_p^V M$ *denote the **horizontal** and **vertical component** of* $X \in T_p M$, *respectively.*

The map $T\pi_{|T^H M}$ thus identifies the horizontal tangent bundle with $\pi^* T\widetilde{M}$. In addition, the metric is used here to choose an embedding of this space into TM. The pullback of a vector $\widetilde{X} \in T\widetilde{M}$ embedded in TM yields a horizontal tangent vector $\widetilde{X}^*$.

Definition 6.3.2 *Associated to* $\widetilde{X} \in T_{\widetilde{p}} \widetilde{M}$ *and* $p \in \pi^{-1}(\widetilde{p})$ *let the **horizontal lift** be the uniquely determined vector* $\widetilde{X}^* \in T_p^H M$ *satisfying* $T\pi(\widetilde{X}^*) = \widetilde{X}$. *A submersion is called a **Riemannian submersion*** $\pi : (M, g) \to (\widetilde{M}, \widetilde{g})$ *if* $T\pi : T^H M \to T\widetilde{M}$ *is everywhere an isometry of Euclidean vector spaces (Fig. 6.4).*

In general, $X \in T_p M$ is a **lift** of $\widetilde{X} \in T_{\widetilde{p}} \widetilde{M}$, if $T\pi(X) = \widetilde{X}$. Of course, this does not uniquely determine X.

Without using Riemannian metrics $T^H M$ can also be constructed as $TM/T^V M$. These definitions already show some fundamental differences to the case of immersions $\widetilde{M} \hookrightarrow M$:

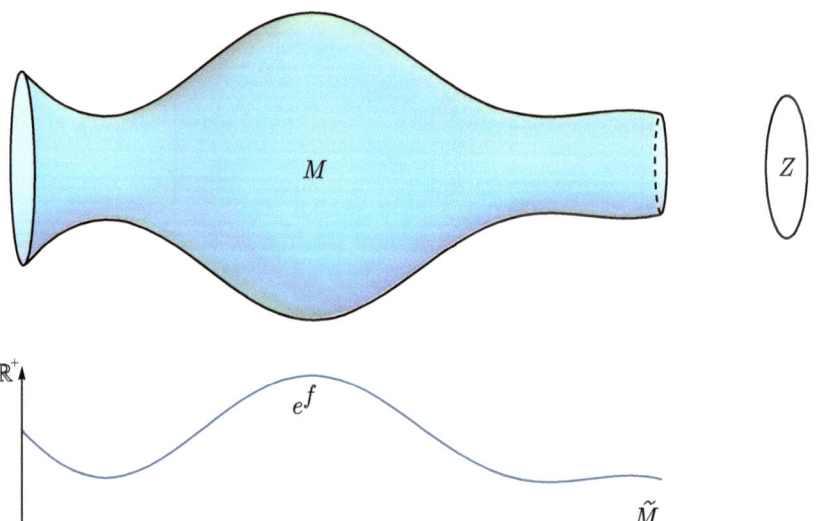

Fig. 6.5 Warped product.

1) The tangent space TM has at every point a decomposition induced by π. In the case of an immersion this is true only for $TM_{|\widetilde{M}}$,

2) therefore a canonical lift $\Gamma(\widetilde{M}, T\widetilde{M}) \to \Gamma(M, TM)$ does not exist for immersions (though non-canonical extensions do).

3) The requirement that the submersion be Riemannian is essentially a requirement on the geometry of M: If p, q lie in a fiber, $T_p^H M \to T_q^H M, X \mapsto (T_p\pi(X))^*$ must be an isometry. For immersions, on the other hand, the requirement "Riemannian" can always be achieved by appropriate choice of the metric on $\widetilde{M}$. In the case of submersions, this is true for the choice of a metric on the fibers and on $\widetilde{M}$ as well as the choice of a decomposition $TM = T^V M \oplus \pi^* T\widetilde{M}$. These structures always exist, as they can be constructed for instance by choosing any Riemannian metrics $\widetilde{g}$, g' on $\widetilde{M}$, M, setting $g^Z := g'_{|T^V M}$, choosing $T^H M$ as the orthogonal complement of $T^V M$ with respect to g' and setting $g := g^Z \oplus \pi^* \widetilde{g}$.

Example $\pi_1 : (\widetilde{M}, g_{\widetilde{M}}) \times (Z, g_Z) \overset{\text{can.}}{\to} (\widetilde{M}, g_{\widetilde{M}})$ and

$$\pi : (\widetilde{M} \times Z, \pi_1^* g_{\widetilde{M}} + \pi_1^* e^{2f} \cdot \pi_2^* g_Z) \to (\widetilde{M}, g_{\widetilde{M}})$$

associated to a function $f \in C^\infty(\widetilde{M}, \mathbb{R})$ are Riemannian submersions. The latter is called the **warped product** (Fig. 6.5).

Lemma 6.3.3 *If $\widetilde{X}, \widetilde{Y}$ are vector fields on $\widetilde{M}$ and U, V are vertical vector fields on M, then $[\widetilde{X}, \widetilde{Y}]^* = [\widetilde{X}^*, \widetilde{Y}^*]^H$ and $[\widetilde{X}^*, U], [U, V] \in \Gamma(M, T^V M)$.*

Proof. Because $T\pi(\widetilde{X}^*) = \widetilde{X}, T\pi(U) = 0$, Lemma 1.4.7 implies that $T\pi([\widetilde{X}^*, \widetilde{Y}^*]) = [\widetilde{X}, \widetilde{Y}]$ and $T\pi[\widetilde{X}^*, U] = 0 = T\pi[U, V]$. $\square$

The latter also follows because U, V are vector fields on the fibers and so are their Lie brackets accordingly. Theorem 2.2.9 shows that the following two operators are tensorial:

Definition 6.3.4 *Suppose E, F are vector fields on M. Let $T, A \in \Gamma(M, T^*M^{\otimes 2} \otimes TM)$ be defined by*

$$T_E F := (\nabla_{E^V} F^V)^H + (\nabla_{E^V} F^H)^V, \qquad A_E F := (\nabla_{E^H} F^V)^H + (\nabla_{E^H} F^H)^V.$$

*A is called the **O'Neill tensor**.*

The tensor T is just the tensor from Definition 5.1.13 associated to the embedding of the fibers $Z_p \subset M$ with normal bundle $T^H M_{|Z_p}$, which extended the 2nd fundamental form skew-symmetrically. Analogous to the role of T in immersions, A, T govern the relations between connections and curvatures on M and $\widetilde{M}$. The decomposition of the tangent vector in whose direction we differentiate also plays a role here, in contrast to the immersion case.

O'Neill's Theorem 6.3.5 ([ON1][2]) *Let $\pi : M \to \widetilde{M}$ be a Riemannian submersion, let X, Y be horizontal and U, V be vertical vector fields and let W be a horizontal lift. Denote the Levi-Civita connections on $M, Z, \widetilde{M}$ by $\nabla, \nabla^Z, \widetilde{\nabla} = \nabla^{T\widetilde{M}}$, respectively. Let II^Z denote the 2nd fundamental form of a fiber $Z \subset M$. Then*

1) $\nabla_X Y = \nabla_X^{\pi^*T\widetilde{M}} Y + \frac{1}{2}[X, Y]^V$,
2) $\nabla_U W = A_W U + T_U W$,
3) $\nabla_W U = A_W U + (T_U W + [W, U])$,
4) $\nabla_U V = II^Z(U, V) + \nabla_U^Z V$ *on each fiber Z.*

In each case the first summand is horizontal and the second is vertical. In $\nabla_X^{\pi^*T\widetilde{M}} Y$ the bundles $\pi^*T\widetilde{M}$ and $T^H M$ are identified: Y is interpreted as an element of $\pi^*T\widetilde{M}$, differentiated, and the result is mapped back to $T^H M$. In particular,

$$\widetilde{\nabla}_{\widetilde{X}} \widetilde{Y} = T\pi(\nabla_X Y)$$

for horizontal lifts X, Y of vector fields $\widetilde{X}, \widetilde{Y}$ on $\widetilde{M}$. The horizontal component of $\nabla_U X$ is no longer tensorial in X for general horizontal vector fields X which are not necessarily lifts.

Proof. 1) Let X, Y, W be horizontal lifts of $\widetilde{X}, \widetilde{Y}, \widetilde{W}$ on $\widetilde{M}$. The Koszul formula implies that

$$
\begin{aligned}
2g(\nabla_X Y, W) &= X.g(Y, W) + \cdots + g([X, Y], W) \\
&= X.(\pi^*\widetilde{g}(\widetilde{Y}, \widetilde{W})) + \cdots + g([X, Y]^H, W) \\
&\overset{6.3.3}{=} \widetilde{X}.\widetilde{g}(\widetilde{Y}, \widetilde{W}) + \cdots + \widetilde{g}([\widetilde{X}, \widetilde{Y}], \widetilde{W}) = 2\widetilde{g}(\widetilde{\nabla}_{\widetilde{X}} \widetilde{Y}, \widetilde{W}).
\end{aligned}
$$

---

[2] 1966, Barrett O'Neill, 1924–2011.

If U is a vertical vector field U, then

$$2g(\nabla_X Y, U) = - \underbrace{U.g(X,Y)}_{=U.\pi^*\widetilde{g}(\widetilde{X},\widetilde{Y})=0} + g([X,Y],U) - \underbrace{g([X,U],Y)}_{=0 \; (6.3.3)} - \underbrace{g([Y,U],X)}_{=0}.$$

Therefore $(\nabla_X Y)^H = \left(\widetilde{\nabla}_{\widetilde{X}}\widetilde{Y}\right)^*$, $(\nabla_X Y)^V = \frac{1}{2}[X,Y]^V$. The latter is tensorial in X,Y, hence the formula follows for any horizontal X,Y.

2),3) Because $\nabla_W U - \nabla_U W = [W,U] \in T^V M$, one gets $(\nabla_U W)^H = (\nabla_W U)^H = A_W U$.

4) is the definition of the 2nd fundamental form of the fiber. □

One can also interpret Theorem 6.3.5 as follows ([BGV, Prop. 10.6]): The Levi-Civita connections on $\widetilde{M}$ and Z induce a connection $\nabla^\oplus := \nabla^{\pi^* T\widetilde{M}} \oplus \nabla^Z$ on $TM = T^H M \oplus T^V M \overset{f}{\cong} \pi^* T\widetilde{M} \oplus TZ$, where the term $\nabla_X^Z U$ not defined by the immersion $Z \hookrightarrow M$ is taken to be $(\nabla_X U)^V$. This connection is metric since f is an isometry. The difference $S := \nabla^\oplus - \nabla$ is skew-symmetric in the last two components. By definition it is given by the vertical terms in (1),(2) and the horizontal terms in (2),(3),(4).

Lemma 6.3.6 *The tensors* $A, T \in T_2^1 M$ *satisfy the following symmetries:*

1) $A_X Y$ *is skew-symmetric in* $X, Y \in T^H M$,
2) $T_U V = II^Z(U,V)$ *is symmetric in* $U, V \in T^V M$,
3) $T_E, A_E \in \operatorname{End} TM$ *are skew-symmetric with respect to* g *for any* $E \in TM$.

According to (3), the right-hand summand in the definition of A (or T) is the negative of the adjoint of the left-hand summand.

Remark 6.3.7 Given vectors $P, Q, R \in TM$ and the form S from Exercise 3.3.14, the symmetries of T and A imply

$$\begin{aligned}
-g(S_P Q, R) &= g\left((\nabla - \nabla^\oplus)_P Q, R\right) \\
&= g(A_P Q + T_P Q, R^V) + g(A_Q P + A_P Q + T_P Q, R^H) \\
&= g(A_P Q^H, R) + g(A_R P^H, Q) + g(A_R Q^H, P) \\
&\quad + g(II(P,Q),R) - g(II(P,R),Q).
\end{aligned}$$

Therefore the connection $\nabla^\oplus$ is torsion free if and only if $A = 0$ and $T = 0$.

Proof. 1) O'Neill's Theorem 6.3.5(1) shows that $A_X Y = \frac{1}{2}[X,Y]^V$. Thus it is skew-symmetric.

2) This is Theorem 5.1.2.

3) If E is horizontal, then

$$0 = Y.\langle V, E\rangle = \langle \nabla_Y V, E\rangle + \langle V, \nabla_Y E\rangle = \langle A_Y V, E\rangle + \langle V, A_Y E\rangle,$$

and A_E vanishes for vertical E. Similarly for T, which has already been shown together with the Weingarten equation 5.1.14. □

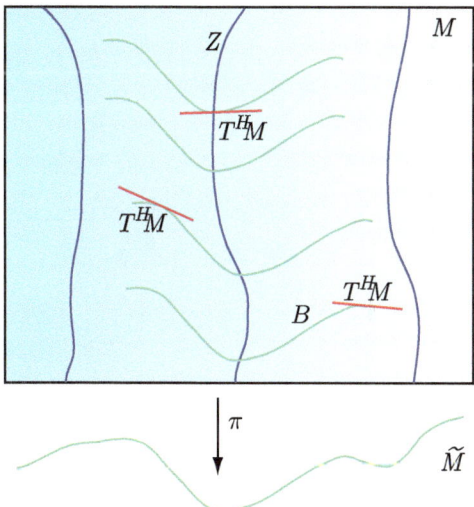

Fig. 6.6 Integrable horizontal distribution.

Remark By Frobenius' Theorem 2.3.10, $A \equiv 0$ if and only if the distribution $T^H M$ is integrable, i.e., at every $p \in M$ there exists a submanifold B_p whose tangent space equals $T^H M$ everywhere. According to Theorem 6.3.5(1), B_p is then totally geodesically embedded (Fig. 6.6).

By the definition in Exercise 5.2.14, $T \equiv 0$ means that the fibers Z are totally geodesically embedded.

Corollary 6.3.8 ([Her])) *1) Let c be a curve on M with $\dot{c} \in T^H M$ and $\widetilde{c} := \pi \circ c$. Then c is geodesic if and only if $\widetilde{c}$ is geodesic.*

2) If a geodesic c on M has horizontal velocity at at least one point p, then its velocity is horizontal everywhere.

Proof. 1) According to Theorem 6.3.5, any horizontal curve satisfies

$$\nabla^{c^* TM}_{d/dt} \dot{c} = \left(\nabla^{\widetilde{c}^* T\widetilde{M}}_{d/dt} \dot{\widetilde{c}} \right)^* + \underbrace{A_{\dot{c}} \dot{c}}_{-0} .$$

2) Without restriction, let I be so small that $\widetilde{c}(I) \subset \widetilde{M}$ is a submanifold. Associated to $\widetilde{c}$ let $\widehat{c}$ be the integral curve of $\dot{\widetilde{c}}^*$ on the submanifold $\pi^{-1}(\widetilde{c}(I)) \subset M$ starting at p. By (1), $\widehat{c}$ is a geodesic, hence the result follows because of the uniqueness of geodesics to an initial velocity (Fig. 6.7).   □

A map is called **proper** if preimages of compacta are compact.

Lemma 6.3.9 *Suppose $\widetilde{M}$ is a connected manifold.*

1) ([Her]) If M is complete, then $\widetilde{M}$ is also complete and π is surjective.

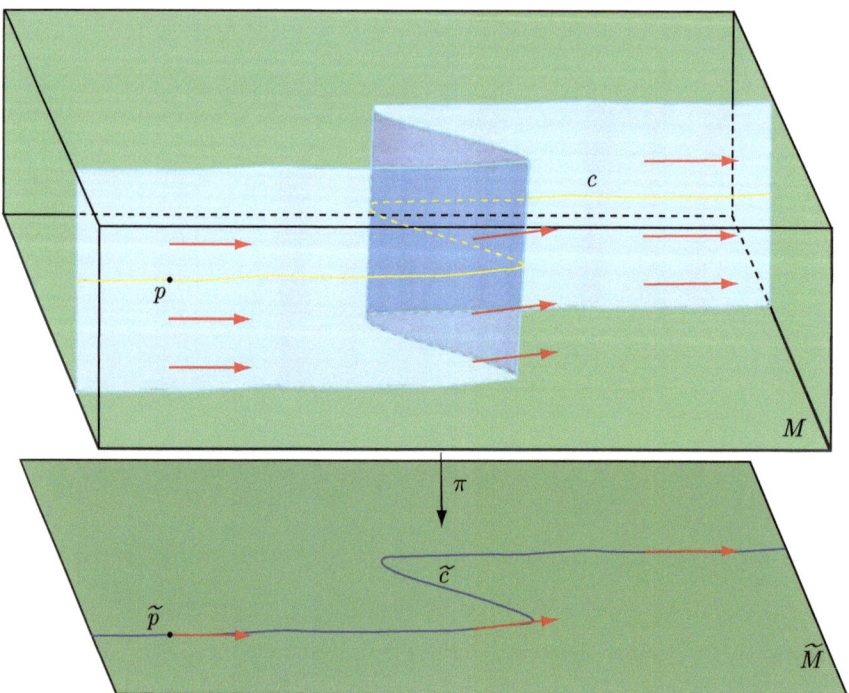

Fig. 6.7 Horizontal lift of a geodesic.

2) Every submersion is open.

3) Every proper submersion $\pi : M \to \widetilde{M}$ is surjective.

In the case of submersions with complete total space, the behavior of geodesics is thus much more straightforward than in the case of immersions: There is a bijection between geodesics with horizontal initial vector at $p \in M$ and geodesics on $\widetilde{M}$ starting at $\pi(p)$.

Proof. 1) Given a fixed point $p \in M$ and any $\widetilde{X} \in T_{\pi(p)}\widetilde{M}$, $\exp_p(t\widetilde{X}^*)$ is defined on all $\mathbb{R}$ and horizontal by Corollary 6.3.8(2). Therefore, by Corollary 6.3.8(1) $\exp_{\pi(p)}(t\widetilde{X}) = \pi\left(\exp_p(t\widetilde{X}^*)\right)$ is also defined on all of $\mathbb{R}$. Hence $\widetilde{M}$ is complete. By the Hopf–Rinow Theorem, every point in $\widetilde{M}$ has the form $\exp_{\pi(p)}\widetilde{X}$, and $\exp_p(t\widetilde{X}^*)$ lies in the preimage under π. Thus π is surjective.

2) Let $U \subset M$ be open. Around $p \in U$ choose a ball-shaped normal neighborhood in U. Then the image of the horizontal geodesic is a geodesic ball in $\widetilde{M}$ around $\pi(p)$ in $\pi(U)$.

3) As a proper map π is closed (Exercise 6.3.18). Therefore $\pi(M)$ as an open (by (2)) and closed set equals $\widetilde{M}$. $\square$

The openness of submersions also follows more elementarily by the Implicit Function Theorem. Analogous to Ehresmann's Fibration Theorem for proper submersions the following holds:

Hermann's Fibration Theorem 6.3.10 ([Her][3]) *Suppose M is a complete manifold and $\widetilde{M}$ is connected. Then every Riemannian submersion $\pi : M \to \widetilde{M}$ is a fiber bundle.*

This thus turns (modulo completeness) the infinitesimal criterion "$T\pi$ surjective" into the criterion "fiber bundle" locally on the base and globally on the fibers.

Proof. Consider $\widetilde{p} \in \widetilde{M}$ and a normal neighborhood $\widetilde{U} \subset \widetilde{M}$. Given $\widetilde{q} \in \widetilde{U}$, set $\widetilde{X}^{\widetilde{q}} := \exp_{\widetilde{p}}^{-1} \widetilde{q}$. Let $\varphi : \widetilde{U} \times Z_{\widetilde{p}} \to \pi^{-1}(\widetilde{U}), (\widetilde{q}, p) \mapsto \exp_p (\widetilde{X}^{\widetilde{q}})^*$. Then

$$\varphi^{-1}(q) = \left(\pi(q), \exp_q \left((T_{\widetilde{X}^{\pi(q)}} \exp_{\widetilde{p}})(-\widetilde{X}^{\pi(q)}) \right)^* \right)$$

holds, as

$$\left((T_{\widetilde{X}^{\pi(q)}} \exp_{\widetilde{p}})(\widetilde{X}^{\pi(q)}) \right)^*$$

is the velocity vector at q of the geodesic $\exp_p t(\widetilde{X}^{\pi(q)})^*$. Hence, the geodesic $\exp_q(\dots)$ in the formula for φ^{-1} runs back on the same path on which $\exp_p t(\widetilde{X}^{\widetilde{q}})^*$ runs from p to q. $\square$

Remark 6.3.7 shows that the curvatures are related by $\nabla^2 + \nabla S + S \wedge S = (\nabla^{\oplus})^2$. Thus, analogous to the role of the 2nd fundamental form in the case of immersions, the curvatures of M, Z and $\widetilde{M}$ can be compared using A and T. Three of the curvature identities describing values of $\Omega(U, V)$ with U, V vertical correspond to the Gauß, Codazzi–Mainardi and Ricci equations for immersions. In addition, three more arise concerning values of $\Omega(X, Y)$ and $\Omega(X, U)$ with X, Y horizontal. In the following theorem, we summarize only the two of these equations that will be used later. See Exercise 6.3.19 for the third and Exercise 6.3.20 for the interpretation of the Ricci equation in terms of A.

Theorem 6.3.11 (O'Neill's[4] formulas) *If $\pi : M \to \widetilde{M}$ is a Riemannian submersion, X, Y, Z, W are horizontal and U, V, U', V' are vertical vectors, then the following identities hold:*

1) The curvature tensor $\widetilde{R}$ of $\widetilde{M}$ satisfies

$$R_p(X, Y, Z, W) = (\pi^* \widetilde{R})_p(X, Y, Z, W) + \langle A_Y Z, A_X W \rangle - \langle A_X Z, A_Y W \rangle$$
$$-2 \langle A_X Y, A_Z W \rangle.$$

In particular, if X, Y are horizontal vector fields, then

$$K_p(X \wedge Y) = \widetilde{K}_{\pi(p)}(T\pi(X) \wedge T\pi(Y)) - \frac{3}{4} \frac{\| [X, Y]^V \|^2}{\| X \wedge Y \|^2}.$$

---

[3] 1960, Robert Hermann, 28.4.1931–10.2.2020.
[4] 1966, Barrett O'Neill.

2) $R(X, U, Y, V) = \langle ((\nabla_X T)_U - (\nabla_U A)_X) V, Y \rangle - \langle T_U X, T_V Y \rangle + \langle A_X U, A_Y V \rangle.$

Proof. 1) Horizontal lifts X, Y, Z, W of vector fields $\widetilde{X}, \widetilde{Y}, \widetilde{Z}, \widetilde{W}$ satisfy

$$
\begin{aligned}
g(\nabla_X \nabla_Y Z, W) &= X.g(\nabla_Y Z, W) - g(\nabla_Y Z, \nabla_X W) \\
&\overset{6.3.5(1)}{=} \widetilde{X}.\widetilde{g}(\widetilde{\nabla}_{\widetilde{Y}} \widetilde{Z}, \widetilde{W}) - \widetilde{g}(\widetilde{\nabla}_{\widetilde{Y}} \widetilde{Z}, \widetilde{\nabla}_{\widetilde{X}} \widetilde{W}) - g(A_Y Z, A_X W) \\
&= \widetilde{g}(\widetilde{\nabla}_{\widetilde{X}} \widetilde{\nabla}_{\widetilde{Y}} \widetilde{Z}, \widetilde{W}) - g(A_Y Z, A_X W).
\end{aligned}
$$

Furthermore,

$$
\begin{aligned}
g(\nabla_{[X,Y]} Z, W) &\overset{6.3.5(1)}{=} g(\nabla_{[X,Y]^H} Z, W) + g(\nabla_{2A_{XY}} Z, W) \\
&\overset{6.3.3, 6.3.5(2)}{=} \widetilde{g}(\widetilde{\nabla}_{[\widetilde{X},\widetilde{Y}]} \widetilde{Z}, \widetilde{W}) + 2g(A_Z A_X Y, W) \\
&\overset{6.3.6}{=} \widetilde{g}(\widetilde{\nabla}_{[\widetilde{X},\widetilde{Y}]} \widetilde{Z}, \widetilde{W}) - 2g(A_Z W, A_X Y).
\end{aligned}
$$

Addition in $R(X, Y, W, Z) = g(-\nabla_X \nabla_Y Z + \nabla_Y \nabla_X Z + \nabla_{[X,Y]} Z, W)$ provides the result we are looking for.

2) One finds

$$
\langle (\nabla_X T)_U V, Y \rangle = \langle \nabla_X (T_U V) - T_{(\nabla_X U)^V} V - T_U (\nabla_X V)^V, Y \rangle
$$

and so

$$
\begin{aligned}
\langle -\nabla_U \nabla_X V &+ \nabla_X \nabla_U V + \nabla_{[U,X]} V, Y \rangle \\
&= \langle (\nabla_X T)_U V - \nabla_U (\nabla_X V)^H + \nabla_X (\nabla_U V)^V + \nabla_{\nabla_U X - (\nabla_X U)^H} V, Y \rangle \\
&= \langle (\nabla_X T)_U V - \nabla_U (A_X V) + A_X (\nabla_U V) + \nabla_{T_U X} V, Y \rangle \\
&= \langle (\nabla_X T)_U V - (\nabla_U A)_X V - A_{\nabla_U X} V, Y \rangle - \langle T_V Y, T_U X \rangle \\
&= \langle (\nabla_X T)_U V - (\nabla_U A)_X V, Y \rangle + \langle A_X U, A_Y V \rangle - \langle T_V Y, T_U X \rangle. \qquad \square
\end{aligned}
$$

Thus, according to (1), $\widetilde{M}$ has sectional curvature at least as large as M. In the case of totally geodesic fibers, i.e. $T \equiv 0$, these formulas provide further estimates on the curvatures of $M, \widetilde{M}$ and Z.

Corollary 6.3.12 *If $\pi : M \to \widetilde{M}$ is a Riemannian submersion, $K > 0$ (or $K \geq 0$) for horizontal planes implies $\widetilde{K} > 0$ (or $\widetilde{K} \geq 0$, respectively).*

Exercises

Exercise* 6.3.13 Let π be the submersion $\pi : S^{2n+1} \to \mathbb{P}^n \mathbb{C}, (x_0, \ldots, x_n) \mapsto [(x_0 : \cdots : x_n)]$, let J be the complex structure on $\mathbb{C}^{n+1} \supset S^{2n+1}$ (i.e. $J : T_p \mathbb{C}^{n+1} \to T_p \mathbb{C}^{n+1}$ equals multiplication by i) and let $\mathfrak{n}$ be the outward pointing normal vector field on S^{2n+1}.

1) Prove that the fibers Z are great circles and determine $T^V S^{2n+1}$ in terms of $J, \mathfrak{n}$.
2) Show that this submersion is Riemannian for a suitable metric on $\mathbb{P}^n \mathbb{C}$.
3) Describe the geodesics on $\mathbb{P}^n \mathbb{C}$.
4) Determine the tensors A, T.
5) Compute the sectional curvature $K(\tilde{X} \wedge \tilde{Y})$ and determine its range.

Exercise 6.3.14 Let $\pi : M \to \tilde{M}$ be a Riemannian submersion and let c be any curve in M. Show that c is at least as long as $\pi \circ c$. Apply this to conclude directly that c is a geodesic if $\pi \circ c$ is a geodesic and $\dot{c} \in T^H M$.

Exercise 6.3.15 Show directly by applying Exercise 6.3.14 and Corollary 5.3.10, but without using A and T, that horizontal lifts $X, Y \in T_p M$ of $\tilde{X}, \tilde{Y}$ and the sectional curvatures $K, \tilde{K}$ of $M, \tilde{M}$ satisfy

$$\tilde{K}(\tilde{X} \wedge \tilde{Y}) \geq K(X \wedge Y).$$

Exercise 6.3.16 Consider a Riemannian submersion $\pi : M \to \tilde{M}$ and a geodesic c on M. Show that $\pi \circ c$ is a geodesic if and only if

$$T_{\dot{c}V} \dot{c}^V + 2A_{\dot{c}H} \dot{c}^V \equiv 0.$$

Exercise 6.3.17 Compute A and T in the case of a **warped product**

$$(B \times Z, \pi_1^* g_B + \pi_1^* e^{2f} \cdot \pi_2^* g_Z)$$

for Riemannian manifolds $(B, g_B), (Z, g_Z), \pi_1 : B \times Z \to B, \pi_2 : B \times Z \to Z$ and $f \in C^\infty(B, \mathbb{R})$.

Exercise* 6.3.18 Show that any proper map $f : M \to \tilde{M}$ between manifolds is closed.

Exercise 6.3.19 Prove the last of O'Neill's six equations about the curvature of submersions:

$$R(X, Y, Z, U) = \langle (\nabla_Z A)_X Y, U \rangle + \langle A_X Y, T_U Z \rangle - \langle A_Y Z, T_U X \rangle - \langle A_Z X, T_U Y \rangle,$$

where X, Y, Z are horizontal lifts and U is vertical.

Exercise 6.3.20 Given a Riemannian submersion $\pi : M \to \tilde{M}$ and a fiber Z, $N := T^H M_{|Z}$ is the normal bundle to $Z \hookrightarrow M$. Show for all vertical U, V and all horizontal lifts X, Y that the curvature of the connection $\nabla_U^N X := \nabla_U X - T_U X$ satisfies

$$g(\Omega^N(U, V)X, Y)$$
$$= g((\nabla_V A)_X Y, U) - g((\nabla_U A)_X Y, V) - g(A_X U, A_Y V) - g(A_X V, A_Y U).$$

6.4 Quotients

principal fiber bundle	chart adapted to the orbits
principal bundle	equivariant trivialization
proper operation	structure group
stabilizer	Maurer–Cartan 1-form
isotropy group	Cartan connection
free operation	frame bundle
equivariant	Kähler manifold

In this section, the Lie group G does not have to be connected. Under certain conditions, a Riemannian manifold has a quotient by an action of G, and $M \to M/G$ is again a fiber bundle. In this case the bundle is called a **principal fiber bundle** or **principal bundle**. In this section, G is assumed to act on the right. Of course, this is just a convention, and all results apply analogously to operations on the left; the quotient is then written as $G \backslash M$.

Crucially, the slice Theorem 6.4.6 ensures the existence of suitable charts on the quotient. Since the definition of the term "smooth manifold" consists of many individual requirements, most of which are topological in nature, one needs a correspondingly large number of conditions on the operation. The first one ensures that the quotient is Hausdorff:

Theorem 6.4.1 *Let a Lie group G act continuously on the right on a manifold M. The operation is **proper** if one of the following three equivalent conditions is satisfied::*

1) The map $f : M \times G \to M \times M, (p, \gamma) \mapsto (p\gamma, p)$ is proper, i.e. the preimage $f^{-1}(K')$ of any compactum is compact.

2) If $K \subset\subset M$ is compact, then $G_K := \{\gamma \in G \mid K\gamma \cap K \neq \emptyset\}$ is compact.

3) Around any point $p, q \in M$ there are neighborhoods U, V, such that $\{\gamma \in G \mid U\gamma \cap V \neq \emptyset\}$ is relatively compact in G.

According to (2), this condition is satisfied in particular if G is compact. Since every Lie subgroup $H \subset G$ is closed, H acts properly if G does. If M is compact and the action is proper, then the compactness of G also follows from (2).

Proof. (1)$\Rightarrow$(2): Let $\pi_2 : M \times G \overset{\text{can.}}{\to} G$. Then

$$\pi_2(f^{-1}(K \times K)) = \pi_2(\{(p, \gamma) \in K \times G \mid p\gamma \in K\}) = G_K,$$

hence G_K is compact.

(2)$\Rightarrow$(3): If U, V are relatively compact neighborhoods of p, q and $K := \overline{U} \cup \overline{V}$, then the closure of

$$\{\gamma \in G \mid U\gamma \cap V \neq \emptyset\} \subset \{\gamma \in G \mid K\gamma \cap K \neq \emptyset\}$$

is compact.

(3)⇒(1): Let $(U_{p,q} \times V_{p,q})_{p,q \in M}$ be a cover of $M \times M$ by neighborhoods as in (3), which themselves are also relatively compact. Let $(U_j \times V_j)_{j \in J}$ be a finite subcover of K'. Then

$$f^{-1}(K') = \{(p, \gamma) \mid (p\gamma, p) \in K'\} \subset \bigcup_{j \in J} \{(p, \gamma) \mid (p\gamma, p) \in U_j \times V_j\}$$

$$= \bigcup_{j \in J} \{(p, \gamma) \mid p \in U_j \gamma^{-1} \cap V_j\} \subset \bigcup_{j \in J} \left(V_j \times \{\gamma \mid U_j \gamma^{-1} \cap V_j \neq \emptyset\} \right)$$

is relatively compact and closed, hence compact.   □

Remark If K is any compactum, then the projection $\pi_2 : K \times G \to G$ is a closed map. Thus the set $G_K = \pi_2(f^{-1}(K \times K))$ is always closed. Therefore, relative compactness also suffices in (2).

Under the assumption (2), the **stabilizer** (or **isotropy group**) $G_p := \{\gamma \in G \mid p \cdot \gamma = p\}$ of any point $p \in M$ is compact. A group G acts **freely** if for all $p \in M, \gamma \in G \setminus \{e\}$ we have $p\gamma \neq p$, i.e. if no $\gamma \neq e$ has any fixed points or, in other words, if $G_{\{p\}} = \{e\}$ for any $p \in M$.

Example 1) The circle S^1 acts isometrically on the Euclidean plane $\mathbb{C}$, and via the radius $\mathbb{C}/S^1$ is identified with $\mathbb{R}_0^+$ as a topological space. Obviously, this quotient is not a topological manifold. On $\mathbb{C}^\times$, on the other hand, S^1 acts freely with quotient $\mathbb{C}^\times/S^1 \cong \mathbb{R}^+$.

2) The subgroup $\mathbb{Q} \subset \mathbb{R}$ acts freely, but not properly on $\mathbb{R}$, for two points $p, q \in \mathbb{R}$ never have neighborhoods U, V satisfying the condition in Theorem 6.4.1(3). Every subset of $\mathbb{Q}$ that contains an interval allows sequences without cluster points and is therefore not relatively compact.

3) If $\alpha \in \mathbb{R} \setminus \pi\mathbb{Q}$, then the operation of $\mathbb{Z}$ on S^1 by $e^{i\beta} * n := e^{i(n\alpha+\beta)}$ is free, but not proper, since S^1 is compact and $\mathbb{Z}$ is not.

A map $f : M \to N$ between manifolds with G-operation is called **(G-)equivariant** if for all $\gamma \in G, p \in M$ we have $f(p \cdot \gamma) = f(p) \cdot \gamma$. Let R denote the group action on the right.

Proposition 6.4.2 (Equivariant Rank Theorem) *Let a Lie group G act on N and transitively on M. Then every equivariant map $f : M \to N$ has constant rank.*

Proof. If $q = p \cdot \gamma$, then $T_q f \circ T_p R_\gamma = T_{f(p)} R_\gamma \circ T_p f$, and R_γ has maximal rank on M and N.   □

Lemma 6.4.3 *Suppose a Lie group G acts C^∞ on a manifold M. Let $p \in M$ and $\mu : G \to M, \gamma \mapsto p\gamma$. Then the stabilizer G_p is a Lie subgroup and* rank $T\mu \equiv$ dim $\mathfrak{g}$ – dim $\mathfrak{g}_p$.

Proof. Set $\rho : \mathfrak{g} \to \Gamma(M, TM), X \mapsto X_{M|q} := \frac{\partial}{\partial t}\big|_{t=0} q \cdot e^{tX}$. With this, X_M has flow $\Phi_t^{X_M}(q) = q \cdot e^{tX}$, because the latter satisfies the condition in Corollary 1.5.2. Let $V \subset T_{e_G} G$ be a neighborhood of 0 and choose $U \subset G$ such that $\exp_G : V \to U$ is a

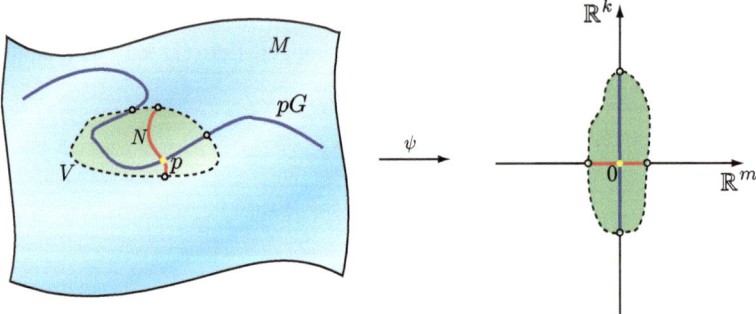

Fig. 6.8 Construction of N in the proof of the Slice Theorem.

diffeomorphism. Then $G_p \cap U = \exp_G(\ker T_e\mu \cap V)$: One finds $X_{M|p} = T_{eG}\mu(X)$. The equation $p \cdot e^{tX} = p$ implies $X_{M|p} = 0$. And if $X_{M|p} = 0$, then $\Phi_t^{X_M}(p) \equiv p$, hence $e^{tX} \in G_p$.

Therefore $G_p \cap U$ is a submanifold and rank $T_{eG}\mu = \dim \mathfrak{g} - \dim T_{eG}G_p$. Because of the equivariance, $T\mu$ has constant rank and $G_p = \mu^{-1}(\{p\})$ is a Lie subgroup. $\qquad\square$

Proposition 6.4.4 *Suppose a Lie group G acts freely and C^∞ on a manifold M. Then the orbits $pG \subset M$ are injective immersions of G. If in addition the operation is proper, then pG is an embedding of G.*

Proof. Set again $\mu : G \to M, \gamma \mapsto p\gamma$. Then μ is injective, because $p\gamma = p\widetilde{\gamma} \Leftrightarrow p\gamma\widetilde{\gamma}^{-1} = p \overset{\text{op. free}}{\Leftrightarrow} \gamma = \widetilde{\gamma}$.

By 6.4.3, rank $T_e\mu \equiv \dim G$ holds, since the operation is free. Thus μ is an immersion.

As a restriction of the map $f : M \times G \to M \times M$ from Theorem 6.4.1 to the closed subset $\{p\} \times G$, μ is proper and thus closed by Exercise 6.3.18. In particular, as a closed injective immersion, μ is an embedding. $\qquad\square$

Definition 6.4.5 *A chart (U, φ) is called **adapted to the orbits** if*

1) $\varphi(U) = \Omega_1 \times \Omega_2 \subset \mathbb{R}^m \times \mathbb{R}^k$ *with* $k := \dim G, m := n - k$,
2) *for all $p \in M$ we have $pG \cap U = \emptyset$ or there exists a $\mathbf{x}_0 \in \mathbb{R}^m$ such that* $\varphi(pG \cap U) = \{\mathbf{x}_0\} \times \mathbb{R}^k$.

Proposition 6.4.6 (Slice Theorem) *Under the same assumptions as in the last proposition, there exists an adapted chart around each $p \in M$.*

Proof. Let (V, ψ) be a chart as in Definition 1.1.1 associated to the submanifold pG, i.e. $\psi(pG \cap V) \subset \{0_{\mathbb{R}^m}\} \times \mathbb{R}^k$ and $\psi(p) = 0_{\mathbb{R}^{m+k}}$. Set $N := \psi^{-1}(\mathbb{R}^m \times \{0_{\mathbb{R}^k}\})$ and $\omega : N \times G \to M, (q, \gamma) \mapsto q\gamma$ (Fig. 6.8). Claim: Locally around (p, e), ω is a diffeomorphism. Indeed, im $T_e(\omega(p, \cdot)) = T(pG)$ is k-dimensional by Proposition 6.4.4, im $T_p(\omega(\cdot, e)) = \operatorname{im} T_p\operatorname{id}_N = \operatorname{im}\operatorname{id}_{T_pN}$ and $T_pM = T_pN \oplus T_p(gG)$. Hence $T_{(p,e)}\omega$ has maximal rank and by the Implicit Function Theorem ω is locally a

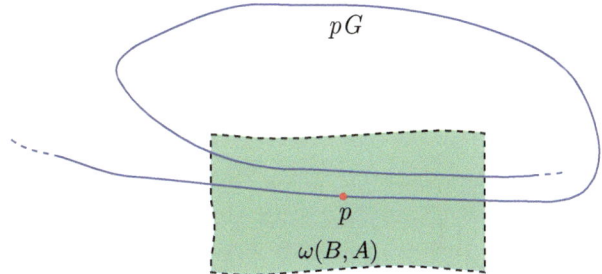

pG

p

$\omega(B,A)$

Fig. 6.9 This is what the Slice Theorem is designed to avoid.

diffeomorphism. Choose neighborhoods A, B of p, e in N, G that are diffeomorphic to balls, with $\overline{A}$ compact, such that ω is bijective on $A \times B$.

First, we show that A can be chosen sufficiently small such that for all $q \in A$ we have $qG \cap A = \{q\}$ (Fig. 6.9): Let $A_j \subset A_{j-1}$ be a sequence of open neighborhoods of p in A such that $\bigcap A_j = \{p\}$. Suppose that in each A_j there are q_j, r_j with $r_j = q_j \gamma_j$ for some $\gamma_j \neq e$. Then $(\gamma_j)_j$ is a sequence in the compactum $G_{\overline{A}}$, so it has a cluster point $\gamma \in G$. Because $q_j, r_j \to p$, it follows that $p = p\gamma$, hence $\gamma = e$ because G acts freely. Thus there is a $\gamma_\ell \in B$, and $\omega(q_\ell, \gamma_\ell) = \omega(r_\ell, e)$. ↯ to the bijectivity of ω.

Now choose A correspondingly small and choose diffeomorphisms $\zeta_1 : A \to B_1(0_{\mathbb{R}^m})$, $\zeta_2 : B \to B_1(0_{\mathbb{R}^k})$. Then $\varphi := (\zeta_1, \zeta_2) \circ \omega^{-1} : AB \to B_1(0_{\mathbb{R}^m}) \times B_1(0_{\mathbb{R}^k})$ is an adapted chart. This is because in each orbit $\widetilde{p}G$ intersecting AB there is exactly one $q \in A$, and $\varphi(qG \cap AB) \subset \{\zeta_1(q)\} \times \mathbb{R}^k$. $\qquad \square$

Proposition 6.4.7 *Suppose that a Lie group G acts continuously on a manifold M. Then $\pi : M \twoheadrightarrow M/G$ is an open map.*

Proof. Let $U \subset M$ be open. Then $\pi^{-1}(\pi(U)) = \bigcup_{\gamma \in G} U\gamma$ is open, therefore by definition of the quotient topology $\pi(U)$ is open. $\qquad \square$

After we have verified these critical technical theorems, the individual axioms from the definition of manifolds can now be checked one by one.

Theorem 6.4.8 [5] *Let a Lie group G act freely, properly and C^∞ on a manifold M. Then M/G has a canonical C^∞-manifold structure and $\pi : M \twoheadrightarrow M/G$ is a fiber bundle with typical fiber G.*

Proof. M/G is a Hausdorff space: the image

$$R := \{(p\gamma, p) \in M \times M \mid \gamma \in G, p \in M\} \subset M \times M$$

of the closed set $M \times G$ under the proper map $M \times G \to M \times M, (p, \gamma) \mapsto (p\gamma, p)$ is closed. Now set $[p] \neq [q] \in M/G$, i.e. $(p, q) \notin R$. If $U \times V$ is a neighborhood

---

[5] 1950, A.M. Gleason, 1921–2008 (G compact); 1953, J.-L. Koszul, 1921–2018 (G non-compact).

of (p, q) in $M \times M \setminus R$, then according to Proposition 6.4.7 $\pi(U), \pi(V)$ are disjoint neighborhoods of $[p], [q]$.

M/G is a Lindelöf space: If $(U_j)_j$ is an open cover of M/G, then $(\pi^{-1}(U_j))_j$ is an open cover of M. Reduce this to a countable cover $(\pi^{-1}(U_{j_n}))_{n \in \mathbb{N}}$. Then $(U_{j_n})_{n \in \mathbb{N}}$ is a countable cover of M/G.

M/G is locally homeomorphic to $\mathbb{R}^m$: Let $\varphi : AB \to \Omega_1 \times \Omega_2$ be an adapted chart around a point $p \in M$. Then $\pi_{|A} : A \to \widetilde{U} := \pi(AB)$ is bijective and continuous. According to Proposition 6.4.7 $\pi_{|A}$ is open: If $A \cap U \overset{\text{open}}{\subset} A$, then $(A \cap U)B \overset{\text{open}}{\subset} M$, because $\omega^{-1}((A \cap U)B) = (A \cap U) \times B \overset{\text{open}}{\subset} M \times G$. Thus $\pi((A \cap U)B) = \pi(A \cap U)$ is open. Therefore $\pi_{|A}$ is a homeomorphism. Set $\pi_1 : \Omega_1 \times \Omega_2 \to \Omega_1$. Then $\pi_1 \circ \varphi_{|A} = \zeta_1 : A \to \Omega_1$ is a homeomorphism, hence also $\widetilde{\varphi} := \pi_1 \circ \varphi \circ (\pi_{|A})^{-1} : \widetilde{U} \to \Omega_1$.

M/G has a canonical C^∞ structure (given by $\{\widetilde{\varphi}_j \mid j \in J\}$ associated to an atlas $\{\varphi_j \mid j \in J\}$ of adapted charts of M): Consider $\widetilde{p} \in M/G$ and adapted charts φ_1, $\varphi_2 : AB \to \Omega_1 \times \Omega_2$ around $p_1 \in \pi^{-1}(\widetilde{p})$, $p_2 = p_1 \gamma \in \pi^{-1}(\widetilde{p})$, respectively. Then

$$\varphi_2' := \varphi_2 \circ R_\gamma : (A\gamma^{-1})(\gamma B \gamma^{-1}) \to \Omega_1 \times \Omega_2$$

is another adapted chart around p_1 such that $\widetilde{\varphi}_2' = \widetilde{\varphi}_2$, because

$$(\varphi_2 \circ R_\gamma) \circ (\pi_{|A\gamma^{-1}})^{-1} = \varphi_2 \circ (\pi_{|A})^{-1}.$$

Since the adapted charts map orbits to constant **x**-coordinates, one gets $(\varphi_1 \circ \varphi_2'^{-1})\begin{pmatrix} \mathbf{x} \\ \mathbf{y} \end{pmatrix} = \begin{pmatrix} \alpha(\mathbf{x}) \\ \beta(\mathbf{x},\mathbf{y}) \end{pmatrix}$ with C^∞-maps α, β. Hence $\widetilde{\varphi}_1 \circ \widetilde{\varphi}_2^{-1} = \widetilde{\varphi}_1 \circ \widetilde{\varphi}_2'^{-1} = \alpha$ is smooth.

π is a fiber bundle: Consider $\widetilde{p} \in M/G$. Choose $p \in \pi^{-1}(\widetilde{p})$ and an adapted chart $\varphi : AB \to \Omega_1 \times \Omega_2$ with ζ_2 as in the proof of the Slice Theorem. Then the $\varphi \circ R_{\widehat{\gamma}}$ provide adapted charts for all $\widehat{\gamma} \in G$, which cover $\pi^{-1}(\widetilde{U})$ for $\widetilde{U} := \pi(A)$. When read through the charts φ and $\widetilde{\varphi}$, $\widetilde{\varphi} \circ \pi \circ \varphi^{-1} : \Omega_1 \times \Omega_2 \to \Omega_1$ equals π_1. Hence π is smooth everywhere on $\pi^{-1}(\widetilde{U})$ (and even a submersion). The set $\widetilde{U}$ is a neighborhood of $\widetilde{p}$, every orbit in $\pi^{-1}(\widetilde{U})$ intersects A in exactly one point q and

$$h : \pi^{-1}(\widetilde{U}) \to \widetilde{U} \times G,$$
$$q\gamma \mapsto (\pi(q), \gamma)$$

is therefore well-defined and bijective. Furthermore, h is smooth, because around each $q' = q\gamma \in \pi^{-1}(\widetilde{U})$ it satisfies

$$(\widetilde{\varphi}, \zeta_2 \circ R_{\gamma^{-1}}) \circ h \circ (\varphi \circ R_{\gamma^{-1}})^{-1} = \mathrm{id}_{\Omega_1 \times \Omega_2}. \qquad \square$$

Remark The trivializations h constructed in the proof satisfy

$$h^{-1}([q], \gamma'\gamma) = h^{-1}([q], \gamma') \cdot \gamma,$$

hence they are equivariant. The fiber bundles $M \to M/G$ together with the operation of G on M are also called (smooth) **principal fiber bundles** with **structure group** G.

Example If $K = \mathbb{R}, \mathbb{C}$ or $\mathbb{H}$, then $K^\times$ acts freely, properly and C^∞ by multiplication on $K^{n+1} \setminus \{0\}$. This yields as the quotient once more the projective space $\mathbb{P}^n K$. $\mathbb{Z}^n$ acts additively freely, properly and C^∞ on $\mathbb{R}^n$, and the quotient is the torus.

As the final structure, the existence of a canonical Riemannian metric on the quotient is checked under an additional condition.

Theorem 6.4.9 *Let a Lie group G operate freely, properly, C^∞ by isometries on a Riemannian manifold (M, g). Then M/G carries a canonical metric $\widetilde{g}$ such that $\pi : M \twoheadrightarrow M/G$ is a Riemannian submersion.*

Proof. Consider $\widetilde{p} \in M/G$ and $X, Y \in T_{\widetilde{p}}(M/G)$. If $p \in \pi^{-1}(\widetilde{p})$, then set $\widetilde{g}_{\widetilde{p}}(X, Y) := g_p(X_p^*, Y_p^*)$. This is independent of the choice of p: If $q \in \pi^{-1}(\widetilde{p})$, then there is exactly one $\gamma \in G$ such that $q = p\gamma$, and hence $g_p = \gamma^* g_q$. In particular, $\gamma^* T_q^H M = T_p^H M$ and $T\gamma(X_p^*) = X_q^*, T\gamma(Y_p^*) = Y_q^*$ hold. $\square$

Example The rotation around the north-south axis on $S^2 \setminus \{N, S\}$ yields as orbits the latitudes and as quotient $]-1, 1[$. This example already shows that even in the case of quotients without further conditions the tensor T does not become all that simple, even if A and T are naturally G-invariant. However, one can assign vertical vector fields X' to the elements $X \in \mathfrak{g}$ as in Exercise 6.4.12, which makes the handling of A, T more straightforward. The associated **Maurer–Cartan 1-form** $\vartheta \in \Gamma(M, T^*M \otimes \mathfrak{g})$ is the map $\vartheta(Y_p) = X$ with $Y_p^V = (X')_p$. This leads to the notion of **Cartan connections** [Car], [KoN, ch. II], which we shall not elaborate on here.

According to a result of Vilms [Vi] there is always a metric on M with respect to which T vanishes.

Example 6.4.10 The round metric on S^{2n+1} induces a metric on $\mathbb{P}^n\mathbb{C}$ via $\mathbb{P}^n\mathbb{C} = (\mathbb{C}^{n+1} \setminus \{0\})/\mathbb{C}^\times = (\mathbb{C}^{n+1} \setminus \{0\}/\mathbb{R}^+)/(\mathbb{C}^\times/\mathbb{R}^+) = S^{2n+1}/S^1$ (see the Exercises 6.3.13, 6.4.14). The group $\mathbb{C}^\times$ itself does not act by isometries on the Euclidean $\mathbb{C}^{n+1}$. The analogous result holds in the case of $\mathbb{P}^n\mathbb{H} = \mathbb{H}^{n+1} \setminus \{0\}/\mathbb{H}^\times = S^{4n+3}/S^3$.

Example 6.4.11 The **orthogonal frame bundle** $O(M)$ of a Riemannian manifold (M, g) is defined as the disjoint union of the sets

$$O(M)_p := \{\varphi : (\mathbb{R}^n, \langle \cdot, \cdot \rangle_{\text{Eucl}}) \to (T_p M, g_p) \text{ is an isometry}\}$$

together with the map $\pi : O(M) \overset{\text{can}}{\to} M$. Each φ corresponds canonically to the orthogonal basis

$$(\varphi(e_1), \ldots, \varphi(e_n)) \in (T_p M)^n,$$

which yields an embedding $O(M)_p \subset TM^{\oplus n}$ as a submanifold. By definition, $O(M)_p \cong O(n)$ and there is a canonical operation of $O(n)$ on the right on $O(M)$ via $O(M) \times O(n) \rightarrow O(M), (\varphi, A) \mapsto \varphi \circ A$. This operation is free and proper, hence $O(M)/O(n) = M$. **Be careful:** $O(M)$ is <u>not</u> the bundle of isometries $T_p M \rightarrow T_p M$.

Exercises

Exercise* 6.4.12 Suppose a Lie group G acts C^∞ on a manifold M via $\rho : \gamma \mapsto (p \mapsto p\gamma)$. Thus, each $X \in \mathfrak{g}$ induces a diffeomorphism $\rho(e^X)$ of M, and $X' := \frac{d}{dt}_{|t=0}\rho(e^{tX})$ is a vector field on M. Show that $[X', Y'] = [X, Y]'$. What is the corresponding identity for the fields $X'' := \frac{d}{dt}_{|t=0}\rho(e^{-tX})$ (or for an operation on the left)?

Exercise* 6.4.13 In the proof of Theorem 6.4.8, replace the proof of the Lindelöf property by deducing the second countability of M/G from that of M.

Exercise 6.4.14 Suppose $\pi : S^{2n+1} \rightarrow \mathbb{P}^n\mathbb{C}$ is the canonical projection from the Example 6.4.10. Let $J : T_p\mathbb{C}^{n+1} \rightarrow T_p\mathbb{C}^{n+1}$ be the multiplication by i. Using Exercise 6.3.13 show that

1) J induces an isometry $J : T_p\mathbb{P}^n\mathbb{C} \rightarrow T_p\mathbb{P}^n\mathbb{C}$ such that $\widetilde{\nabla}J = 0$ (a Riemannian manifold M is a **Kähler manifold** if and only if there exists such an automorphism satisfying $J^2 = -1$).
2) Determine the lengths of all geodesics on $\mathbb{P}^n\mathbb{C}$.
3) Consider two geodesics on $\mathbb{P}^n\mathbb{C}$ with different paths, but starting at the same point. After which distance do they meet for the first time (proof by cases)?
4) Compute the metric of $\mathbb{P}^n\mathbb{C}$ pulled back by the chart φ_0 from Section 1.2.

Exercise 6.4.15 Let G be a Lie group and let G_0 be the connected component of e_G.

1) Show that G_0 is a normal subgroup of G.
2) Conclude that G/G_0 is a discrete Lie group and that all connected components are diffeomorphic.

6.5 Discrete Fibers

Riemannian covering	simply connected
degree of a covering	homotopy groups
leaf	Hadamard–Cartan Theorem
properly discontinuous action	Klein's bottle
group of deck transformations	Möbius strip
universal covering	lense space
fundamental group	

In this section, the results of the last two chapters are specialized to the case of discrete fibers and discrete groups. This leads in particular to the classification of spaces of constant curvature.

Definition 6.5.1 *A map $\pi : M \to \widetilde{M}$ is called a **Riemannian covering** if for all $p \in \widetilde{M}$ there exist a neighborhood $U \subset \widetilde{M}$ and a discrete set Z (i.e., a 0-dimensional manifold) for which $\pi^{-1}(U) \cong U \times Z$ as Riemannian manifolds.*

Since $\#Z$ is locally constant, for any p one can choose the same set Z and π becomes a fiber bundle with typical fiber Z. The number $\#Z$ is called the **degree** of π or **number of leaves**. Thus, a Riemannian covering is a fiber bundle with discrete fiber which is Riemannian as a submersion.

Example $\mathbb{R} \to S^1$ is an ∞-fold covering.

Lemma 6.5.2 *Suppose $\pi : M \to \widetilde{M}$ is a covering and $\widetilde{g}$ is a Riemannian metric on $\widetilde{M}$. Then π induces a metric on M such that π is a Riemannian covering.*

Proof. Choose $g_p := \pi^* \widetilde{g}_{\pi(p)}$. As π is local diffeomorphism, $T\pi$ has maximal rank and one finds $g > 0$. □

Conversely, not every local diffeomorphism is a covering. For example, the map $]0, 4\pi[\to S^1, t \mapsto e^{it}$ is a local isometry with compact fibers, but $0 \in S^1$ has one preimage 2π, while every other point has two preimages (Fig. 6.10).

Lemma 6.5.3 *Let $\pi : M \to \widetilde{M}$ be a Riemannian covering. Then M is complete if and only if $\widetilde{M}$ is complete.*

Proof. "$\Rightarrow$" is a special case of Lemma 6.3.9.

"$\Leftarrow$": Suppose $\widetilde{M}$ is complete and $c :]a, b[\to M$ is a geodesic which cannot be extended to any $t > b$. Then $\widetilde{c} := \pi(c)$ is a geodesic, hence defined on $\mathbb{R}$. On a sufficiently small neighborhood U of $\widetilde{c}(b)$, the fiber bundle π is trivial. If $\widetilde{c}([b - \varepsilon, b + \varepsilon]) \subset U$, then U is isometric to a neighborhood of $c(b - \varepsilon)$. Thus c can be continued to $b + \varepsilon$. ↯ □

Lemma 6.5.4 *Suppose M is a complete manifold and $\widetilde{M}$ is connected. Then every map $\pi : M \to \widetilde{M}$ whose derivative $T_p\pi$ is an isometry at every point is a Riemannian covering.*

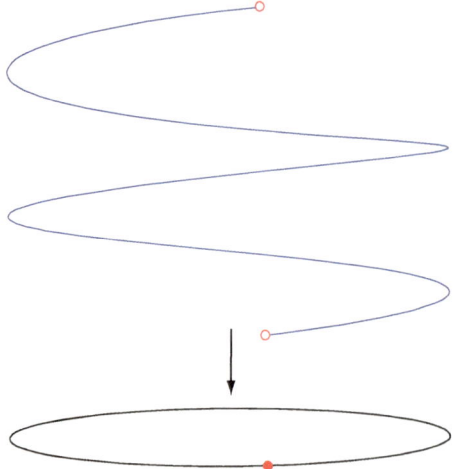

Fig. 6.10 A local isometry that is not a covering.

Proof. By Hermann's Fibration Theorem 6.3.10, M is a fiber bundle. Since the isometry $T_p\pi$ is invertible, the fibers are discrete.   □

A proper operation of a discrete group is called **properly discontinuous**. The criteria (2),(3) from Theorem 6.4.1 then apply with "finite" in place of "(relatively) compact".

Example $\pi : S^n \to \mathbb{P}^n\mathbb{R}$ induces a metric on $\mathbb{P}^n\mathbb{R}$.

Lemma 6.5.5 *Suppose $\varphi, \psi : M \to \widetilde{M}$ are two local isometries between Riemannian manifolds and M is connected. If there is a point $p \in M$ such that $T_p\varphi = T_p\psi$, then $\varphi = \psi$ holds.*

Proof. Set $N := \{q \in M \mid T_q\varphi = T_q\psi\}$ ($\ni p$, hence $\neq \emptyset$). Then N is closed. If $q \in N$, then $\varphi(\exp_q X) = \exp_{\varphi(q)} T_q\varphi(X) = \exp_{\psi(q)} T_q\psi(X) = \psi(\exp_q X)$ holds, because φ, ψ map geodesics to geodesics and each geodesic is uniquely determined by its initial vector. Therefore, the normal neighborhood of each $q \in N$ is a subset of N and thus N is open. As a nonempty open closed subset, $N = M$.   □

Theorem 6.5.6 *Suppose $\pi : M \to \widetilde{M}$ is a Riemannian covering and M is connected. Then the **group of deck transformations***

$$\Gamma := \{\varphi : M \to M \text{ diffeomorphism} \mid \pi \circ \varphi = \pi\}$$

of π acts freely on each fiber, properly discontinuously, isometrically and C^∞ on the left on M. If Γ acts transitively on a fiber, it follows that $\widetilde{M} \cong \Gamma\backslash M$.

Proof. As $T\pi$ is pointwise an isometry, because $T_p\varphi = (T_{\varphi(p)}\pi)^{-1} \circ T_p\pi$, every $\varphi \in \Gamma$ is also an isometry. If $\varphi(p) = p$ for a $p \in M$, then it follows that $T_p\varphi =$ id. On the other hand $T_p \mathrm{id}_M = \mathrm{id}$ holds and by Lemma 6.5.5 one thus gets $\varphi = \mathrm{id}_M$. Hence Γ acts freely.

Because φ maps normal neighborhoods to normal neighborhoods, Γ acts properly discontinuously: Given $p, q \in M$ with $\pi(p) \neq \pi(q)$, choose normal neighborhoods U, V with $\pi(U), \pi(V)$ disjoint. Then $\{\varphi \in \Gamma \mid \varphi(U) \cap V \neq \emptyset\} = \emptyset$. In the case $\pi(p) = \pi(q)$ choose normal neighborhoods U, V such that $\pi(U) = \pi(V) = $ normal neighborhood of $\pi(p)$. Because of the free operation, there is at most one $\varphi \in \Gamma$ satisfying $\varphi(p) = q$, thus $\#\{\varphi \in \Gamma \mid \varphi(U) \cap V \neq \emptyset\} \leq 1$.

Therefore $\tilde{\pi} : \Gamma \backslash M \to M$ is a covering. If Γ acts transitively on $\pi^{-1}(p)$, $\tilde{\pi}$ is 1-fold. $\qquad \square$

The topological covering theory (see e.g. [Ha]) shows this fact more generally and further the following statements about coverings, which will not be proved here:

Theorem and Definition 6.5.7 *Any manifold $\tilde{M}$ has a uniquely determined maximal covering M which has no further nontrivial coverings. M is called the **universal covering** of $\tilde{M}$. The associated group of deck transformations is called the **fundamental group** $\pi_1(\tilde{M}) := \Gamma$ of $\tilde{M}$. A connected manifold $\tilde{M}$ is called **simply connected** if $\pi_1(\tilde{M}) = 0$.*

In particular, the universal covering of $\tilde{M}$ is simply connected.

Remark In Algebraic Topology it is further shown that every covering $\overline{M}$ of $\tilde{M}$ corresponds to a subgroup $H \subset \pi_1(\tilde{M})$ which becomes the group of deck transformations of $M \to \overline{M}$. The group of deck transformations of $\overline{M} \to \tilde{M}$ is then (normalizer of H in $\pi_1(\tilde{M}))/H$. If H is a normal subgroup, the group of deck transformations of $\overline{M} \to \tilde{M}$ acts transitively on the fibers.

Given any covering $\pi : M \to \tilde{G}$ of a Lie group, M can always be given a Lie group structure for which π becomes a homomorphism by lifting the multiplication (see [War, 3.24]). In particular, by Exercise 6.5.13 $\pi_1(\tilde{G})$ is always abelian. In general, given a fiber bundle, in Topology a long exact sequence of **homotopy groups** π_j is constructed ([Stee, §17]), which in the case of a quotient $M \to M/G$ takes the form

$$\cdots \longrightarrow \pi_2(G) \longrightarrow \pi_2(M) \longrightarrow \pi_2(M/G) \rightsquigarrow$$
$$\rightsquigarrow \pi_1(G) \longrightarrow \pi_1(M) \longrightarrow \pi_1(M/G) \rightsquigarrow$$
$$\rightsquigarrow \pi_0(G) \longrightarrow \pi_0(M) \longrightarrow \pi_0(M/G) \longrightarrow 0$$

(where all maps except the last three are group homomorphisms). If G is connected, $\pi_0(G)$ vanishes. Moreover for compact groups one can show $\pi_2(G) = 0$ ([BtD, Lemma V.7.5]). Using the Iwasawa decomposition (see [Hel, ch. VI §5]), this follows for arbitrary Lie groups G, and the above sequence implies for G connected an exact sequence

$$0 \to \pi_2(M) \to \pi_2(M/G) \to \pi_1(G) \to \pi_1(M) \to \pi_1(M/G) \to 0. \quad (6.1)$$

In particular, for a simply connected G the homotopy groups π_0, π_1, π_2 of the total space and base coincide. Conversely, if M is connected and simply connected, then

$$0 \to \pi_1(M/G) \to \pi_0(G) \to 0. \quad (6.2)$$

Thus in this case $\pi_1(M/G)$ equals the group of connected components of G.

Theorem 6.5.8 *Suppose $\tilde{M}$ is complete and connected with sectional curvature $K = -1, 0$ or 1. Then $\tilde{M}$ is isometric to*

$$\Gamma \backslash H^n, \ \Gamma \backslash \mathbb{R}^n \quad or \quad \Gamma \backslash S^n.$$

Proof. By (the remark following) Cartan's Theorem 6.2.1 there exists a map $\varphi : H^n, \mathbb{R}^n, S^n \setminus \{p\} \to \tilde{M}$ which is a local isometry (for any $p \in S^n$). In the case $K = 1$, let $q \in S^n \setminus \{p, -p\}$. Then Cartan's Theorem also provides a local isometry $\psi : S^n \setminus \{-q\} \to \tilde{M}$ to $T_q \psi := T_q \varphi$. By Lemma 6.5.5 it follows that $\psi = \varphi$ holds on the common domain of definition, hence φ can be continued to p by $\psi(p)$ as a map $\varphi : S^n \to \tilde{M}$. By Lemma 6.5.4, φ is a Riemannian covering. In particular, for any covering $M \to H^n, \mathbb{R}^n$ or S^n, again $M = H^n, \mathbb{R}^n$ or S^n, respectively. Thus, by Theorem 6.5.7, the covering is universal, and by Theorem 6.5.6, $\tilde{M}$ is of the stated form. $\qquad \square$

Remark More elegantly, the case of positive constant curvature can be solved by applying the stronger Cartan–Ambrose–Hicks Theorem 7.3.2.

Hadamard–Cartan Theorem 6.5.9 [6] *Suppose M is a complete manifold with $K \leq 0$ everywhere and $p \in M$. Then $\exp_p : T_p M \to M$ is a covering. In particular, M is diffeomorphic to $T_p M \cong \mathbb{R}^n$ if and only if $\pi_1(M) = 0$, and in that case there is exactly one geodesic that passes through any two points.*

This generalizes an axiom of Euclidean geometry: Exactly one straight line passes through 2 points.

Proof. Let c be a geodesic, $\|\dot{c}\| = 1$ and let $Y = T_{tX} \exp_p tV$ be a Jacobi field along c satisfying $0 \neq V \perp \dot{c}$. Then

$$\frac{\partial^2}{\partial t^2} \|Y_t\|^2 \quad = \quad 2\|\nabla_{\partial/\partial t} Y\|^2 + 2g((\nabla_{\partial/\partial t})^2 Y, Y)$$

$$\overset{\text{Jacobi ODE}}{=} \quad \underbrace{2\|\nabla_{\partial/\partial t} Y\|^2}_{\geq 0} \underbrace{-2K(\dot{c} \wedge Y) \cdot \|Y\|^2}_{\geq 0}.$$

---

[6] 1928, Élie Joseph Cartan, 1869–1951; for surfaces in 1881 by Hans Carl Friedrich von Mangoldt, 1854–1925. Jacques Salomon Hadamard, 1865–1963, had no new contribution to it.

Hence $\|Y\|^2$ is convex. Because $Y_0 = 0, Y_0' \neq 0$, it follows that $\|Y_t\|^2 > 0$ for any $t > 0$. The velocity field of the geodesic likewise satisfies $\|t\dot{c}\|^2 \neq 0$ there. Therefore $T_{tX} \exp_p tV \neq 0$ for any $X, V \in T_pM, V \neq 0$, and $\exp_p$ is a local diffeomorphism everywhere. When choosing the metric $\exp_p^* g$ on T_pM, $\exp_p$ becomes a local isometry, and the radial straight lines are geodesics. Hence T_pM is complete. According to Lemma 6.5.4 $\exp_p$ is a covering. $\qquad\square$

Exercises

Exercise 6.5.10 Let $(v_1, \ldots, v_n)$ be a basis of $\mathbb{R}^n$ and $\Lambda := \mathbb{Z}v_1 + \cdots + \mathbb{Z}v_n$. Show that Λ as a subgroup of $M = \mathbb{R}^n$ acts freely and properly discontinuously on M.

Exercise 6.5.11 Show that the following Riemannian manifolds are well-defined:

1) **Klein's bottle** $\{\pm 1\}\backslash(S^1)^2$, where -1 acts as $(e^{i\varphi}, e^{i\psi}) \mapsto (-e^{i\varphi}, e^{-i\psi})$.
2) **Möbius strip** $\{\pm 1\}\backslash S^1 \times \mathbb{R}$, where -1 acts as $(e^{i\varphi}, x) \mapsto (-e^{i\varphi}, -x)$.
3) **Lens space** $L(m; k_1, \ldots, k_n) := (\mathbb{Z}/m\mathbb{Z})\backslash S^{2n-1}$ with the operation

$$(\ell, (z_1, \ldots, z_n)) \mapsto (\zeta^{\ell k_1} z_1, \ldots, \zeta^{\ell k_n} z_n)$$

for a fixed primitive mth root of unity ζ and $k_1, \ldots, k_n$ relatively prime to m.

Exercise* 6.5.12 Prove that the following maps are well-defined and isometries (possibly up to a rescaling):

1) $\varphi : \{\pm 1\}\backslash S^3 \xrightarrow{\cong} SO(3), q \mapsto (v \mapsto qvq^{-1})$ (the latter being understood as an endomorphism of $\mathbb{R}^\perp \subset \mathbb{H}$),
2) $\psi : \{\pm 1\}\backslash(S^3 \times S^3) \xrightarrow{\cong} SO(4) \subset \text{End}(\mathbb{H}), (q, \widetilde{q}) \mapsto (v \mapsto qv\widetilde{q}^{-1})$.

Exercise 6.5.13 Let $\pi : G \to \widetilde{G}$ be a Lie group homomorphism and a universal covering with group of deck transformations $\Gamma = \pi_1(\widetilde{G})$. Suppose G is connected. Conclude from $R_{\gamma(e_G)}e_G = L_{\gamma(e_G)}e_G = \gamma(e_G)$ for any $\gamma \in \Gamma$ that Γ lies in the center of G. Hence in particular it is abelian.
  (*Remark. If $M \to \widetilde{G}$ is any covering, then M can always be given such a Lie group structure by lifting the multiplication $\widetilde{G} \times \widetilde{G} \to \widetilde{G}$. In particular $\pi_1(\widetilde{G})$ is always abelian.*)

Exercise 6.5.14 Suppose a finite group Γ acts freely and C^∞ on a manifold M. Given a Γ representation V, set $V^\Gamma := \{v \in V \,|\, \forall \gamma \in \Gamma : \gamma v = v\}$. Show that $\pi^* : \mathfrak{A}^\bullet(\Gamma\backslash M) \to \mathfrak{A}^\bullet(M)^\Gamma$ is an isomorphism. (*Hint: To show surjectivity, use the local trivialization.*)

Exercise 6.5.15 With the notations of Exercise 6.5.14, let $\pi^\Gamma : V \to V^\Gamma, v \mapsto \frac{1}{\#\Gamma} \sum_\gamma \gamma v$ be the projection onto V^Γ.

1) Prove that $\pi^* : H^\bullet(\Gamma\backslash M) \to H^\bullet(M)$ is injective.

2) Prove that $\pi^*(H^\bullet(\Gamma\backslash M)) = H^\bullet(M)^\Gamma$.

3) Show that $H^\bullet(\mathbb{Z}^n\backslash\mathbb{R}^n) \neq H^\bullet(\mathbb{R}^n)^{\mathbb{Z}^n}$ for $n > 0$.

In particular one finds $H^\bullet(\Gamma\backslash M) \cong H^\bullet(M)^\Gamma$.

Exercise 6.5.16 Compute $H^n(\mathbb{P}^n\mathbb{R})$ using $\mathbb{P}^n\mathbb{R} = (\mathbb{Z}/2\mathbb{Z})\backslash S^n$ under the assumption that $H^n(S^n) = \mathbb{R}$.

6.6 Left-Invariant Metrics on Lie Groups

biinvariant metric	Killing form
right-invariant metric	simple Lie algebra
left-invariant metric	

As a preparation for the study of the geometry of homogeneous spaces, in this section the geometry of Lie groups equipped with a left-invariant metric is studied. Further results for this situation in a more general context will be given in later sections.

Definition 6.6.1 *Suppose G is a Lie group. A metric on G is called **biinvariant** if $L_h : a \mapsto ha, R_h : a \mapsto ah$ are isometries for any $h \in G$ (analogously **left-**, **right-invariant**).*

Example Every metric g_e on $\mathfrak{g}$ induces a left-invariant metric $g_h := L_{h^{-1}}^* g_e$ on G, and vice versa. By Exercise 3.1.23, $SO(n)$ and hence any Lie subgroup $G \subset SO(n)$ carries a biinvariant metric.

Given a scalar product g on $\mathfrak{g}$ and $X \in \mathfrak{g}$, let $\mathrm{ad}_X^* : \mathfrak{g} \to \mathfrak{g}$ be the adjoint to ad_X, i.e. $g([X, Y], Z) = g(Y, \mathrm{ad}_X^* Z)$. Let U be the symmetric bilinear form

$$U : \mathfrak{g}^2 \to \mathfrak{g}, (X, Y) \mapsto (\mathrm{ad}_X^* Y + \mathrm{ad}_Y^* X)/2.$$

Lemma 6.6.2 *Let g be a left-invariant metric on G and let X, Y be left-invariant vector fields. Then $\nabla_X Y = \frac{1}{2}[X, Y] - U(X, Y)$ (in particular it is left-invariant).*

Proof. Because of the left-invariance of g, $g(X, Y) \equiv$ const. holds. The Koszul formula thus yields

$$2g(\nabla_X Y, Z) = g([X, Y], Z) - g(Y, [X, Z]) - g(X, [Y, Z])$$
$$= g([X, Y] - \mathrm{ad}_X^* Y - \mathrm{ad}_Y^* X, Z). \qquad \square$$

Theorem 6.6.3 *Suppose G is a Lie group and X, Y are left-invariant. The curvature of a left-invariant metric is given by*

$$R(X,Y,X,Y) = -g(U(X,X),U(Y,Y)) + \|U(X,Y)\|^2 - \frac{3}{4}\|[X,Y]\|^2$$
$$-\frac{1}{2}g([[Y,X],X],Y) - \frac{1}{2}g([[X,Y],Y],X).$$

Proof. By Lemma 6.6.2 one finds

$$R(X,Y,X,Y) = g(\nabla_Y \nabla_X X - \nabla_X \nabla_Y X - \nabla_{[Y,X]}X, Y)$$
$$= -g(\nabla_X X, \nabla_Y Y) + g(\nabla_Y X, \nabla_X Y) - g(\nabla_{[Y,X]}X, Y)$$
$$= -g(U(X,X), U(Y,Y))$$
$$+ g(\frac{1}{2}[Y,X] - U(X,Y), \frac{1}{2}[X,Y] - U(X,Y))$$
$$- \frac{1}{2}g([[Y,X],X] - \text{ad}^*_{[Y,X]}X - \text{ad}^*_X[Y,X], Y)$$
$$= -g(U(X,X), U(Y,Y)) + \left(\|U(X,Y)\|^2 - \frac{1}{4}\|[X,Y]\|^2\right)$$
$$- \frac{1}{2}g([[Y,X],X],Y) - \frac{1}{2}g([[X,Y],Y],X) - \frac{1}{2}\|[X,Y]\|^2. \quad \square$$

In the next section, it will be useful to have a description of the Levi-Civita connection also in terms of right-invariant vector fields.

Theorem 6.6.4 *Let g be a left-invariant metric on G and let $X_e, Y_e \in T_eG = \mathfrak{g}$ have right-invariant extensions $\widehat{X}, \widehat{Y}$ on G. Then*

$$(\nabla_{\widehat{X}}\widehat{Y})|_\gamma = -\frac{1}{2}R_{\gamma*}[X_e,Y_e] - L_{\gamma*}(U(\text{Ad}_{\gamma^{-1}}X_e, \text{Ad}_{\gamma^{-1}}Y_e))$$
$$= \frac{1}{2}[\widehat{X},\widehat{Y}]|_\gamma - L_{\gamma*}(U(\text{Ad}_{\gamma^{-1}}X_e, \text{Ad}_{\gamma^{-1}}Y_e)).$$

Remark As always the first Lie bracket on T_eG is meant here as the Lie bracket induced by the left-invariant vector fields. The value of U is taken as an element of T_eG. The formula makes clear that for right-invariant fields the derivative $\nabla_{\widehat{X}}\widehat{Y}$ does not have to be right-invariant, i.e., except at single points like e, it does not correspond to an element of $\mathfrak{g}$.

Proof. Let $\widetilde{Z}$ be the left-invariant vector field associated to $Z \in T_eG$. Then

$$\widehat{X}_\gamma = TR_\gamma X_e = TL_\gamma \text{Ad}_{\gamma^{-1}} X_e = (\widetilde{\text{Ad}_{\gamma^{-1}}X_e})|_\gamma$$

holds. Since for all $a, b \in G$ we have $L_a R_b = R_b L_a$, left- and right-invariant vector fields $\widetilde{X}, \widehat{Y}$ commute. By Lemma 6.6.2 one thus obtains

$$
\begin{aligned}
\nabla_{\widehat{X}} \widehat{Y}|_\gamma &= \nabla_{\widehat{Ad_{\gamma^{-1}X_e}} \widehat{Y}|_\gamma} = \nabla_{\widehat{Y}} (\widehat{Ad_{\gamma^{-1}X}})|_\gamma \\
&= \nabla_{\widehat{Ad_{\gamma^{-1}Y}}} (\widehat{Ad_{\gamma^{-1}X}})|_\gamma \\
&= \frac{1}{2} \underbrace{[\widehat{Ad_{\gamma^{-1}Y}}, \widehat{Ad_{\gamma^{-1}X}}]}_{=TL_\gamma \circ Ad_{\gamma^{-1}}[Y_e, X_e]} - U(\widehat{Ad_{\gamma^{-1}X}}, \widehat{Ad_{\gamma^{-1}Y}}) \\
&= \frac{1}{2} R_{\gamma*}[Y_e, X_e] - L_{\gamma*}(U(\widehat{Ad_{\gamma^{-1}X}}, \widehat{Ad_{\gamma^{-1}Y}})).
\end{aligned}
$$

Exercise 1.6.29 shows that $[\widehat{X}, \widehat{Y}]_{|e} = -[X_e, Y_e]$.   □

Alternatively, one can carry out this proof only at $g = e$ and left-translate afterwards.

Exercises

Exercise 6.6.5 Let G be the Heisenberg group (cf. Exercise 1.6.28), parameterized by

$$
\begin{pmatrix} 1 & x & z \\ 0 & 1 & y \\ 0 & 0 & 1 \end{pmatrix}.
$$

1) Show that the vector fields $A := \frac{\partial}{\partial x}, B := \frac{\partial}{\partial y} + x \frac{\partial}{\partial z}, C := \frac{\partial}{\partial z}$ are left-invariant.
2) Let g be the metric for which A, B, C form an orthonormal basis. Is g right-invariant?
3) Calculate the Levi-Civita connection on (G, g).
4) Show that almost all geodesics on G through id are given by

$$
\begin{aligned}
x(t) &= \cot \varphi \cdot (\sin(t \sin \varphi + \vartheta) - \sin \vartheta) \\
y(t) &= -\cot \varphi \cdot (\cos(t \sin \varphi + \vartheta) - \cos \vartheta) \\
z(t) &= \frac{t}{2} (\sin \varphi + \frac{1}{\sin \varphi}) + \frac{1}{2} \cot^2 \varphi \cdot \Big(2 \cos(t \sin \varphi + \vartheta) \cdot \sin \vartheta \\
&\quad - \cos(t \sin \varphi) \cdot \sin(t \sin \varphi + 2\vartheta) \Big)
\end{aligned}
$$

(up to reparametrization).
5) Determine the remaining geodesics through id.
6) Compare the geodesics in (4) to the Lie group exponential map $\exp_G$.

Exercise* 6.6.6 Let $\mathfrak{g}$ be a finite dimensional Lie algebra and

$$B : \mathfrak{g} \times \mathfrak{g} \to \mathbb{R}$$
$$(X, Y) \mapsto \mathrm{Tr}\,(\mathrm{ad}_X \circ \mathrm{ad}_Y)$$

be the **Killing form**.

1) Let $\mathfrak{h} \subset \mathfrak{g}$ be an ideal (cf. Exercise 1.6.35). Prove that then $\mathfrak{h}^\perp$ is also an ideal ($\perp$ with respect to the Killing form).
2) Assume $B < 0$. Using (1), show that $\mathfrak{g}$ can be decomposed as a sum $\mathfrak{g}_1 \oplus \cdots \oplus \mathfrak{g}_k$, where the $\mathfrak{g}_j$ are **simple**, i.e. they are not abelian and have no nontrivial ideals.

Exercise 6.6.7 Let G be a connected abelian Lie group equipped with a left-invariant metric. Show that $K \equiv 0$ and conclude that there exists a discrete subgroup $\Gamma \subset \mathbb{R}^n$ for which G is isometric to $\Gamma \backslash \mathbb{R}^n$.

Exercise 6.6.8 Equip $\mathrm{SU}(3)$ with the metric induced by the embedding $\mathrm{SU}(3) \subset \mathrm{SO}(6)$ (with $a + ib \mapsto \left(\begin{smallmatrix} a & b \\ -b & a \end{smallmatrix} \right)$). Determine by Exercise 6.6.7 a lattice Γ for which the torus of diagonal matrices in $\mathrm{SU}(3)$ is isometric to $\Gamma \backslash \mathbb{R}^2$.

Exercise 6.6.9 (cf. Exercise 1.6.26) ([Miln]) Let H^{n+1} be the upper half-space model of hyperbolic space and suppose G equals the subgroup $\mathbb{R}^+ \ltimes \mathbb{R}^n$ of affine transformations of Euclidean $\mathbb{R}^n$ with the operation

$$G \times \mathbb{R}^n \to \mathbb{R}^n,$$
$$\begin{pmatrix} a_0 \\ a \end{pmatrix} \cdot x \mapsto a_0 x + a.$$

1) Determine the Lie bracket of G.
2) Show that $\rho_{\left(\begin{smallmatrix} a_0 \\ a \end{smallmatrix} \right)} : H^{n+1} \to H^{n+1}$, $\begin{pmatrix} x_0 \\ x \end{pmatrix} \mapsto \begin{pmatrix} a_0 x_0 \\ x+a \end{pmatrix}$ is an isometry.
3) Consider the standard scalar product g_{eG} on $T_{eG} G = \mathbb{R}^{n+1}$. Show that G with the associated left-invariant metric is isometric to H^{n+1}. Find an explicit isometry.

6.7 Existence of Homogeneous Metrics

homogeneous space	faithful operation
Klein geometry	effective operation
homogeneous metric	almost faithful operation
Myers–Steenrod Theorem	core
Gleason Theorem	reductive

In this section existence criteria and descriptions are worked out for Riemannian metrics which admit a transitive isometric operation of a Lie group.

Here and in the following chapters, properties and objects which actually belong to G are often given in terms of quotients G/H. This leads to the fact that for two diffeomorphic spaces $G/H \cong G'/H'$ some formulas are valid for G, H which are not true for G', H'. One can avoid misunderstandings by emphasizing the pair (G, H) in place of G/H, but this will make the notation more cumbersome. Anyway, one should keep in mind that properties like "reductive" or "normal" are properties of pairs or triples and not of the space, and the later theorems about symmetric spaces assume special choices of G and H.

Definition 6.7.1 *A* ***(Riemannian) homogeneous space*** *(or* ***Klein geometry***) *M is a Riemannian manifold M on which a Lie group G acts transitively, C^∞ and isometrically. The Riemannian metric is then called a* ***homogeneous metric***.

Example $SO(n) \rtimes \mathbb{R}^n$ acts transitively and isometrically on $\mathbb{R}^n$, $SO(n+1)$ likewise on S^n, $SO(n, 1)$ on H^n, $\mathbb{R}^n$ on T^n. Any Lie group equipped with a left-invariant metric acts transitively isometrically on itself.

Remark According to the **Myers–Steenrod Theorem** (1939) [MySt], [Hel, ch. IV, Th. 2.5], the isometry group G of a manifold M equipped with the compact-open topology is a (continuous) Lie group. The compact-open topology is generated by the sets $\{f \in C(M, M) \mid f(K) \subset U\}$ for $K \subset M$ and $U \subset M$ open. In particular, it is uniquely induced by M. According to the **Gleason–Montgomery–Zippin Theorem** (1952) [Gl], [MoZi] a continuous Lie group carries a uniquely determined C^∞-structure by which it becomes a C^∞-Lie group. Thus, viewed in this way, a homogeneous space is a Riemannian manifold with a transitively operating isometry group. The group G is not necessarily closed in the isometry group, hence it is not necessarily a Lie subgroup.

Theorem 6.7.2 *Every homogeneous space M is complete.*

Proof. Suppose a geodesic c of speed 1 is defined at most on $]t_0, t_1[$ for $t_0 \in \mathbb{R} \cup \{-\infty\}$, $t_1 \in \mathbb{R}$. Let $p \in M$ be arbitrary and $\exp_p$ be defined on $B_{2\varepsilon}(0)$. Let γ be an isometry with $\gamma(p) = c(t_1 - \varepsilon)$. Then $\exp_{c(t_1-\varepsilon)} t\dot{c}(t_1 - \varepsilon)$ is an extension of c to $t_1 + \varepsilon \nmid$ (Fig. 6.11).   □

In particular, any left-invariant metric on a Lie group G is metrically complete.

Lemma 6.7.3 *Let H be a (not necessarily connected) Lie subgroup of a Lie group G. Then $G/H := \{aH \mid a \in G\}$ is a manifold, $G \to G/H$ is a fiber bundle and G acts by left multiplication smoothly and transitively on G/H.*

Proof. H acts freely on G because $ah = a \Leftrightarrow h = e$. H acts properly, because for $K \subset\subset G$ one gets

$$H_K = \{h \in H \mid Kh \cap K \neq \emptyset\} = \{h \in H \mid \exists k, k' \in K : kh = k'\}$$
$$= H \cap K^{-1}K.$$

Since H is closed and $K^{-1}K$ is compact, H_K is compact.

According to Theorem 6.4.8, G/H is a manifold and $G \to G/H$ is a fiber bundle. If $\pi_1 : G \to G/H$, $\pi_2 : G \times G \to G \times (G/H)$, then π_2 has local sections

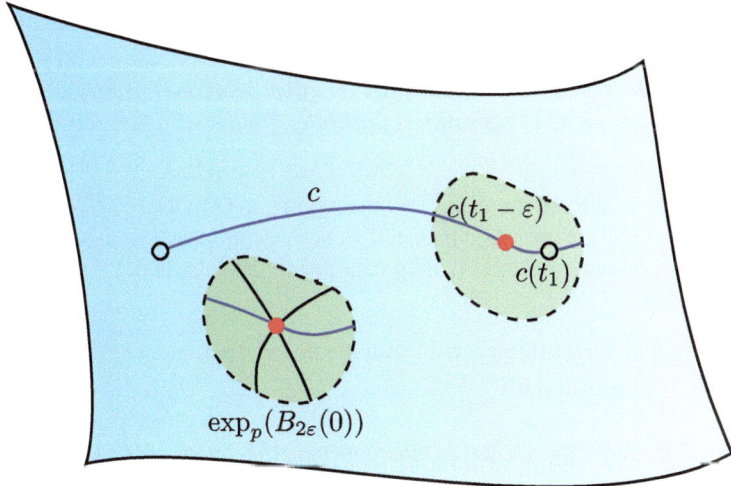

Fig. 6.11 Completeness of homogeneous spaces.

$\sigma : U \to G \times G, U \subset G \times G/H$. Therewith the operation from G to G/H is locally equal to

$$U \xrightarrow{\sigma} G \times G \xrightarrow{\text{group action}} G \xrightarrow{\pi_1} G/H,$$
$$(\gamma, aH) \mapsto (\gamma, a) \mapsto \gamma a \mapsto \gamma aH,$$

hence C^∞. Because for all $a, b \in G$ we have $ba^{-1} \cdot aH = bH$, G acts transitively on G/H. $\qquad\square$

If H is a normal divisor, an analog of the last argument shows that G/H is a Lie group, since the multiplication is smooth as required in Definition 1.6.1.

Lemma 6.7.4 *Suppose a Lie group G operates transitively and C^∞ on a manifold M. Consider $p \in M$. Then M is equivariantly diffeomorphic to G/G_p.*

Proof. By Lemma 6.4.3, G_p is a Lie subgroup of G. Then $\varphi : G/G_p \to M, aG_p \mapsto a \cdot p$ is a G-equivariant diffeomorphism:

Well-defined: For all $h \in G_p$ we have $\varphi(ahG_p) = ah \cdot p = a \cdot p = \varphi(aG_p)$.

Equivariant: $\varphi(abG_p) - a \cdot \varphi(bG_p)$.

Surjective: Because of the transitivity, for every $q \in M$ there is an $a \in G$ such that $\varphi(aG_p) = a \cdot p = q$.

Injective: $\varphi(aG_p) = \varphi(bG_p) \Leftrightarrow b^{-1}aG_p = G_p \Leftrightarrow b^{-1}a \in G_p$.

C^∞: Set $\mu : G \to M, a \mapsto a \cdot p$. Locally on G/G_p choose a C^∞ section $\sigma : U \to G$ of π. Because $\varphi \circ \pi = \mu$, one gets $\varphi = \mu \circ \sigma$, thus φ is C^∞. For $T_{[e]}\varphi : \mathfrak{g}/\mathfrak{g}_p \to T_pM$, $X + \mathfrak{g}_p \mapsto \frac{\partial}{\partial t}|_{t=0} e^{tX} \cdot p$ it follows that rank $T_{[e]}\varphi = \operatorname{rank} T_e\mu = \dim \mathfrak{g}/\mathfrak{g}_p$ by Lemma 6.4.3. Because of the equivariance, φ has constant rank, so by the Inverse Function Theorem φ^{-1} is also C^∞. $\qquad\square$

This diffeomorphism maps the chosen point p to the equivalence class $[e_G] = G_p$ of e_G.

Definition 6.7.5 *A group G acts **faithfully** (or **effectively**) on a set M if only $e_G \in G$ acts as id_M. A Lie group G acts **almost faithfully** if only a discrete subset of G acts as id_M.*

In other words, effectivity means that the map from G to the bijective maps from M to M induced by the operation is injective, because its kernel vanishes. Therefore, given a homogeneous space, G is then canonically an (algebraic) subgroup of the isometry group.

Example $SO(2)$ acts faithfully on S^1. But the induced action on $\mathbb{P}^1\mathbb{R}$ is not faithful, because $\left(\begin{smallmatrix} -1 & 0 \\ 0 & -1 \end{smallmatrix}\right)$ acts trivially.

Lemma 6.7.6 *Every space G/H is diffeomorphic to a quotient G'/H' on which G' acts faithfully.*

Proof. Let $N := \bigcap_{a \in G} aHa^{-1} \overset{a=e}{\subset} H$ be the **(normal) core** of H. Every normal subgroup of G contained in H is also contained in aHa^{-1}. On the other hand, clearly $N \triangleleft G$, hence N is the largest normal subgroup of G contained in H. The Lie algebra $\mathfrak{n} = \bigcap_{a \in G} \mathrm{Ad}_a \mathfrak{h}$ shows as in Lemma 6.4.3 that N is a Lie subgroup. An element $b \in G$ acts trivially if and only if for all $a \in G$ we have $b \cdot aH = aH$, i.e. $a^{-1}ba \in H$, i.e. $b \in N$.

Thus the core is the obstruction to faithfulness. If $G' := G/N, H' := H/N$, then G' acts faithfully on G'/H' and $r : G'/H' \to G/H, (\gamma \cdot N) \cdot H' \mapsto \gamma H$ is bijective. This map is a diffeomorphism: Consider the maps

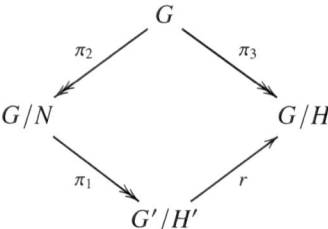

and local sections σ_j of π_j. Then $r = \pi_3 \circ \sigma_2 \circ \sigma_1, r^{-1} = \pi_1 \circ \pi_2 \circ \sigma_3$ are smooth. □

The action is almost faithful if and only if the core is discrete.

Theorem 6.7.7 *Consider a Riemannian manifold M^n, a fixed point $p \in M$ and a Lie group G acting faithfully by isometries on M. Then $\dim G \leq \frac{n(n+1)}{2}$. The isotropy group G_p is an immersed subgroup of $O(T_pM)$, hence in particular $\dim G_p \leq \frac{n(n-1)}{2}$.*

Proof. By Lemma 6.5.5 isometries are uniquely determined by their derivative at a point. Consequently, G acts freely on the frame bundle $O(M) \subset TM^n$ of orthonormal bases of the tangent bundle (Example 6.4.11). The orbit of an orthonormal basis $(e_j)_j$ of T_pM is thus an injective immersion of G of dimension $\leq \dim M + \dim O(n) = n + \frac{n(n-1)}{2} = \frac{n(n+1)}{2}$. Because $T_p\gamma \in O(T_pM)$ for $\gamma \in G_p$, G_p is a subgroup of $O(T_pM)$ in a canonical way. $\qquad\square$

Here $O(T_pM)$ is the isometry group of T_pM rather than the fiber $O_p(M)$ of the frame bundle. Any choice of an orthonormal basis $(e_j)_j$ of T_pM identifies these spaces, but G_p is canonically embedded into $O(T_pM)$.

Example Because $SO(n+1)/SO(n) \cong S^n$, the bounds in Theorem 6.7.7 are sharp.

As an immersed subgroup of $O(T_pM)$, G_p carries a biinvariant metric, thus it follows that:

Corollary 6.7.8 *Let a Lie group G act faithfully by isometries on M. Then the isotropy group G_p of each point $p \in M$ can be equipped with a biinvariant metric.*

If G is a closed subgroup of the isometry group (e.g., if G is the isometry group itself), then G_p is also closed in $O(T_pM)$, i.e., compact, by Lemma 6.7.4.

Remark The Myers–Steenrod Theorem and the Gleason–Montgomery–Zippin Theorem thus imply: Every homogeneous space M is diffeomorphic to a quotient G/H for a Lie group G and a compact Lie subgroup $H \subset G$ such that the canonical operation of G is isometric with respect to the metric induced by the diffeomorphism. In this situation M is compact if and only if G is compact. For if M is compact, then the preimage G under the proper map $\pi : G \to G/H \cong M$ is also compact. Conversely, if G is compact, then $\pi(G)$ is also compact.

Having thus found that homogeneous spaces are diffeomorphic to quotients G/H of certain Lie groups, the question arises whether there is a Riemannian metric on G such that the quotient metric yields the Riemannian metric of M. More generally, one can ask whether there are G-invariant metrics on a quotient G/H at all, and if so, how to describe them. This problem will be addressed in the next theorems.

Theorem 6.7.9 *Consider a Lie subgroup H of a Lie group G and $h \in H, Z \in \mathfrak{h}$. Then $\mathrm{Ad}_h, \mathrm{ad}_Z$ induce operations on $\mathfrak{g}/\mathfrak{h} = T_{[e]}G/H$, and there is a bijection between*

1) G-invariant metrics g on G/H (for which, in particular, G/H is a homogeneous space),

2) inner products $g_{[e]}$ on $T_{[e]}(G/H)$, which are Ad_h-invariant for all $h \in H$.

The latter implies that

3) for all $Z \in \mathfrak{h}$ the endomorphism ad_Z is skew-symmetric with respect to $g_{[e]}$

and if H is connected, the converse is also true.

Proof. If $X \in \mathfrak{h}$, then $\mathrm{Ad}_h X, \mathrm{ad}_Z X \in \mathfrak{h}$, hence these act on $\mathfrak{g}/\mathfrak{h}$.

(1) implies (2): Since for all $X \in \mathfrak{g}$ we have $L_h e^{tX} H = C_h(e^{tX})H$, one finds that $T_{[e]}L_h = \mathrm{Ad}_h$ holds. Therefore $g_{[e]}$ is Ad_h-invariant for all $h \in H$ and for any G-invariant metric g.

Inverse map from (2) to (1): Set $g_{[a]} := L^*_{a^{-1}} g_{[e]}$. Then

$$g_{[ah]} = L^*_{h^{-1}a^{-1}} g_{[e]} = L^*_{a^{-1}} L^*_{h^{-1}} g_{[e]} = L^*_{a^{-1}} g_{[e]} (\mathrm{Ad}_{h^{-1}} \cdot, \mathrm{Ad}_{h^{-1}} \cdot) = g_{[a]}.$$

(2)$\Rightarrow$(3): Differentiating $g_{[e]}(\mathrm{Ad}_h X, \mathrm{Ad}_h Y) = g_{[e]}(X, Y)$ at $h = e$ in the direction $Z \in \mathfrak{h}$ yields $g_{[e]}(\mathrm{ad}_Z X, Y) + g_{[e]}(X, \mathrm{ad}_Z Y) = 0$.

(3)$\Rightarrow$(2): Corollary 1.6.25 shows that

$$g_{[e]}(\mathrm{Ad}_{\exp Z} X, \mathrm{Ad}_{\exp Z} Y) \quad = \quad g_{[e]}(\underbrace{\exp(\mathrm{ad}_Z)}_{:=\sum \frac{(\mathrm{ad}_Z)^m}{m!}} X, \exp(\mathrm{ad}_Z) Y)$$

$$\overset{\text{Assumption}}{=} g_{[e]}(\exp(-\mathrm{ad}_Z) \exp(\mathrm{ad}_Z) X, Y) = g_{[e]}(X, Y).$$

Thus, Ad_h-invariance holds for all h in a neighborhood of e in H. By Theorem 1.6.19, this neighborhood generates all of H if H is connected, so the ad-invariance holds in general. $\square$

Corollary 6.7.10 *If H is connected, then the existence of a G-invariant metric on G/H depends only on the operation of $\mathfrak{h}$ on $\mathfrak{g}/\mathfrak{h}$.*

Definition 6.7.11 *The quotient G/H is called **reductive** if there exists a vector subspace $\mathfrak{m} \subset \mathfrak{g}$ such that $\mathfrak{g} = \mathfrak{m} \oplus \mathfrak{h}$ and $\mathrm{Ad}_H \mathfrak{m} \subset \mathfrak{m}$. Any $X \in \mathfrak{g}$ is decomposed into the summands $X^\mathfrak{h}, X^\mathfrak{m}$ accordingly.*

This implies by differentiating $[\mathfrak{h}, \mathfrak{m}] \subset \mathfrak{m}$. On the other hand, for H connected it follows that $\mathrm{ad}_Z \mathfrak{m} \subset \mathfrak{m} \Rightarrow e^{\mathrm{ad}_Z} \mathfrak{m} \subset \mathfrak{m} \Rightarrow \mathrm{Ad}_{e^Z} \mathfrak{m} \subset \mathfrak{m} \Rightarrow \mathrm{Ad}_H \mathfrak{m} \subset \mathfrak{m}$.

Example 6.7.12 The following examples are discussed in more detail in the next section.

1) Choose $H = \{e\}$. Then $G/H = G$ is reductive with $\mathfrak{h} = 0, \mathfrak{m} = \mathfrak{g}$ and the operation of G on the homogeneous space $G/\{e\}$ is the left multiplication.

2) Consider a Lie group G, let $G' := G \times G$ and $H := \{(a, a) \mid a \in G\} \subset G'$. Then $\varphi : G'/H \overset{\cong}{\to} G, [(a, b)] \mapsto ab^{-1}$ is a diffeomorphism. Via φ, $(a, b) \in G \times G$ acts on $c \in G$ as

$$(a, b) \cdot \varphi^{-1}(c) = (a, b) \cdot [(c, e)] = [(ac, b)] = \varphi^{-1}(acb^{-1}),$$

hence as left multiplication by a and right multiplication by b^{-1}. This space is reductive with $\mathfrak{m} := \{(X, -X) \mid X \in \mathfrak{g}\}$ because

$$\mathrm{Ad}_{(a,a)}(X, -X) = (\mathrm{Ad}_a X, -\mathrm{Ad}_a X) \in \mathfrak{m}.$$

This is an example of a natural operation that does not need to be faithful: The action of $G \times G$ on G'/H is faithful if and only if for all $(a, b) \in G \times G \setminus \{(e, e)\}$ there exists a point $[(\gamma, e)] = \varphi^{-1}(\gamma) \in G'/H$ such that $[(a\gamma, b)] \neq [(\gamma, e)]$. If $a \neq b$, then this is always true because, for example, $[(e, e)]$ is not a fixed point. But for $a = b$ this condition becomes: There exists a $\gamma \in G$ such that for all $c \in G$ we have

$(a\gamma, a) \neq (\gamma c, c)$ or there exists a $\gamma \in G$ such that $\gamma^{-1}a\gamma \neq a$. Thus, exactly the pairs (a, a) with $a \in Z(G)$ operate trivially. The operation is faithful if $Z(G) = \{e\}$.

Theorem 6.7.13 *Consider a Lie subgroup $H \subset G$. Then the following implications hold:*

(1) H is compact.
$\Rightarrow$ *(2) G carries a G-left-invariant, H-biinvariant metric.*
$\Rightarrow$ *(3) G/H is reductive and $\mathfrak{m}$ carries an Ad_H-invariant scalar product.*
$\Rightarrow$ *(4) G/H carries a G-invariant metric.*

If G acts faithfully on G/H, then (2)–(4) are equivalent.

If in (4) G is closed subgroup of the isometry group of G/H, then by the Myers–Steenrod Theorem (4)$\Rightarrow$(1) follows and all four conditions become equivalent. By applying Theorem 7.4.13 to the implication (1)$\Rightarrow$(2) (with the Lie group G there in the role of H here), this statement can be further refined.

Proof. (1)$\Rightarrow$(2): Let $\langle \cdot, \cdot \rangle$ be any scalar product on $\mathfrak{g}$ with volume form $d\mathrm{vol}_e$. Let $d\mathrm{vol}_h := R^*_{h^{-1}} d\mathrm{vol}_e$ be the associated right-invariant volume form on G. Setting

$$g_e(X, Y) := \int_H \langle \mathrm{Ad}_h X, \mathrm{Ad}_h Y \rangle \, d\mathrm{vol}_h$$

one obtains for any $a \in H$

$$g_e(\mathrm{Ad}_a X, \mathrm{Ad}_a Y)$$
$$= \int_H \langle \underbrace{\mathrm{Ad}_{ha} X}_{= \mathrm{Ad}_h \mathrm{Ad}_a X}, \mathrm{Ad}_{ha} Y \rangle \, d\mathrm{vol}_h = \int_H \left(R^*_a \langle \mathrm{Ad}. X, \mathrm{Ad}. Y \rangle \right)_h d\mathrm{vol}_h$$
$$= \int_H R^*_a \left(\langle \mathrm{Ad}. X, \mathrm{Ad}. Y \rangle d\mathrm{vol} \right)_h = \int_H \langle \mathrm{Ad}. X, \mathrm{Ad}. Y \rangle \, d\mathrm{vol} = g_e(X, Y).$$

Therefore g_e is Ad_a-invariant. Thus the left-invariant metric $h \mapsto L^*_{h^{-1}} g_e$ on G is H-biinvariant, because for all $h \in H, a \in G$

$$R^*_h(L^*_{a^{-1}} g_e) = L^*_{a^{-1}} L^*_h \qquad \underbrace{L^*_{h^{-1}} R^*_h g_e}_{= g_e(\mathrm{Ad}_{h^{-1}}\cdot, \mathrm{Ad}_{h^{-1}}\cdot) = g_e} = L^*_{ha^{-1}} g_e.$$

(2)$\Rightarrow$(3): Set $\mathfrak{m} := \mathfrak{h}^{\perp}$ with the scalar product $g_{e|\mathfrak{m}}$.
(3)$\Rightarrow$(4): Theorem 6.7.9 with the canonical identification $\mathfrak{m} \cong \mathfrak{g}/\mathfrak{h}$.
(4)$\Rightarrow$(2): Let $n := \dim G/H$, $p \in G/H$ and $(e_j)_j$ be an orthonormal basis of $T_p M$. By Theorem 6.7.7 G immerses as the orbit of $(e_j)_j$ into $O(TM)$. Now $O(TM) \subset TM^{\oplus n}$ carries a G-left-invariant metric with isometric right action of $O(n)$. Thus, via the immersion of H into $O(n)$ induced by $(e_j)_j$, the metric on G induced by the immersion is G-left, H-right invariant. $\qquad \square$

6.8 Geometry of Homogeneous Spaces

Ricci curvature	semisimple
Killing form	naturally reductive
center	polar decomposition

In the study of connections, curvature and geodesics on homogeneous spaces, in this section we will predominantly use a metric on G as in Theorem 6.7.13(2). For such a metric, we set $\mathfrak{m} := \mathfrak{h}^{\perp}$ accordingly. At the end of the section, we consider the more special class of naturally reductive spaces.

Remark 6.8.1 Because of the Ad_H-invariance, ad_X is skew-symmetric with respect to g on all $\mathfrak{g}$ for all $X \in \mathfrak{h}$, as in the proof of Theorem 6.7.9.

Lemma 6.8.2 *Consider a homogeneous space G/H and let $\pi : G \to G/H$ and G carry a metric as in Theorem 6.7.13(2). Then the following hold:*

1) *The left-invariant vector fields $\mathfrak{h}$ are vertical.*
2) *The left-invariant vector fields $\mathfrak{m}$ are horizontal (but in general they are no lifts).*
3) *The right-invariant vector fields $\widehat{X}$ associated to $X_e \in T_e G$ are lifts (but not necessarily horizontal). Their images under $T\pi$ are Killing fields on G/H.*
4) $\rho(X_e) := (\widehat{X})^{\mathfrak{m}}$ *is the horizontal lift of the Killing field $T\pi(\widehat{X})$ on G/H.*

Proof. 1) Holds because $\pi^{-1}([a]) = aH = L_a H$.

2) Because of the left-invariance of the metric, $\mathfrak{m} \perp \mathfrak{h}$ holds everywhere.

3) $\pi(ah) = \pi(a)$ implies $\pi \circ R_h = \pi$ and thus $T\pi \circ TR_h = T\pi$.

4) The right-invariant vector fields are Killing fields on G by Theorem 1.6.13, since their flow acts by left-multiplication. Because G acts isometrically on G/H by left-multiplication, their images under $T\pi$ are also Killing. $\qquad\square$

The horizontal lift of $TR_a X_e + \mathfrak{h}$ is thus $h \mapsto (TR_{ah} X_e)^{\mathfrak{m}}$ for $h \in H$ (Fig. 6.12). Because of the left-invariance of $\mathfrak{h}$, the lift can also be written as

$$(TR_a X_e)^{\mathfrak{m}} = L_{a*}(L_{a^{-1}*} R_{a*} X_e)^{\mathfrak{m}} = L_{a*}(\mathrm{Ad}_{a^{-1}} X_e)^{\mathfrak{m}}.$$

The elements of $\mathfrak{h}$ are mapped to Killing fields that vanish at $[e]$.

Remark 6.8.3 If X is a horizontal lift to $\pi : G \to G/H$ and V is a vertical vector field, then $[V, X]$ is vertical. But here $[\mathfrak{h}, \mathfrak{m}] \subset \mathfrak{m}$ holds. This fits since vector fields in $\mathfrak{m}$ are not horizontal lifts in general. Because $L_a R_b = R_b L_a$, left- and right-invariant vector fields $V, \widehat{X}$ commute, hence $[V, (\widehat{X})^{\mathfrak{m}}] = -[V, (\widehat{X})^{\mathfrak{h}}] \in T^V G$ also follows directly.

Of course, G/H may carry other Killing fields; for example, in the case of a biinvariant metric on G, the left-invariant vector fields on G/e are Killing fields.

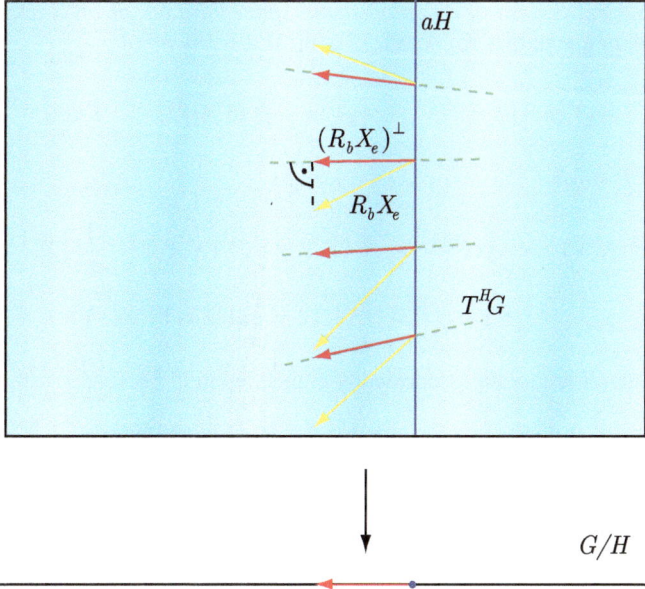

Fig. 6.12 Horizontal lift in the case of homogeneous spaces.

Lemma 6.8.4 *The Lie bracket on the Killing fields is given by $\rho([X_e, Y_e]) = -[\rho(X_e), \rho(Y_e)]$.*

Proof. The Lie bracket of right-invariant vector fields equals $-[\cdot, \cdot]$ at e_G. The assertion follows by Lemma 1.4.7.   □

Example The group $SO(3)$ acts transitively on S^2, and the isotropy group H at the north pole is isomorphic to S^1. The fields $T\pi(TR.X_e)$ associated to $X_e \in \mathfrak{h}$ are Killing fields on S^2, which vanish at the poles because $(X_e)^{\mathfrak{m}} = 0$. Therefore they are the Killing fields corresponding to rotations around the north-south axis.

Corollary 6.8.5 *Given a metric on G as in Theorem 6.7.13(2), then for all $A, B \in \mathfrak{g}$ we have $U(A, B) \in \mathfrak{m}$ and for all $V, W \in \mathfrak{h}$ the Lie algebra element $U(V, W)$ vanishes.*

Proof. According to Remark 6.8.1, for all $V \in \mathfrak{h}$ we have $\mathrm{ad}_V^* = -\mathrm{ad}_V$. Thus one obtains

$$g(2U(A, B), V) = g(\mathrm{ad}_A^* B + \mathrm{ad}_B^* A, V)$$

$$= g(B, [A, V]) + g(A, [B, V]) \overset{\substack{\mathrm{ad}_V \ \mathrm{skew-} \\ \mathrm{symmetric}}}{=} 0$$

and $2U(V, W) = -[V, W] - [W, V] = 0$.   □

Theorem 6.8.6 *Let G/H be a homogeneous space for a metric on G as in Theorem 6.7.13(2). With $\mathfrak{g} = \mathfrak{m} \oplus \mathfrak{h}$, for $X \in \mathfrak{g}$, $Y \in \mathfrak{m}$, $V \in \mathfrak{h}$ one obtains*

$$A_X Y = \frac{1}{2}[X,Y]^{\mathfrak{h}}, \qquad A_X V = -\frac{1}{2}(\mathrm{ad}_X^* V)^{\mathfrak{m}}, \qquad T \equiv 0.$$

For X, Y orthonormal the curvature is given by

$$K^{G/H}(X \wedge Y) = -g(U(X,X), U(Y,Y)) + \|U(X,Y)\|^2 - \frac{3}{4}\|[X,Y]^{\mathfrak{m}}\|^2$$
$$-\frac{1}{2}g([[Y,X],X],Y) - \frac{1}{2}g([[X,Y],Y],X).$$

The fibers aH are totally geodesically embedded in G because $T \equiv 0$.

Proof. Given $X \in \mathfrak{m}$, the first formula for A follows from Lemma 6.6.2, because the left-invariant vector fields $\mathfrak{h}$ are vertical and the values of U lie in $\mathfrak{m}$. Alternatively, O'Neill's Theorem can be used. Because $[\mathfrak{h}, \mathfrak{m}] \subset \mathfrak{m}$, the formula is also valid for $X \in \mathfrak{h}$. In the same way, $T_V W = -U(V,W) \overset{6.6.2}{=} 0$. Because of the skew-symmetry of T from Lemma 6.3.6(3), T vanishes entirely. The skew-symmetry of A from Lemma 6.3.6(3) implies the second formula for A. The formula for the curvature follows from O'Neill's Formula and Theorem 6.6.3:

$$K^G(X \wedge Y) = -g(U(X,X), U(Y,Y)) + \|U(X,Y)\|^2 - \frac{3}{4}\|[X,Y]\|^2$$
$$-\frac{1}{2}g([[Y,X],X],Y) - \frac{1}{2}g([[X,Y],Y],X). \qquad \square$$

In the case of formulas concerning the connection, in contrast to the tensorial curvature, one has to make a choice in terms of which vector fields to express it. In contrast to G, G/H in general has a nontrivial tangent bundle. In the next theorem the projections of right-invariant vector fields are used for this purpose, which we know to be Killing fields.

Theorem 6.8.7 *Consider $X_e, Y_e \in \mathfrak{g}$ with right-invariant extensions $\widehat{X}, \widehat{Y}$ to G. Then the associated Killing vector fields $\widetilde{X} = T\pi(\widehat{X}), \widetilde{Y} = T\pi(\widehat{Y})$ on G/H with a metric as in Theorem 6.7.13(2) satisfy*

$$(\nabla_{\widetilde{X}}\widetilde{Y})_{|[\gamma]} = T_\gamma \pi\Big(\frac{1}{2}[\widehat{X}, \widehat{Y}]$$
$$-\frac{1}{2}L_{\gamma*}\big(\mathrm{ad}^*_{\mathrm{Ad}_{\gamma^{-1}}X_e}(\mathrm{Ad}_{\gamma^{-1}}Y_e)^{\mathfrak{m}} + \mathrm{ad}^*_{\mathrm{Ad}_{\gamma^{-1}}Y_e}(\mathrm{Ad}_{\gamma^{-1}}X_e)^{\mathfrak{m}}\big)\Big)$$
$$= T_\gamma \pi\Big(-\frac{1}{2}L_{\gamma*}[(\mathrm{Ad}_{\gamma^{-1}}X_e)^{\mathfrak{m}}, (\mathrm{Ad}_{\gamma^{-1}}Y_e)^{\mathfrak{m}}]$$
$$-L_{\gamma*}[(\mathrm{Ad}_{\gamma^{-1}}X_e)^{\mathfrak{m}}, (\mathrm{Ad}_{\gamma^{-1}}Y_e)^{\mathfrak{h}}]$$
$$-L_{\gamma*}\big(U((\mathrm{Ad}_{\gamma^{-1}}X_e)^{\mathfrak{m}}, (\mathrm{Ad}_{\gamma^{-1}}Y_e)^{\mathfrak{m}})\big)\Big).$$

In particular, if $X_e, Y_e \in \mathfrak{m}$, then

$$(\nabla_{\widetilde{X}}\widetilde{Y})_{|[e]} = T_e\pi\left(-\frac{1}{2}[X_e, Y_e] - U(X_e, Y_e)\right).$$

Remark 1) $\nabla_{\widetilde{X}}\widetilde{Y}$ does not have to be a Killing field, so except for single points like e it does not have to correspond to an element of $\mathfrak{g}$. As always, the Lie bracket on $T_e G$ is the Lie bracket induced by the left-invariant vector fields.
2) The second summand lies in $\mathfrak{m}$ according to Corollary 6.8.5 in both formulas.
3) By left-invariance, the value at $[\gamma]$ can also be calculated from the value at $[e]$ thanks to

$$(L_{\gamma*}\widehat{X})_{|a} = L_{\gamma*}R_{\gamma^{-1}a*}X_e = R_{\gamma^{-1}a*}R_{\gamma*}\mathrm{Ad}_\gamma X_e = \widehat{\mathrm{Ad}_\gamma X_e}_{|a}.$$

Proof. By O'Neill's Theorem one gets

$$\nabla_{\widetilde{X}}^{G/H}\widetilde{Y} = T\pi(\nabla_{\widehat{X}^{\mathfrak{m}}}^G\widehat{Y}^{\mathfrak{m}})$$
$$= T\pi(\nabla_{\widehat{X}}^G\widehat{Y} - \nabla_{\widehat{X}^{\mathfrak{m}}}^G\widehat{Y}^{\mathfrak{h}} - \nabla_{\widehat{X}^{\mathfrak{h}}}^G\widehat{Y}^{\mathfrak{m}} - \nabla_{\widehat{X}^{\mathfrak{h}}}^G\widehat{Y}^{\mathfrak{h}})$$
$$= T\pi(\nabla_{\widehat{X}}^G\widehat{Y} - A_{\widehat{X}^{\mathfrak{m}}}\widehat{Y}^{\mathfrak{h}} - A_{\widehat{Y}^{\mathfrak{m}}}\widehat{X}^{\mathfrak{h}} - II(\widehat{X}^{\mathfrak{h}}, \widehat{Y}^{\mathfrak{h}})).$$

According to Theorem 6.6.4 one finds

$$\nabla_{\widehat{X}}^G\widehat{Y}_{|\gamma} = \frac{1}{2}[\widehat{X}, \widehat{Y}] - L_{\gamma*}(U(\mathrm{Ad}_{\gamma^{-1}}X_e, \mathrm{Ad}_{\gamma^{-1}}Y_e))$$
$$= \frac{1}{2}[\widehat{X}, \widehat{Y}] - \frac{1}{2}L_{\gamma*}(\mathrm{ad}^*_{\mathrm{Ad}_{\gamma^{-1}}X_e}\mathrm{Ad}_{\gamma^{-1}}Y_e + \mathrm{ad}^*_{\mathrm{Ad}_{\gamma^{-1}}Y_e}\mathrm{Ad}_{\gamma^{-1}}X_e).$$

Theorem 6.8.6 provides the values of A and $II = 0$: Since $\widehat{X}_\gamma = L_{\gamma*}\mathrm{Ad}_{\gamma^{-1}}X_e$, one gets

$$T\pi(A_{\widehat{X}}\widehat{Y}^{\mathfrak{h}}) = T\pi(-L_{\gamma*}\mathrm{ad}^*_{\mathrm{Ad}_{\gamma^{-1}}X_e}(\mathrm{Ad}_{\gamma^{-1}}Y_e)^{\mathfrak{h}}).$$

This gives the first equation. By Lemma 1.4.7, $[\widetilde{X}, \widetilde{Y}] = T\pi([\widehat{X}, \widehat{Y}])$ and at e this is equal to $T\pi(-[X_e, Y_e])$. If $Z \in \mathfrak{h}$, then $\mathrm{ad}^*_Z = -\mathrm{ad}_Z$. Thus one gets

$$T_\gamma\pi\left(-\frac{1}{2}R_{\gamma*}[X_e, Y_e]\right.$$
$$-\frac{1}{2}L_{\gamma*}\left(\mathrm{ad}^*_{(\mathrm{Ad}_{\gamma^{-1}}X_e)^{\mathfrak{h}}}(\mathrm{Ad}_{\gamma^{-1}}Y_e)^{\mathfrak{m}} + \mathrm{ad}^*_{(\mathrm{Ad}_{\gamma^{-1}}Y_e)^{\mathfrak{h}}}(\mathrm{Ad}_{\gamma^{-1}}X_e)^{\mathfrak{m}}\right)\right)$$
$$= T_\gamma\pi\left(-\frac{1}{2}L_{\gamma*}[\mathrm{Ad}_{\gamma^{-1}}X_e, \mathrm{Ad}_{\gamma^{-1}}Y_e]\right.$$
$$\left. +\frac{1}{2}L_{\gamma*}\left([(\mathrm{Ad}_{\gamma^{-1}}X_e)^{\mathfrak{h}}, (\mathrm{Ad}_{\gamma^{-1}}Y_e)^{\mathfrak{m}}] - [(\mathrm{Ad}_{\gamma^{-1}}X_e)^{\mathfrak{m}}, (\mathrm{Ad}_{\gamma^{-1}}Y_e)^{\mathfrak{h}}]\right)\right).$$

Using $[\mathfrak{h}, \mathfrak{h}] \subset \mathfrak{h}$, the second equation follows by splitting into the components. $\square$

Lemma 6.8.8 *Every geodesic $c : \mathbb{R} \to M$ on a homogeneous space M is either injective or simply periodic (i.e., injective on $\mathbb{R}/\text{period}\cdot\mathbb{Z}$).*

Proof. Let without restriction $eH = c(0) = c(t_0)$ and $X_a := T\pi(T_e R_a \dot{c}(0))$ be the Killing field at $\dot{c}(0)$. Now, in general, Killing fields along a geodesic satisfy

$$\frac{\partial}{\partial t} g(X_{c(t)}, \dot{c}(t)) \overset{c \text{ geod.}}{=} g(\nabla^{c^* TM}_{\partial/\partial t} X_{c(t)}, \dot{c}(t)) = g(\nabla^{TM}_{\dot{c}(t)} X_{c(t)}, \dot{c}(t)) \overset{X \text{ Killing}}{=} 0.$$

Therefore

$$g(\dot{c}(0), \dot{c}(t_0)) = g(X_{c(0)}, \dot{c}(t_0)) = g(X_{c(t_0)}, \dot{c}(t_0)) = g(X_{c(0)}, \dot{c}(0)) = \|\dot{c}(0)\|^2$$

and $\dot{c}(0) = \dot{c}(t_0)$ by Cauchy–Schwarz.   □

This argument works in general for geodesics where $\dot{c}(0)$ belongs to a Killing field.

Theorem 6.8.9 *Every homogeneous space M with $K \leq 0, \text{Ric} < 0$ is simply connected, for all $p \in M$ the map $\exp_p : T_p M \to M$ is a diffeomorphism and there is exactly one geodesic going through any two given points.*

Proof. Suppose $\pi_1(M) \neq 0$, i.e., there exists a nontrivial covering $\pi : \widetilde{M} \to M$. Consider $p \neq q \in \widetilde{M}$ such that $\pi(p) = \pi(q) = [e]$. By Lemma 6.5.3 there is a geodesic $\widetilde{c}$ from p to q. Then by Lemma 6.8.8, $c := \pi(\widetilde{c})$ is a periodic geodesic on M. Because $\text{Ric} < 0$, there exists an $X_e \in \mathfrak{m}$ such that $R(X_e, \dot{c}(0), X_e, \dot{c}(0)) < 0$. Let $X_a := T\pi(T_e R_a X_e)$ be the associated Killing field. By Exercise 5.3.14, $X_{|c}$ is a Jacobi field, but $t \mapsto \|X_{c(t)}\|^2$ is periodic. $\notz$ to the convexity in the proof of the Hadamard–Cartan Theorem.

Because of the completeness, $\exp_p$ is a diffeomorphism according to the Hadamard–Cartan Theorem 6.5.9.   □

Definition 6.8.10 *Consider a Lie algebra $\mathfrak{g}$. The **Killing form** on $\mathfrak{g}$ is defined as the symmetric bilinear form*

$$B : \mathfrak{g} \times \mathfrak{g} \to \mathbb{R}$$
$$(X, Y) \mapsto \text{Tr}(\text{ad}_X \circ \text{ad}_Y).$$

*The **center** $\mathfrak{z}$ of $\mathfrak{g}$ is $\mathfrak{z} := \{X \in \mathfrak{g} \mid \text{ad}_X = 0\}$. The Lie algebra $\mathfrak{g}$ and any associated Lie group G is called **semisimple** if B is non-degenerate.*

Given a Lie group G with Lie algebra $\mathfrak{g}$, by Exercise 1.6.33 $\mathfrak{z}$ is the Lie algebra of $Z(G)$.

Example For the compact Lie group $G = \mathbb{R}^n/\mathbb{Z}^n$ the canonical metric is biinvariant, the Killing form vanishes and $\mathfrak{z} = \mathfrak{g}$.

Proposition 6.8.11 *If $\mathfrak{g}$ is semisimple, then $\mathfrak{z}$ is trivial.*

Proof. Suppose there exists an $X \in \mathfrak{z} \setminus \{0\}$, i.e. $\mathrm{ad}_X = 0$. Then $B(X, \cdot) = 0$, hence B is degenerate. □

Lemma 6.8.12 *Let $H \subset G$ be a Lie subgroup and let G carry a metric as in Theorem 6.7.13(2). Then $B^{\mathfrak{h}} \leq 0$ and $B_{|\mathfrak{h} \times \mathfrak{h}} \leq 0$ follow for the Killing forms $B^{\mathfrak{h}}$, B of $\mathfrak{h}$ and $\mathfrak{g}$, respectively. If G acts almost faithfully on G/H or if $\mathfrak{h} \cap \mathfrak{z} = 0$, it follows that $B^{\mathfrak{h}} < 0$ and $B_{|\mathfrak{h} \times \mathfrak{h}} < 0$.*

Proof. Because ad_X is skew-symmetric with respect to g for $X \in \mathfrak{h}$, the eigenvalues lie in $i\mathbb{R}$. Thus $(\mathrm{ad}_X)^2$ has eigenvalues in $\mathbb{R}_0^-$. If G acts almost faithfully, by Lemma 6.7.6 the intersection of the center of G with H is discrete (since this intersection lies in the normal subgroup N) and $\mathfrak{h} \cap \mathfrak{z} = 0$. From $\mathfrak{h} \cap \mathfrak{z} = 0$ it follows that $\mathrm{ad}_X \neq 0$ in $\mathrm{End}(\mathfrak{g})$ for every $X \in \mathfrak{h} \setminus \{0\}$, so $\mathrm{Tr}_{\mathfrak{g}}(\mathrm{ad}_X)^2 \neq 0$. □

Definition 6.8.13 *A reductive homogeneous space G/H together with a choice of an Ad_H-invariant splitting $\mathfrak{g} = \mathfrak{m} \oplus \mathfrak{h}$ is called **naturally reductive** if the metric $g^{G/H}$ on $T_{[e]}G/H \cong \mathfrak{m} \subset \mathfrak{g}$ satisfies*

$$g^{G/H}([X,Y]^{\mathfrak{m}}, Z) = g^{G/H}(X, [Y,Z]^{\mathfrak{m}}) \qquad \text{for all } X, Y, Z \in \mathfrak{m}.$$

Remark Thus, this property depends not only on the space G/H, but also on the choice of G and, in addition, on the chosen splitting $\mathfrak{g} = \mathfrak{m} \oplus \mathfrak{h}$.

Lemma 6.8.14 *Suppose g is a G-left-invariant, H-biinvariant metric on G. Then G/H with the induced metric is naturally reductive if and only if $U \equiv 0$.*

Proof. In terms of g on $\mathfrak{g} = \mathfrak{m} \oplus \mathfrak{h}$, naturally reductive means that

$$\text{for all } X, Y, Z \in \mathfrak{m} \text{ we have } g([X,Y], Z) = g(X, [Y,Z])$$

or $U(X,Z)^{\mathfrak{m}} = 0$. By Corollary 6.8.5 this is equivalent to the vanishing of $U(X,Z)$ for all $X, Z \in \mathfrak{m}$. □

Lemma 6.8.15 *Consider $X_e, Y_e \in \mathfrak{g}$ with right-invariant extensions $\widehat{X}, \widehat{Y}$ to G. Given a naturally reductive space G/H and $\widetilde{X} = T\pi(\widehat{X}), \widetilde{Y} = T\pi(\widehat{Y})$, one obtains*

$$(\nabla_{\widetilde{X}}\widetilde{Y})_{|[\gamma]} = T_\gamma \pi \Big(-\frac{1}{2} L_{\gamma *}[(\mathrm{Ad}_{\gamma^{-1}}X_e)^{\mathfrak{m}}, (\mathrm{Ad}_{\gamma^{-1}}Y_e)^{\mathfrak{m}}]$$

$$- L_{\gamma *}[(\mathrm{Ad}_{\gamma^{-1}}X_e)^{\mathfrak{m}}, (\mathrm{Ad}_{\gamma^{-1}}Y_e)^{\mathfrak{h}}] \Big).$$

In particular, for $X_e, Y_e \in \mathfrak{m}$ it follows that

$$(\nabla_{\widetilde{X}}\widetilde{Y})_{|[e]} = T_e \pi (\frac{1}{2}[\widehat{X},\widehat{Y}]) = \frac{1}{2}[\widetilde{X},\widetilde{Y}].$$

The geodesics are the curves $c : t \mapsto \pi(ae^{tX})$ for $X \in \mathfrak{m}, a \in G$.

Remark In particular, in this case $\exp_{[e_G]} = \pi \circ \exp_G$ holds.

Proof. The formula for the connection follows from Theorem 6.8.7 with $U \equiv 0$. On the Lie group G, by Lemma 6.6.2 for all $X, Y \in \mathfrak{m}$ the equation

$$\nabla_X^G Y = \frac{1}{2}[X, Y] - U(X_e, Y_e) = \frac{1}{2}[X, Y]$$

holds. In particular $c(t) := e^{tX}$ is a geodesic on G, because $\nabla_X X = 0$. The geodesic c is horizontal, so the projection is geodesic by Corollary 6.3.8. The same is true for any left translation of c because of the left-invariance of the metric on G.   □

Corollary 6.8.16 *Let G/H be naturally reductive and $H \subset K \subset G$ be a Lie subgroup. Then K/H is a totally geodesic submanifold of G/H and it is naturally reductive with the induced metric.*

Remark Conversely, according to Kobayashi–Nomizu ([KoN, Ch. VII, Cor. 8.10]), every totally geodesic submanifold of a homogeneous space is homogeneous.

Proof. The tangent space of K/H at e is identified with $\mathfrak{k} \cap \mathfrak{m}$, and if $X_e, Y_e, Z_e \in \mathfrak{k} \cap \mathfrak{m}$, then $[X_e, Y_e] \in \mathfrak{k}$ holds. So for a G-left-invariant, H-biinvariant extension of the metric on G one obtains

$$g([X, Y]^{\mathfrak{k} \cap \mathfrak{m}}, Z) = g([X, Y]^{\mathfrak{m}}, Z) = g(X, [Y, Z]^{\mathfrak{m}}) = g(X, [Y, Z]^{\mathfrak{k} \cap \mathfrak{m}})$$

and K/H is naturally reductive. Hence, by Lemma 6.8.15, the geodesics have the form $\pi(ae^{tX})$ for $a \in K$, $X \in \mathfrak{k} \cap \mathfrak{m}$, and these are also geodesics on G/H.   □

Alternatively, the formula for the connection in Lemma 6.8.15 at $\gamma = e_G$ shows that $II = 0$ holds there for the embedding $K/H \subset G/H$.

Theorem 6.8.17 *Given a naturally reductive homogeneous space G/H and orthonormal $X, Y \in \mathfrak{m}$, one gets*

$$K^{G/H}(X \wedge Y) = \frac{1}{4}\|[X, Y]^{\mathfrak{m}}\|^2 + g([[X, Y]^{\flat}, X], Y).$$

Proof. According to the general formula from Theorem 6.8.6, with an appropriate extension of the metric to G and $U_{|\mathfrak{m} \times \mathfrak{m}} \equiv 0$,

$$K(X \wedge Y) = -\frac{3}{4}\|[X, Y]^{\mathfrak{m}}\|^2$$
$$-\frac{1}{2}g([[Y, X], X], Y) - \frac{1}{2}g([[X, Y], Y], X).$$

Now because of the natural reductivity

$$-\frac{1}{2}g([[Y, X], X], Y) = -\frac{1}{2}g([Y, X]^{\mathfrak{m}}, [X, Y]) - \frac{1}{2}g([[Y, X]^{\flat}, X], Y),$$
$$-\frac{1}{2}g([[X, Y], Y], X) = -\frac{1}{2}g([X, Y]^{\mathfrak{m}}, [Y, X]) - \frac{1}{2}g([[X, Y]^{\flat}, Y], X).$$

Because of the biinvariance of the metric with respect to H one further finds

$$g([[X,Y]^\flat,Y],X) = -g(Y,[[X,Y]^\flat,X]). \qquad \qquad \square$$

As an application, one finds the following nice qualitative statement about the topology of noncompact Lie groups: They are diffeomorphic to a product of a vector space with a compact group if, for example, there is an associated homogeneous space of negative curvature as in Lemma 6.8.9.

Theorem 6.8.18 *Given a naturally reductive homogeneous space G/H, suppose $\exp_{[eG]} : \mathfrak{m} \to G/H$ is injective. Then $\varphi : \mathfrak{m} \times H \to G$, $(X,h) \mapsto e^X h$ is a diffeomorphism.*

Proof. Since $\exp_{[eG]}$ is a local diffeomorphism and is also surjective because of the completeness of G/H, injectivity implies that $\exp_{[eG]}$ is a diffeomorphism. By Lemma 6.8.15, $\exp_{[eG]} X = \pi(e^X)$. Set $f := e^{(\exp_{[eG]})^{-1}} \circ \pi : G \to G$. If $a \in G$, then there is an $X \in \mathfrak{m}$ such that $\pi(a) = \exp_{[eG]} X = \pi(e^X)$. Hence $\pi(f(a)) = \pi(a)$, i.e. there exists an $h \in H$ such that $f(a) = ah$. Therefore

$$\psi : G \to \mathfrak{m} \times G$$
$$a \mapsto (\exp^{-1}_{[e]} \pi(a), f(a)^{-1} \cdot a) = (\exp^{-1}_{[e]} \pi(a), e^{-\exp^{-1}_{[e]} \pi(a)} \cdot a)$$

indeed takes on values in $\mathfrak{m} \times H$. Because $\psi \circ \varphi(X,h) = (X, e^{-X} \cdot e^X h) = (X,h)$, φ, ψ are diffeomorphisms. $\qquad \square$

In particular, in this case, if G is connected, then H is also connected and H, G have the same fundamental group.

Example The hyperbolic space $SO_0(1,n)/SO(n) = H^n$: Using the Gram matrix

$$A := \begin{pmatrix} 1 & 0 & 0 & 0 \\ 0 & -1 & 0 & 0 \\ 0 & 0 & \ddots & \vdots \\ 0 & 0 & \cdots & -1 \end{pmatrix} \text{ of the Minkowski form, one finds}$$

$$\mathfrak{so}(1,n) = \{X \in \mathbb{R}^{n \times n} \mid X^t A + AX = 0\}.$$

That is, the elements are skew-symmetric in almost all entries, only the first column and row are equal and the top left entry is 0. For the L^2-metric on the $n \times n$-matrices, $\mathfrak{m} := \mathfrak{h}^\perp$ consists of the symmetric matrices in $\mathfrak{so}(1,n)$, so for any $X,Y \in \mathfrak{m}$ one gets

$$[X,Y]^t = (XY - YX)^t = -[X,Y] \qquad \text{or} \qquad [X,Y] \in \mathfrak{h}.$$

In particular, H^n is naturally reductive. Because $\exp_{[eG]} : \mathbb{R}^n \overset{\text{diffeom.}}{\to} H^n$, one gets

$$SO_0(1,n) \overset{\text{diffeom.}}{\cong} \mathbb{R}^n \times SO(n).$$

Exercises

Exercise 6.8.19 Suppose M is a homogeneous space, $p \in M$ and $X \in T_p M$. Show that there is a Killing vector field Y on M satisfying $Y_p = X$.

Exercise* 6.8.20 Suppose an m-dimensional Lie group G acts faithfully and isometrically on G/H. Deduce from Theorem 6.7.7 an upper bound for $\dim H$ as a function of m which is sharp for infinitely many values of m. What do you obtain for $m = 2, 3, 4, 5$?

Exercise 6.8.21 Let $p, q \in \mathbb{Z}^+$, $n := p + q$ and $G_{\mathbb{R}}(p, q)$ be the set of p-dimensional $\mathbb{R}$-vector subspaces in $\mathbb{R}^n$.

1) Show that $G := SO(n)$ acts transitively on $G_{\mathbb{R}}(p, q)$.
2) Determine an isotropy group H and use it to provide $G_{\mathbb{R}}(p, q)$ with the structure of a Riemannian homogeneous space.
3) Conclude $G_{\mathbb{R}}(p, q) \cong G_{\mathbb{R}}(q, p)$ and determine $\dim G_{\mathbb{R}}(p, q)$.
4) Does G act effectively?

Exercise 6.8.22 1) Show that $G' := U(n + 1)$ acts transitively on $\mathbb{P}^n \mathbb{C}$ by

$$
A \cdot (z_0 : \cdots : z_n) := \left[A \begin{pmatrix} z_0 \\ \vdots \\ z_n \end{pmatrix} \right].
$$

2) Compute the isotropy group H' of $(1 : 0 : \cdots : 0)$.
3) Find the maximal connected normal subgroup N of G' in H' and identify $G := G'/N, H := H'/N$ with subgroups of G'.
4) Determine the orthogonal complement $\mathfrak{m} := \mathfrak{h}^\perp$ in $\mathfrak{g}$ with respect to the standard L^2-metric on $\mathbb{C}^{n \times n}$.
5) Show that $[\mathfrak{h}, \mathfrak{m}] \subset \mathfrak{m}$.
6) Determine explicitly the Lie bracket $[X, Y]$ for two vectors $X, Y \in \mathfrak{m}$. Does it lie in a special subspace of $\mathfrak{g}$?
7) Calculate the norm $\|[X, Y]\|^2$ for $X, Y \in \mathfrak{m}$ with $\|X\| = \|Y\| = 1$, $X \perp Y$.

Exercise* 6.8.23 Consider an ideal $\mathfrak{m} \subset \mathfrak{g}$. Consider the Killing forms of $\mathfrak{m}$ and $\mathfrak{g}$ and show that

$$
B^{\mathfrak{m}} = B^{\mathfrak{g}}_{|\mathfrak{m} \times \mathfrak{m}}.
$$

Exercise 6.8.24 Consider a Riemannian manifold M and Killing fields X, Y, Z (cf. Exercise 3.3.10).

1) Prove that

$$
2g(\nabla_X Y, Z) = g([X, Y], Z) + g([X, Z], Y) + g(X, [Y, Z]).
$$

2) Use this to find another proof of the formula at $[e_G]$ in Theorem 6.8.7.
3) Show that $[X, Y]$ is again a Killing field.

Exercise 6.8.25 Let $M := GL(n)/SO(n)$ (with the canonical embedding of $SO(n)$ in $GL(n)$ and the standard L^2-metric on $\mathfrak{gl}(n)$).

1) Find an $Ad_{SO(n)}$-invariant $\mathfrak{m}$ and show that M is a naturally reductive space.
2) Prove that $\exp_{[e_G]}$ is injective on $\mathfrak{m}$. Hint: Eigenspace decomposition and the map $[A] \mapsto AA^t$.
3) Deduce that $GL(n)$ is diffeomorphic to $O(n) \times \mathbb{R}^m$ (and give m explicitly). This is the **polar decomposition**.
4) Prove analogously $SL(n, \mathbb{R}) \cong SO(n) \times \mathbb{R}^{m'}$.

Exercise 6.8.26 Suppose G/H is a homogeneous space and G is compact. Prove that every closed form $\alpha \in \mathfrak{A}(G/H)$ is equal to a closed G-invariant form up to an exact form (i.e., the cohomology can be represented by G-invariant forms).

Chapter 7
Symmetric Spaces

Between the constant curvature requirement and the transitive isometry group of the last chapter, there is a property of the isometry group that provides a rather large class of spaces, that are however similar to the spaces of constant curvature. For these symmetric spaces, the isometry group is required to contain a geodesic point reflection at each point $p \in M$. In contrast to the unmanageable abundance of subgroups defining homogeneous spaces, there are exactly 22 families and 34 sporadic cases of symmetric spaces up to products (which is not shown in this book, please see [Hel, ch. 10], [Wolf1, Sect. 8.11]). Many of these spaces play a central role in other areas of mathematics because they parameterize solutions to problems from those areas; for example, in algebraic geometry and number theory.

In the first section, we once again construct more special metrics with a particularly large isometry group on Lie groups. This leads to a class of symmetric spaces. After the definitions in the second section, the symmetric spaces are characterized by an infinitesimal property in the third section. This is followed in the fifth section by an exact characterization in terms of Lie algebras. In the penultimate section it is shown that the set of symmetric spaces can be divided into pairs of mutually dual spaces, one compact and one noncompact. Finally, some results on the irreducibility of the isotropy group operation on the tangent space are given.

Even on the fundamental foundations of symmetric spaces, much more can be said than can be accommodated in this book, which is designed as a two-semester course. For further reading on the subject of symmetric spaces, the following books are recommended: Helgason [Hel], Wolf [Wolf1] and Loos [Loos]. An investigation using more differential geometric methods can be found in [KoN, ch. XI], [Kl2, ch. 2.2], [ChEb], [ON2, ch. 11].

© The Editor(s) (if applicable) and The Author(s), under exclusive license to Springer-Verlag GmbH, DE, part of Springer Nature 2024
K. Köhler, *Differential Geometry and Homogeneous Spaces*, Universitext,
https://doi.org/10.1007/978-3-662-69721-4_7

7.1 Biinvariant Metrics on Lie Groups

normal homogeneous space   │ Weyl group

In this section, as in Example 6.7.12, we consider the case G'/H with $G' := G \times G$, G connected and $H := G$ embedded diagonally. This investigation is continued at the end of Section 7.4. For non-abelian G, $H \subset G \times G$ is not a normal subgroup and $G \times G/H$ has no induced group structure. The diffeomorphism $\varphi : G \times G/H \cong G, [(a, b)] \mapsto ab^{-1}$ is then not a group isomorphism.

By Example 6.7.12, $G \times G$ acts on the left and right on $G'/H \cong G$. Homogeneous metrics on $(G \times G)/G$ thus induce biinvariant metrics on G, and conversely $G \times G$ acts isometrically on G if it carries a biinvariant metric. As said in Example 6.7.12, this operation is not faithful if G has a nontrivial center. On the level of Lie algebras, $\mathfrak{h} = \{(X, X) \mid X \in \mathfrak{g}\} \cong \mathfrak{g}$ and $\mathfrak{m} := \{(X, -X) \mid X \in \mathfrak{g}\}$. The derivative of φ yields the vector space isomorphism $T_{[e]}\varphi : \mathfrak{g} \times \mathfrak{g}/\mathfrak{g} \to \mathfrak{g}, [(X, Y)] \mapsto X - Y$, in particular $\alpha : \mathfrak{m} \cong T_{[e]}(G'/H) \cong \mathfrak{g}, (X, -X) \mapsto 2X$. The Lie bracket is not preserved by α since $[\mathfrak{m}, \mathfrak{m}] \subset \mathfrak{h}$. Theorem 6.7.9 implies with the identifications $\varphi, T_{[e]}\varphi$:

Corollary 7.1.1 *There is a bijection between biinvariant metrics g on G and Ad_G-invariant scalar products g_e on $\mathfrak{g}$ via*

$$g \mapsto g_e, \qquad g_e \mapsto g_h := L^*_{h^{-1}} g_e.$$

A scalar product g_e on $\mathfrak{g}$ is Ad_h-invariant for all $h \in G \Leftrightarrow \mathrm{ad}_X$ is skew-symmetric with respect to g_e for all $X \in \mathfrak{g}$. Thus, the existence of a biinvariant metric on G depends only on $\mathfrak{g}$.

Theorem 6.7.13 shows in this case:

Corollary 7.1.2 *Any compact Lie group G carries a biinvariant metric.*

Because $[\mathfrak{m}, \mathfrak{m}] \subset \mathfrak{h}$, G'/H is naturally reductive. From here on, it is more convenient to study the Lie group $G = G/e_G$ using the theorems from Chapter 6.6, to express the results directly in terms of the Lie bracket on G. Therefore the

Change of convention: For the rest of the chapter, the homogeneous space G is usually considered as the quotient G/e_G.

By Corollary 7.1.1, the biinvariant metric satisfies $\mathrm{ad}^*_X = -\mathrm{ad}_X$ for all $X \in \mathfrak{g}$, i.e. $U \equiv 0$. So G/e_G is naturally reductive also for the choice $H = e_G, \mathfrak{h} = 0, \mathfrak{m} = \mathfrak{g}$. By Lemma 6.6.2 it follows that:

Corollary 7.1.3 *Consider a biinvariant metric g on G and left-invariant vector fields X, Y. Then $\nabla_X Y = \frac{1}{2}[X, Y]$ and $\exp_e A = e^A$ for all $A \in \mathfrak{g}$.*

Corollary 7.1.4 *G carries a biinvariant metric $\Rightarrow A \mapsto e^A$ is surjective and $B \leq 0$.*

Proof. The first part follows by Corollary 7.1.2 from the completeness of the homogeneous space G/e. Lemma 6.8.12 applied to the space $(G \times G)/G$ yields $B^{\mathfrak{g}} \leq 0$ for the isotropy group G.   □

Example According to Corollary 5.4.6, $SL_2(\mathbb{R})$ cannot carry a biinvariant metric. Consequently, no group containing $SL_2(\mathbb{R})$ can carry a biinvariant metric either; e.g. $SL_n(\mathbb{R})$ or $GL_n(\mathbb{R})$.

Corollary 7.1.5 *Each Lie subgroup H of a Lie group G with biinvariant metric is totally geodesically embedded, i.e. $II = 0$.*

Proof. If $X, Y \in \mathfrak{h}$, then $\nabla^G_X Y = \frac{1}{2}[X, Y] \in \mathfrak{h}$ holds. □

Lemma 7.1.6 *Let G be a Lie group, X, Y, Z, W left-invariant and let $(e_j)_j$ be an orthonormal basis of $\mathfrak{g}$. The curvature of a biinvariant metric satisfies*

1) $\Omega(X, Y)Z = \frac{1}{4}[Z, [X, Y]]$,
2) $R(X, Y, Z, W) = \frac{1}{4}g([X, Y], [Z, W])$,
3) $K(X \wedge Y) = \frac{1}{4}\frac{\|[X,Y]\|^2}{\|X \wedge Y\|^2} \geq 0$ *for X, Y linearly independent,*
4) $\mathrm{Ric}(X, X) = \frac{1}{4}\sum_j \|[X, e_j]\|^2 \geq 0$, *i.e.* $\mathrm{Ric} = -\frac{1}{4}B$,
5) $s = \frac{1}{4}\sum_{j,k}\|[e_j, e_k]\|^2 \geq 0$.

Proof. Theorem 6.8.17 with $\mathfrak{h} = 0$ implies immediately (3) and thus (4),(5) (instead one could also use Theorem 6.6.3). Since for all $X \in \mathfrak{g}$ we have $\mathrm{ad}^*_X = -\mathrm{ad}_X$, $\mathrm{Ric}(X, X) = -\sum_j g(\mathrm{ad}_X \mathrm{ad}_X e_j, e_j) = -B(X, X)$, hence the assertion follows thanks to the polarization identity. The uniqueness of R for given K under all symmetric bilinear forms on $\Lambda^2 TM$ satisfying the 1st Bianchi identity provides (1) and (2). □

Remark 7.1.7 Just as in Exercise 3.4.10, this can also be computed directly using the formula for ∇. The scalar curvature vanishes if and only if for all j, k the Lie bracket $[e_j, e_k]$ vanishes, i.e. if G is abelian. Thus, for a biinvariant metric, each of the five types of curvature in Lemma 7.1.6 vanishes if and only if G is abelian.

Lemma 7.1.8 *Suppose G is a Lie group with biinvariant metric. Then the curvature tensor $R \in \Gamma(G, T^*G^{\otimes 4})$ satisfies*

$$\nabla^{T^*G^{\otimes 4}} R = 0.$$

Remark As will become apparent in the next chapter, this formula is much stronger than the 2nd Bianchi identity where $\nabla^{\mathrm{End}\, TG}\Omega = 0$ holds. According to Corollary 3.3.6, the operator $\nabla^{\mathrm{End}\, TG} : \mathfrak{A}^2(G, \mathrm{End}\, TG) \to \mathfrak{A}^3(G, \mathrm{End}\, TG)$ involves averaging $\nabla^{T^*G^{\otimes 4}}$ over the symmetric group $\mathfrak{S}_3$ with 6 elements.

Proof. Consider left-invariant vector fields X, Y, Z. Then with $\mathrm{ad}_{|h} \in T^*_h G^{\otimes 2} \otimes T_h G$,

$$(\nabla_X \mathrm{ad})(Y, Z) = \nabla_X(\mathrm{ad}_Y Z) - \mathrm{ad}_{\nabla_X Y} Z - \mathrm{ad}_Y \nabla_X Z$$

$$= \frac{1}{2}\left([X, [Y, Z]] - [[X, Y], Z] - [Y, [X, Z]]\right) \overset{\text{Jacobi}}{=} 0.$$

Thus $R_h = \frac{1}{4}g(\mathrm{ad}\,\cdot, \mathrm{ad}\,\cdot)$ implies both $\nabla R = 0$ and $\nabla^{\Lambda^2 T^*G \otimes \mathrm{End}\, TG}\Omega = 0$. □

The following theorem is actually a statement about Lie algebras, and the proof is carried out only using these. Nevertheless, for the sake of clarity, the formulation uses Lie groups.

Theorem 7.1.9 *Given a Lie group G, the following are equivalent:*

1) B is negative definite,
2) $\mathfrak{z} = 0$ and there exists a biinvariant metric on G,
3) there exists a biinvariant metric on G such that Ric > 0.

In Theorem 7.4.13 this statement will be further enhanced.

Proof. $(1)\Rightarrow(2)$: Let $g := -B$. If $X, Y, Z \in \mathfrak{g}$, the Jacobi equation implies

$$\text{ad}_{[X,Z]}Y = [X, [Z,Y]] - [Z, [X,Y]] = (\text{ad}_X\text{ad}_Z - \text{ad}_Z\text{ad}_X)Y.$$

Therefore, because $\text{Tr}\,AB = \text{Tr}\,BA$, it follows that $B(\text{ad}_X Y, Z) = -B(Y, \text{ad}_X Z)$. Thus, by Corollary 7.1.1, g induces a biinvariant metric on G. The result about $\mathfrak{z}$ follows from Proposition 6.8.11.

$(2)\Rightarrow(3)$: Lemma 6.8.12 applied to the space $(G \times G)/G$ shows (since $\mathfrak{z} = 0$) that $B^{\mathfrak{g}} < 0$, hence Ric $= -B^{\mathfrak{g}}/4 > 0$.

$(3)\Rightarrow(1)$: According to Lemma 7.1.6, Ric $= -\frac{1}{4}B$ holds.   □

Remark Alternatively, one can see $(2)\Rightarrow(3)$ in a more direct way: Consider $X \in \mathfrak{g} \setminus \{0\}$. If $\mathfrak{z} = 0$, then $\text{ad}_X \neq 0$, so there exists a $Y \in \mathfrak{g}$ such that $[X,Y] \neq 0$. Set $e_1 := \frac{Y}{\|Y\|}$ and extend this to an orthonormal basis $(e_j)_j$, then one finds

$$\text{Ric}(X, X) = \frac{1}{4} \sum_j \|[X, e_j]\|^2 > 0.$$

Proposition 7.1.10 *Assume $\mathfrak{g}$ to be simple (cf. Exercise 6.6.6). Then every biinvariant metric g on $\mathfrak{g}$ is proportional to the Killing form B.*

Proof. Define $g^{-1}B \in \text{End}(\mathfrak{g})$ as that vector space endomorphism such that $g(X, (g^{-1}B)Y) = B(X,Y)$; $g^{-1}B$ has at least one eigenvalue λ. Then given any eigenvector $Y \in \text{Eig}_{g^{-1}B}(\lambda)$ and $X, Z \in \mathfrak{g}$, one obtains

$$g(X, (g^{-1}B)[Y,Z]) = B(X, [Y,Z]) = B([Z,X],Y)$$
$$= g([Z,X], (g^{-1}B)Y) = \lambda g([Z,X],Y) = \lambda g(X, [Y,Z]),$$

thus $[Y,Z] \in \text{Eig}_{g^{-1}B}(\lambda)$. Therefore $\text{Eig}_{g^{-1}B}(\lambda)$ is a nontrivial ideal, hence equal to all of $\mathfrak{g}$ and $B = \lambda g$.   □

Remark This is essentially **Schur's Lemma**[1]: Every homomorphism between two irreducible group representations is 0 or invertible.

---
[1] Issai Schur, 1875–1941.

Corollary 7.1.11 *If $\mathfrak{z} = 0$, then every biinvariant metric can be decomposed as a direct sum of multiples of Killing forms.*

Proof. Use the decomposition in Exercise 6.6.6.   □

Definition 7.1.12 *A homogeneous space G/H is called* **normal** *if the metric can be extended to a biinvariant metric on G.*

Corollary 7.1.13 *Any normal homogeneous space is naturally reductive.*

Proof. This follows from $U \equiv 0$.   □

Remark 7.1.14 Berestovskii and Nikonorov show in [BereNi, Th. 25] that for any compact naturally reductive space (M, g) with positive Euler characteristic there exist Lie groups $H \subset G$ such that $M = G/H$ is normal homogeneous.

Corollary 7.1.15 *Consider a Lie subgroup $H \subset G$ of a Lie group with biinvariant metric g. Then the normal space G/H satisfies $K^{G/H}(X \wedge Y) = \frac{1}{4}\|[X,Y]\|^2 + \frac{3}{4}\|[X,Y]^\flat\|^2 \geq 0$ for any $X, Y \in T_e(G/H) \cong \mathfrak{h}^\perp \subset \mathfrak{g}$.*

Proof. Theorem 6.8.17 implies

$$
K^{G/H}(X \wedge Y) \quad = \quad \frac{1}{4}\|[X,Y]^\mathfrak{m}\|^2 + g([[X,Y]^\flat, X], Y)
$$

$$
\overset{\underset{\text{ad skew-}}{\text{symmetric}}}{=} \quad \frac{1}{4}\|[X,Y]\|^2 + \frac{3}{4}\|[X,Y]^\flat\|^2.   \qquad □
$$

Exercises

Exercise 7.1.16 Prove that for $n > 1$ there is no metric on $\mathrm{SL}(n+1, \mathbb{R})/\begin{pmatrix} 1 & 0 \\ 0 & \mathrm{SL}(n,\mathbb{R}) \end{pmatrix}$ for which $\mathrm{SL}(n+1, \mathbb{R})$ acts by isometries. Is this space reductive?

Exercise 7.1.17 Let $\mathrm{SL}(2, \mathbb{R})$ operate on the hyperbolic plane H^2 as in Exercise 3.1.22. Let $\mathrm{PSL}(2, \mathbb{Z})$ denote the quotient of $\mathrm{SL}(2, \mathbb{Z})$ by the normal subgroup $\pm \mathrm{id}_{\mathbb{R}^2}$. Does $\mathrm{PSL}(2, \mathbb{Z})$ act freely? Does $\mathrm{PSL}(2, \mathbb{Z})$ act faithfully?

Exercise* 7.1.18 Consider a Lie algebra $\mathfrak{g}$ with $B < 0$. Show in a direct way that for any $h \in G$, Ad_h is an isometry of $-B$.

Exercise 7.1.19 Using the Jordan normal form, show that

$$
\mathfrak{gl}(n) \to \mathrm{GL}(n, \mathbb{C}), \quad A \mapsto e^A
$$

is surjective (although $\mathrm{GL}(n, \mathbb{C})$ cannot carry a biinvariant metric).

Exercise 7.1.20 Consider

$$
G := \left\{ \left. \begin{pmatrix} 1/a & 0 & 0 \\ 0 & a & b \\ 0 & 0 & 1 \end{pmatrix} \right| a > 0, b \in \mathbb{R} \right\} \subset SL_3(\mathbb{R}).
$$

1) Prove that G is isomorphic to a connected component of the group of affine transformations of real lines from Exercise 1.6.26.
2) Show that $\mathfrak{g} \to G, A \mapsto e^A$ is surjective.
3) Calculate the Killing form associated to G. Can G carry a biinvariant metric?
4) Determine the sectional curvature of the left-invariant metric on G induced by the Euclidean metric of $\mathbb{R}^{3 \times 3}$ on $T_e G$ and find an isometry to a space you already know.

Exercise 7.1.21 Compute the Killing form B on $SO(n)$ and compare it to the metric induced by the Euclidean metric of $\mathbb{R}^{n \times n}$.

Exercise 7.1.22 Determine the sectional curvature of $SU(2)$ with respect to $-B$.

Exercise 7.1.23 Let G be a compact Lie group equipped with a biinvariant metric, let $T \subset G$ be a maximal torus (i.e. there is no torus in G that contains T properly), and $M := G/T$.

1) Let $N := \{\gamma \in G | \gamma T \gamma^{-1} = T\}$ be the normalizer of T in G and let $W_G := N/T$ be the **Weyl group** of G. Show that $W_G \subset M^\gamma$ for any $\gamma \in T$. Conclude by Exercise 5.2.16 that W_G is finite.
2) Prove that for almost all $\gamma \in T$ the fixed point set is $M^\gamma = W_G$.

7.2 Definition of Symmetric Spaces

symmetric space	Lorentzian metric
locally symmetric	

In this section, symmetric spaces are described as homogeneous spaces with certain point reflections. This property is associated with the parallelism of curvature.

Definition 7.2.1 *A Riemannian manifold M is called a **symmetric space** :$\Leftrightarrow$ For every $p \in M$ there exists an isometry $\sigma_p : M \to M$ such that $\sigma_p(p) = p, T_p \sigma_p = -\mathrm{id}_{T_p M}$. M is called **locally symmetric** if such a σ_p exists on a neighborhood around each $p \in M$.*

By Theorem 6.7.7 the point reflection σ_p is uniquely determined and given on every ball-shaped normal neighborhood by $\exp_p X \mapsto \exp_p(-X)$ (Fig. 7.1). Via the exponential map the condition on σ_p becomes equivalent to: locally $\sigma_p^2 = \mathrm{id}$ holds and p is an isolated fixed point of σ_p.

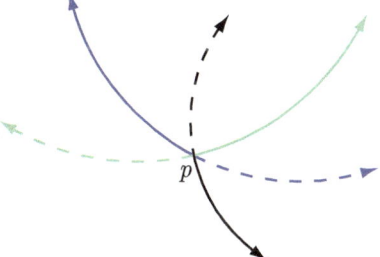

Fig. 7.1 Geodesic point reflection.

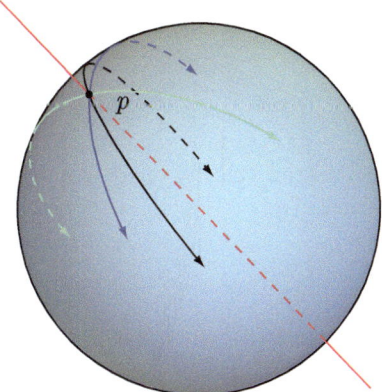

Fig. 7.2 Reflection of the sphere at an axis of rotation.

Remark Because of its transitive group of isometries, a homogeneous space is symmetric if there exists at least one point p at which there is such a point reflection σ_p.

Example $\mathbb{R}^n$ via the multiplication by -1, S^n via the reflection at an axis of rotation through p (Fig. 7.2), analogously H^n with the hyperboloid model, $\mathbb{P}^n\mathbb{C}$ with their standard metrics respectively are symmetric. The point reflection of $\mathbb{P}^n\mathbb{C}$ at $(1 : 0 : \cdots : 0)$ is given by

$$\sigma_{(1:0:\cdots:0)} : (z_0 : z_1 : \cdots : z_n) \mapsto (-z_0 : z_1 : \cdots : z_n).$$

Lie groups G with biinvariant metric are symmetric with $\sigma_p(a) := pa^{-1}p$. In particular, the flat tori $\mathbb{R}^n/\Gamma$ are symmetric.

Every open subset of these spaces is locally symmetric.

Lemma 7.2.2 *Every symmetric space is complete.*

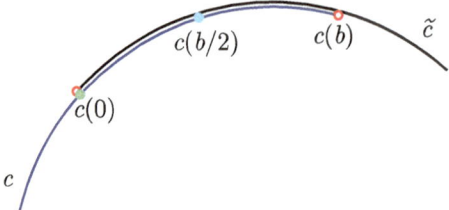

Fig. 7.3 Point reflection of a geodesic.

Proof. Let c be a geodesic and $b > 0$ maximal such that c is defined on $]a, b[$ for an $a < 0$. Then $c(\frac{b}{2} - t) = \exp_{c(\frac{b}{2})}(-t\dot{c}(\frac{b}{2}))$ is thus defined on $] - \frac{b}{2}, \frac{b}{2} - a[$, so also

$$\tilde{c}(t) := \sigma_{c(\frac{b}{2})}(c(\frac{b}{2} - t)) = \exp_{c(\frac{b}{2})} T\sigma(-t\dot{c}(\frac{b}{2})) = \exp_{c(\frac{b}{2})}(t\dot{c}(\frac{b}{2}))$$

(Fig. 7.3). Since $\tilde{c}$ at 0 has the same velocity vector as $c(\frac{b}{2} + t)$, it is equal to $c(\frac{b}{2} + t)$. Hence c is defined on $]a, b - a[$ ↯.   □

Lemma 7.2.3 *M is symmetric $\Rightarrow$ for all $p \in M$ we have $\sigma_p^2 = $ id, i.e. σ_p is an involution.*

Proof. This follows by Theorem 6.7.7, since $T_p\sigma_p^2 = $ id $= T_p$id.   □

Lemma 7.2.4 *Every symmetric space M is homogeneous.*

For the proof of this lemma we need the Myers–Steenrod Theorem and the Gleason–Montgomery–Zippin Theorem: The isometry group G of M is a Lie group. A relatively short proof for the case of symmetric spaces can be found, for example, in [Hel, ch. IV, Th. 2.5] and [Hel, ch. IV, Lemma 3.2]. Alternatively, one can assume homogeneity in the definition of a symmetric space. This would make no difference to the rest of the proofs in this book.

Proof. Given arbitrary points $p, q \in M$, q can be written as $q = \exp_p X$ because of the completeness. By Lemma 7.2.2 p, q are mapped to each other by $\sigma_{\exp_p(X/2)}$. Therefore the isometry group G acts transitively.   □

Lemma 7.2.5 *M is locally symmetric $\Leftrightarrow \nabla^{T^*M^{\otimes 4}} R \equiv 0$.*

For Lie groups with biinvariant metric, "$\Rightarrow$" was Lemma 7.1.8. As noted there, this equation is stronger than the 2nd Bianchi identity.

Proof. "$\Rightarrow$" σ_p is an isometry, hence $\sigma_p^*(\nabla^{T^*M^{\otimes 4}} R) = \nabla^{T^*M^{\otimes 4}} R$. Given $A, X, Y, Z, W \in T_pM$, one thus obtains

$$(\nabla_A^{T^*M^{\otimes 4}} R)(X, Y, Z, W) = \sigma_p^*((\nabla_A^{T^*M^{\otimes 4}} R)(X, Y, Z, W))$$
$$= (\nabla_{-A}^{T^*M^{\otimes 4}} R)(-X, -Y, -Z, -W)$$
$$= -(\nabla_A^{T^*M^{\otimes 4}} R)(X, Y, Z, W).$$

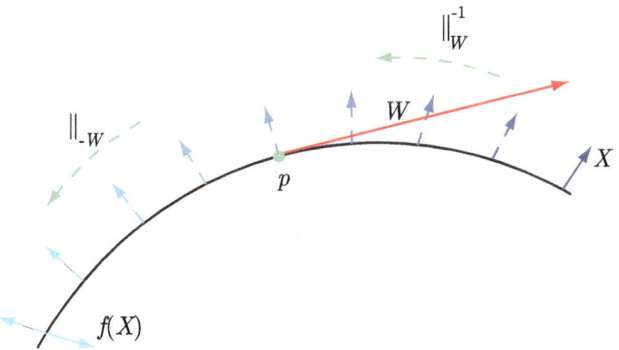

Fig. 7.4 Cartan's Theorem in the case of locally symmetric spaces.

"$\Leftarrow$" We need to show that on a ball-shaped normal neighborhood σ_p : $\exp_p(W) \mapsto \exp_p(-W)$ is an isometry. Because $\nabla^{T^*M^{\otimes 4}} R = 0$, Ω is parallel along any curve, i.e., for X, Y, Z parallel, $\Omega(X, Y)Z$ is parallel. With

$$f : T_{\exp_p wM} \to T_{\exp_p(-W)}M,$$

$$X \mapsto \|{}_{T_p\sigma_p W} \circ T_p\sigma_p \circ \|{}_W^{-1} X = -\|{}_{-W} \circ \|{}_W^{-1} X$$

(Fig. 7.4) and X, Y, Z parallel along $\exp_p tW$, the two sides in Cartan's Theorem 6.2.1

$$f(\Omega_{\exp W}(X, Y)Z) = -\Omega_{\exp(-W)}(X, Y)Z$$

and

$$\Omega_{\exp(-W)}(f(X), f(Y))f(Z) = \Omega_{\exp(-W)}(-X, -Y)(-Z)$$

are equal. So σ_p is a local isometry.   □

Likewise, any other tensor of odd degree determined canonically by g (or more generally being σ_p-invariant) vanishes.

Remark The Ambrose–Singer Theorem [AmSi] generalizes this result to (locally) homogeneous spaces by means of a more complicated differential equation.

Exercises

Exercise 7.2.6 Let $G := \{(a, b) \,|\, a \in \mathbb{R}^+, b \in \mathbb{R}\}$ be the Lie group from Exercise 1.6.26 and $M := \left\{ \begin{pmatrix} x \\ y \end{pmatrix} \in \mathbb{R}^2 \,\middle|\, x + y > 0 \right\}$ with the Lorentzian metric $g = dx \otimes dx - dy \otimes dy$ (similar to a Riemannian metric, but not positive definite). Connections, curvature, geodesics, isometries are defined for g in the same way as for a Riemannian metric.

1) Show that G acts on M transitively, effectively, and isometrically via

$$(a,b) \cdot \begin{pmatrix} x \\ y \end{pmatrix} = \begin{pmatrix} \frac{a+1/a}{2}x + \frac{1/a-a}{2}y + b \\ \frac{1/a-a}{2}x + \frac{a+1/a}{2}y - b \end{pmatrix}.$$

2) Determine the isotropy group of $\begin{pmatrix} 1 \\ 1 \end{pmatrix}$ and find a diffeomorphism $G \to M$.
3) Prove that the curvature of M vanishes (a fortiori thus $\nabla R \equiv 0$), and find a local reflection at any point $\begin{pmatrix} x \\ y \end{pmatrix}$.
4) Show that M is not geodesically complete.

7.3 The Cartan–Ambrose–Hicks Theorem

| Cartan–Ambrose–Hicks Theorem | broken geodesic |
| Whitehead Theorem | |

The Cartan–Ambrose–Hicks Theorem is a global version of Cartan's Theorem 6.2.1. In particular, it will prove here that complete locally symmetric spaces are quotients of symmetric spaces by discrete groups. First we will need a weak form of convexity of normal neighborhoods.

Proposition 7.3.1 *For every point $p \in M$ and sufficiently small $\delta > 0$, there exists an $\varepsilon > 0$ such that for all $q \in B_\varepsilon(p)$, $B_\varepsilon(p) \subset B_\delta(q)$ holds and $B_\delta(q)$ is a normal neighborhood. In particular, every normal neighborhood U of p contains a neighborhood $\tilde{U}$ of p such that all shortest paths between any points in $\tilde{U}$ lie in U.*

Proof. As in Definition 5.2.5, choose an open neighborhood $\Omega \subset TM$ of $0 \in T_pM$ on which $\exp : \Omega \to M$ is defined. By Lemma 5.2.6, the map $f : \Omega \to M \times M$, $X \in T_qM \mapsto (q, \exp_q X)$ has a derivative of the form $\begin{pmatrix} \mathrm{id} & * \\ 0 & \mathrm{id} \end{pmatrix}$, so f is a diffeomorphism on a neighborhood $V \subset \Omega$ of 0. If we choose $\delta > 0$ with $B_{2\delta}(0) \subset T_pM \cap V$, then there is an open neighborhood $W \subset M$ of p such that for all $q \in W$ we have $B_\delta(0) \subset T_qM \cap V$. Choose $\varepsilon > 0$ with $B_\varepsilon(p) \times B_\varepsilon(p) \subset f(W)$.

Thus, for every normal neighborhood U of p there is a $\delta > 0$ and a subset $\tilde{U} := B_\varepsilon(p)$ such that $\bigcup_{q \in B_\varepsilon(p)} B_\delta(q) \subset U$. Because $B_\delta(q)$ by Theorem 5.2.11 contains all shortest paths from q to points in $B_\varepsilon(p)$, these lie in particular in U. □

Remark The **Whitehead Theorem** [Whi], [ChEb, p. 103] states that there are even normal neighborhoods U which contain all shortest paths between any points in U.

The proof of the following theorem uses broken geodesics, since the last step of the proof requires curves that do not always exist as geodesics. Let a **broken geodesic** be a curve $\gamma :\,]t_0, t_m[\to M$ with $t_1 < \cdots < t_m$ and $\gamma_{|]t_j, t_{j+1}[}$ geodesic. Given an isometry $A : T_pM \to T_{\tilde{p}}\tilde{M}$ and $p = \gamma(t_0)$, set $\tilde{\gamma}_{|]t_0, t_1[}(t) := \exp_{\tilde{p}} tA\dot{\gamma}(t_0)$

and $\gamma_j := \gamma_{|]t_0,t_j[}$. Successively, using the parallel transport $||_{\gamma_j}$ along γ_j this is extended as

$$\widetilde{\gamma}_{|]t_j,t_{j+1}[}(t) := \exp_{\lim_{t\,\nearrow t_j}\widetilde{\gamma}(t)} ||_{\widetilde{\gamma}_j} \circ A \circ ||_{-\gamma_j}(\lim_{t\,\searrow t_j} \dot{\gamma}(t))$$

(where at a breaking point the parallel transport contains the corresponding rotation).

Cartan–Ambrose–Hicks Theorem 7.3.2 ([Am][2]) *Let* (M,g), $(\widetilde{M},\widetilde{g})$ *be complete Riemannian manifolds, let M be simply connected, $p \in M$, $\widetilde{p} \in \widetilde{M}$ and $A : T_p M \to T_{\widetilde{p}}\widetilde{M}$ be an isometry. For all broken geodesics γ, $X,Y,Z \in T_{\gamma(t_m)}M$ and*

$$f_\gamma := ||_{\widetilde{\gamma}} \circ A \circ ||_{-\gamma} : T_{\gamma(t_m)}M \to T_{\widetilde{\gamma}(t_m)}\widetilde{M},$$

assume that

$$f_\gamma(\Omega(X,Y)Z) = \widetilde{\Omega}(f_\gamma(X),f_\gamma(Y))f_\gamma(Z).$$

Then for all broken geodesics γ, γ', $\gamma(t_m) = \gamma'(t'_m)$ implies $\widetilde{\gamma}(t_m) = \widetilde{\gamma}'(t'_m)$. Therefore there is a C^∞-map $\varphi : M \to \widetilde{M}, \gamma(t_m) \mapsto \widetilde{\gamma}(t_m)$. φ is a Riemannian covering.

Proof. Rescale γ, γ' and add breaking points at smooth points such that both have $t_1, \ldots, t_m$ as breaking points.

1) First, suppose that for all j, $\gamma(t_{j+1})$, $\gamma'(t_{j+1})$, $\gamma'(t_{j+2}) \in \widetilde{U}$ holds for a ball-shaped normal neighborhood U of $\gamma(t_j)$ and $\widetilde{U}$ as in Proposition 7.3.1. The proof of $\widetilde{\gamma}(t_m) = \widetilde{\gamma}'(t_m)$, $f_\gamma = f_{\gamma'}$ is carried out by induction over m:

Initial case: If γ, γ' lie in a normal neighborhood of p, the statement follows using the local isometry φ from Cartan's Theorem.

Induction step: Let τ be the shortest path from $\gamma(t_{m-2})$ to $\gamma'(t_{m-1})$ (Fig. 7.5). By induction hypothesis, $\widetilde{\gamma_{m-2} \cup \tau}(t_{m-1}) = \widetilde{\gamma}'(t_{m-1})$ and $f_{\gamma_{m-2}\cup\tau} = f_{\gamma'_{m-1}}$. For broken geodesics c starting at $\gamma(t_{m-2})$, let $\overline{c}$ be the geodesic starting at $\widetilde{\gamma}(t_{m-2})$ with initial vector $f_{\gamma_{m-2}}(\dot{c})$. By Cartan's Theorem, since the curves involved lie in a normal neighborhood of $\gamma(t_{m-2})$, one finds with $c := \gamma_{|]t_{m-2},t_m[}$, $c' := \gamma'_{|]t_{m-1},t_m[}$ that

$$\widetilde{\gamma}(t_m) = \overline{c}(t_m) \overset{\text{Cartan}}{=} \widetilde{\tau \cup c'}(t_m) = \widetilde{\gamma_{m-2} \cup \tau \cup c'}(t_m)$$

and

$$f_\gamma = f_c = f_{\tau \cup c'} = f_{\gamma_{m-2}\cup\tau\cup c'}.$$

Thus it follows that

$$f_\gamma = ||_{\widetilde{c}'} \circ f_{\gamma_{m-2}\cup\tau} \circ ||_{-c'} \overset{\substack{\text{induction}\\\text{hypothesis}}}{=} ||_{\widetilde{c}'} \circ f_{\gamma_{m-1}} \circ ||_{-c'} = f_{\gamma'}$$

and $\widetilde{\gamma}(t_m) = \widetilde{\gamma}'(t_m)$.

---

[2] Élie Joseph Cartan, 1869–1951; 1956, Warren Ambrose, 1914–1995; 1966, N. Hicks for general connections.

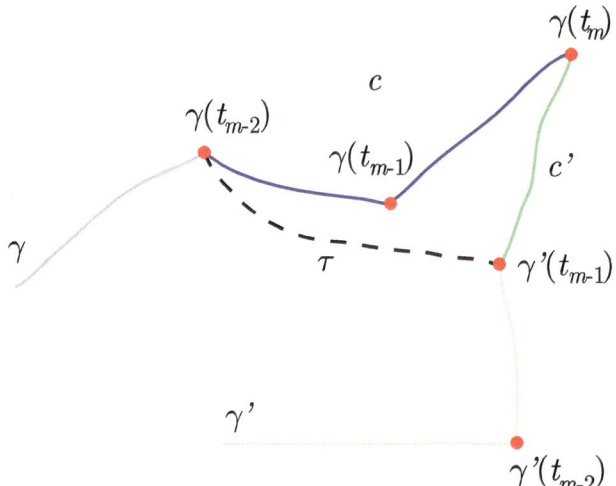

Fig. 7.5 Geodesic in the proof of the Cartan–Ambrose–Hicks Theorem.

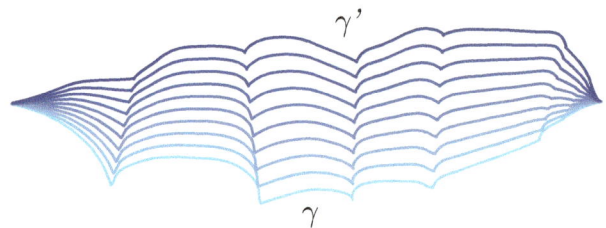

Fig. 7.6 Step 2 in the proof of the Cartan–Ambrose–Hicks Theorem.

2) Now let γ, γ' be any geodesics with the same starting point and the same ending point. Because M is simply connected, there exists a homotopy γ^s from γ to γ' by broken geodesics with breaking points at $t_1 < \cdots < t_m$ for all s (Fig. 7.6). Let $0 < s_1 < \cdots < s_\ell < 1$ and $t_1 < \cdots < t_m$ be sufficiently fine that $\gamma^{s_k}(t_{j+1}), \gamma^{s_{k+1}}(t_{j+1}), \gamma^{s_{k+1}}(t_{j+2})$ lie in $\widetilde{U}$ for a normal neighborhood U of $\gamma^{s_k}(t_j)$ for all j, k. Then, according to (1),

$$\widetilde{\gamma}(t_m) = \widetilde{\gamma}^{s_1}(t_m) = \cdots = \widetilde{\gamma}^{s_\ell}(t_m) = \widetilde{\gamma}'(t_m).$$

On a normal neighborhood of each point φ is the map from Cartan's Theorem, i.e. a local isometry, so by Lemma 6.5.4 it is a Riemannian covering.   □

Corollary 7.3.3 *M is locally symmetric, complete and simply connected $\Rightarrow$ M is symmetric.*

Proof. As in Lemma 7.2.5, $\nabla R = 0$ and simply connected implies that σ can be extended globally.   □

Compare [Hel, ch. IV §5] for a Lie-algebra-theoretic proof.

7.4 Symmetric Spaces and Group Involutions

Ado's Theorem	diameter
shearing	real structure
transvection	Graßmann manifolds

In this section, symmetric spaces are characterized using involutions on Lie groups and Lie algebras, curvature in terms of the Lie algebra is determined, and some geometric consequences are drawn. At the end, the groups carrying a biinvariant metric are determined in more detail. Still valid is the remark from Section 6.7 that quite a few properties assigned to G/H in this chapter are to be understood as properties of the pair (G, H), in some cases additionally including a decomposition $\mathfrak{g} = \mathfrak{h} \oplus \mathfrak{m}$.

Lemma 7.4.1 *Consider a Lie group G and a Lie subgroup $\mathfrak{h} \subset \mathfrak{g}$. Then there is an immersion $H \subset G$ of a Lie group with Lie algebra $\mathfrak{h}$ as algebraic subgroup.*

However, H is not necessarily closed, thus not necessarily a Lie subgroup of G.

Proof. By Frobenius' Theorem 2.3.10, there is a submanifold $\widehat{H}_a$ around any $a \in G$ which is tangent to the left-invariant subbundle $\mathfrak{h} \subset TM$. On $\bigcup_{a \in G} \widehat{H}_a$, choose the topology where U is open if $U \cap H_a$ is open in H_a for all $a \in G$. Let H be the connected component of e in $\bigcup_{a \in G} \widehat{H}_a$. With the charts of H_a, H becomes a manifold immersed in G. For $h' \in H$ and a path h_t in H from e to $h \in H$, $\gamma(t) := h' h_t^{-1}$ is a path from h' to $h' h^{-1}$ with $\dot{\gamma} \in (L_{h'})^* \mathfrak{h} = \mathfrak{h}$, that is, in H. Thus H is subgroup. In particular the multiplication in H is smooth since it is induced by that of G. $\square$

The condition of the version 1.5.9 of Frobenius' Theorem cannot be satisfied by left-invariant vector fields $(X_1, \ldots, X_k)$ in general: If the vector fields commute, H is abelian.

Remark Ado's Theorem[3]. states that for every finite-dimensional Lie algebra $\mathfrak{g}$ there exist an $n \in \mathbb{N}$ and an embedding $\mathfrak{g} \hookrightarrow \mathfrak{gl}(\mathbb{R}^n)$ ([Varad] or [Bour]). Lemma 7.4.1 then shows that for every finite-dimensional Lie algebra $\mathfrak{g}$ there is a Lie group $G \subset \mathrm{GL}_n(\mathbb{R})$. However, there are also finite-dimensional Lie groups which are not subgroups of any $\mathrm{GL}_n(\mathbb{R})$ (e.g., by Exercise 7.4.16 the universal covering of $\mathrm{SL}_2(\mathbb{R})$).

Theorem 7.4.2 *Given any Lie algebra homomorphism $A : \mathfrak{g} \to \mathfrak{h}$, there is a unique local Lie group homomorphism $\varphi : U \to H$ on a neighborhood $U \subset G$ of e. If G is simply connected, a unique $\varphi : G \to H$ exists globally.*

---

[3] 1935, Igor Dmitrievich Ado, 1910–1983

Proof. Consider the graph $\mathfrak{k} := \{(X, AX) \mid X \in \mathfrak{g}\} \subset \mathfrak{g} \times \mathfrak{h}$ of A. $\mathfrak{k}$ is a Lie subalgebra, because

$$[(X, AX), (Y, AY)] = ([X, Y], [AX, AY]) = ([X, Y], A[X, Y]) \in \mathfrak{k}.$$

So, according to Lemma 7.4.1, there exists an immersion as a subgroup of a Lie group $K \subset G \times H$. With $\pi_1, \pi_2 : G \times H \to G, H$ set $\pi := \pi_{1|K} : K \to G$. Then $T_e \pi : \mathfrak{k} \to \mathfrak{g}$ is the projection onto the first factor, so it is bijective. Thus, on a sufficiently small neighborhood U of e, $\pi_{|U}$ is a diffeomorphism. Set $\varphi := \pi_{2|K} \circ \pi_{|U}^{-1}$. Because $T_e \varphi \circ T_e \pi_1 = T_e \pi_2$, it follows that $T_e \varphi(X) = AX$.

Equip G, K with left-invariant metrics such that $T_e \pi$ is an isometry. Because of the equivariance, $T\pi$ is everywhere an isometry and π is a covering by Lemma 6.5.4. If G is simply connected, π must be an isomorphism, hence φ is globally defined. By Lemma 1.6.19, φ is uniquely defined. $\square$

In the following theorem, criterion (2) gives a description of symmetric spaces, and assuming the Myers–Steenrod Theorem and the Gleason–Montgomery–Zippin Theorem, by (0) every symmetric space has this form. (5) gives a particularly elegant characterization in terms of Lie algebras, the first two relations having already appeared in the chapter on homogeneous spaces: $[\mathfrak{h}, \mathfrak{h}] \subset \mathfrak{h}$ means that $\mathfrak{h}$ is subalgebra, and $[\mathfrak{h}, \mathfrak{m}] \subset \mathfrak{m}$ was a consequence of reductivity. Crucially, we have the new relation $[\mathfrak{m}, \mathfrak{m}] \subset \mathfrak{h}$. Inevitably this can describe symmetric spaces only up to covering, therefore the additional conditions for the inversion are necessary.

Theorem 7.4.3 (Élie Cartan) *For Lie groups G, H, G connected, and $M = G/H$ with faithful G-operation, the following holds (and if the Killing form of G is negative definite and $M = G/H$ is normal, the properties in parentheses follow automatically from the statements immediately before):*

(0) M is symmetric and G is the connected component of e of the isometry group,

$\Rightarrow$ *(1) M is symmetric with isometric G-operation and at least one σ_p is induced by an involution $\sigma \in \mathrm{Aut}(G)$,*

$\Rightarrow$ *(2) there exists an involution $\sigma \in \mathrm{Aut}(G)$ whose fixed point set G^σ and whose connected components G_0^σ of e_G satisfy $G_0^\sigma \subset H \subset G^\sigma$ (and $\sigma : G/H \to G/H$ induces an isometry),*

$\Rightarrow$ *(3) $M = G/H$ homogeneous is symmetric and each σ_p is induced by an involution $\sigma \in \mathrm{Aut}(G)$,*

$\Rightarrow$ *(4) $\mathfrak{g}$ has a Lie algebra homomorphism A which is an involution with $\mathfrak{g}^A = \mathfrak{h}$ (and an A-invariant scalar product for which ad_X is skew-symmetric for all $X \in \mathfrak{h}$),*

$\Leftrightarrow$ *(5) $\mathfrak{g}$ has a vector space decomposition $\mathfrak{g} = \mathfrak{h} \oplus \mathfrak{m}$ such that*

$$[\mathfrak{h}, \mathfrak{h}] \subset \mathfrak{h}, \ [\mathfrak{h}, \mathfrak{m}] \subset \mathfrak{m}, \ [\mathfrak{m}, \mathfrak{m}] \subset \mathfrak{h} \tag{7.1}$$

hold (and a scalar product with $\mathfrak{h} \perp \mathfrak{m}$ for which ad_X is skew-symmetric for all $X \in \mathfrak{h}$).

If G is simply connected and H is connected, then (2)–(5) are equivalent.

Proof. $(0) \Rightarrow (1)$ Given the isometry group $\text{Isom}(M)$ set $\sigma : \text{Isom}(M) \to \text{Isom}(M), k \mapsto \sigma_p k \sigma_p$. Then σ leaves the connected component G of e invariant and one gets $\sigma^2 = \text{id}$. The isotropy group H of p in G satisfies $\sigma_p(aH) = \sigma(a)H$ because $\sigma_p H = H$.

$(1) \Rightarrow (2)$: One obtains $H \subset G^\sigma$, since because of the effectiveness $h \in H$ is uniquely determined by $T_p L_h$ and $T_p(\sigma_p \circ L_h \circ \sigma_p) = -(-T_p L_h) = T_p L_h$.

Conversely, for $k \in G^\sigma$ one gets $k \cdot p = \sigma_p(k \cdot p)$. But in a ball-shaped normal neighborhood V of p, p is the only fixed point of σ_p, so $k \cdot p \in V \Rightarrow k \in H$ holds. Thus there exists a neighborhood U of e_G in G with $G^\sigma \cap U = H \cap U$. Since by Theorem 1.6.19 G_0^σ, H_0 are generated by elements of this neighborhood, it follows that $G_0^\sigma = H_0$.

Additional statement in brackets: If $B < 0$, by Corollary 7.1.11 σ is automatically an isometry, because $T_e\sigma$ is a Lie algebra isomorphism, i.e., an isometry of the Killing form on the simple components of $\mathfrak{g}$ (since B is uniquely determined by the Lie algebra structure).

$(2) \Rightarrow (3)$: Let $\sigma_{aH} := L_a \circ \sigma \circ L_{a^{-1}}$.

$(3) \Rightarrow (4)$: follows by differentiating with $A := T_e\sigma$. By Theorem 6.7.13 the extension of the metric to G is H-biinvariant with $\mathfrak{h}^\perp = \mathfrak{m}$. So $\text{ad}_{|\mathfrak{h}}$ is skew-symmetric and, because $A_{|\mathfrak{h}} = \text{id}$ and $A_{|\mathfrak{m}} = -\text{id}$, A is an isometry.

Additional statement in brackets: If the metric on G/H is normal homogeneous, ad_X is skew-symmetric for all $X \in \mathfrak{g}$ with respect to the metric on $\mathfrak{g}$. Because $\mathfrak{g}^A = \mathfrak{h}$, it follows that $A_{|\mathfrak{h}} = \text{id}$, $A_{|\mathfrak{m}} = -\text{id}$ and that A is an isometry.

$(4) \Rightarrow (5)$: Set $\mathfrak{m} = \text{Eig}_A(-1)$. If $X, Y \in \mathfrak{m}, V \in \mathfrak{h}$, then

$$A([X,V]) = [AX, AV] = -[X,V], \quad \text{thus} \quad [\mathfrak{m}, \mathfrak{h}] \subset \mathfrak{m},$$
$$A([X,Y]) = [AX, AY] = [X,Y], \quad \text{thus} \quad [\mathfrak{m}, \mathfrak{m}] \subset \mathfrak{h}.$$

The A-invariance of the scalar product implies $\mathfrak{m} = \mathfrak{h}^\perp$.

Additional statement in brackets: In the case of the Killing form,

$$B(X,V) = \text{Tr}\,[X, [V, \cdot]] = 0$$

already follows from the first half because $[X, [V, \cdot]] : \begin{smallmatrix} \mathfrak{h} \to \mathfrak{m} \\ \mathfrak{m} \to \mathfrak{h} \end{smallmatrix}$.

$(5) \Rightarrow (4)$: Set $A_{|\mathfrak{h}} = \text{id}_\mathfrak{h}, A_{|\mathfrak{m}} = -\text{id}_\mathfrak{m}$. Because of the commutator relations in (5), A is a Lie algebra homomorphism.

$(5) \Rightarrow (2)$ for simply connected G: By Theorem 7.4.2 there is an involution $\sigma : G \to G$ such that $T_e\sigma = A$. Because $A_{|\mathfrak{h}} = \text{id}_\mathfrak{h}$, for H connected $\sigma_{|H} = \text{id}_H$ holds by Theorem 1.6.19. $\square$

Remark The last step follows (locally) also from the Baker–Campbell–Hausdorff formula: For $\sigma(e^X) := e^{AX}$ one gets

$$\sigma(e^X e^Y) = \sigma(e^{X+Y+[X,Y]/2+\cdots}) = e^{AX+AY+A[X,Y]/2+\cdots}$$
$$= e^{AX+AY+[AX,AY]/2+\cdots} = \sigma(e^X)\sigma(e^Y).$$

Theorem 7.4.4 *Let $M = G/H$ be a symmetric space and $\mathfrak{g} = \mathfrak{m} \oplus \mathfrak{h}$ as in Theorem 7.4.3(5). Then M is naturally reductive, in particular $U \equiv 0$. The O'Neill tensors satisfy $T \equiv 0$ and for all $X, Y, Z \in \mathfrak{m}$ we have $A_X Y = \frac{1}{2}[X, Y]$,*

$$\Omega^{G/H}(X, Y)Z = -[[X, Y], Z] \qquad and \qquad \mathrm{Ric}(X, Y) = -\frac{1}{2}B(X, Y).$$

The formula for Ricci curvature is a bit more subtle than it looks at first glance, because the corresponding trace is taken over $\mathfrak{m}$, whereas for B it is taken over $\mathfrak{g}$.

Proof. Because $[X, Y]^{\mathfrak{m}} = 0$, M is naturally reductive, so $U \equiv 0$. By Theorem 6.8.6 it follows that $A_X Y = \frac{1}{2}[X, Y]^{\mathfrak{h}} = \frac{1}{2}[X, Y]$ and Theorem 6.8.17 implies for $X, Y \in \mathfrak{m}$ orthonormal that

$$K^{G/H}(X \wedge Y) = \frac{1}{4}\|[X, Y]^{\mathfrak{m}}\|^2 + g([[X, Y]^{\mathfrak{h}}, X], Y)$$
$$= g([[X, Y], X], Y).$$

As in Lemma 7.1.6, the formula for Ω follows because of the uniqueness of Ω to given K. Thus it follows further that

$$\mathrm{Ric}(X, X) = -\mathrm{Tr}\, \mathrm{ad}_X \mathrm{ad}_{X|\mathfrak{m}} \overset{(7.1)}{=} -\mathrm{Tr}\, \mathrm{ad}_{X|\mathfrak{h}} \mathrm{ad}_{X|\mathfrak{m}}$$
$$= -\mathrm{Tr}\, \mathrm{ad}_{X|\mathfrak{m}} \mathrm{ad}_{X|\mathfrak{h}} \overset{(7.1)}{=} -\mathrm{Tr}\, \mathrm{ad}_X \mathrm{ad}_{X|\mathfrak{h}},$$

and hence $\mathrm{Ric} = -\frac{1}{2}B_{|\mathfrak{m} \times \mathfrak{m}}$ by the polarization identity.   □

Remark At first sight this formula seems to contradict the one from Lemma 7.1.6 for the symmetric space G equipped with a biinvariant metric (by a factor $\frac{1}{4}$). But there the point reflection at e is equal to $\sigma_e : G \mapsto G, a \mapsto a^{-1}$ and for G non-abelian it is not a group homomorphism. Since G acts on the left and right by isometries, the isometry group of G actually contains $G \times G$. Condition (5) from Theorem 7.4.3 is not satisfied for $\mathfrak{m} := \mathfrak{g}, \mathfrak{h} := \mathfrak{e} = 0$, but is satisfied for $\mathfrak{h}' := \{[(X, X)] \mid X \in \mathfrak{g}\}, \mathfrak{m}' := \{[(-X, X)] \mid X \in \mathfrak{g}\}$. The representation from Theorem 7.4.3 in this case is the isometry $\varphi : G \times G/G \cong G, [(a, b)] \mapsto ab^{-1}$ with G embedded diagonally. Then $\sigma : G \times G \to G \times G, (a, b) \mapsto (b, a)$ is an automorphism. This leads to the factor, because due to $T_e \varphi[(X, -X)] = 2X$ it follows that

$$-[[T_e\varphi^{-1}X, T_e\varphi^{-1}Y], T_e\varphi^{-1}Z] = T_e\varphi^{-1}(-\frac{1}{4}[[X, Y], Z]).$$

Definition 7.4.5 *An isometry φ of a Riemannian manifold M is called a **shearing** (or **transvection**) along a geodesic c if*

1) there exists a t_0 such that $\varphi(c(t)) = c(t + t_0)$,
2) $T\varphi_{|c}$ is the parallel transport along c, denoted by $T_{c(t)}\varphi = \|_{c(t)}^{c(t+t_0)}$.

Lemma 7.4.6 *If M is symmetric and c a geodesic then $\varphi_{t_0} := \sigma_{c(t_0/2)}\sigma_{c(0)}$ is a shearing that translates c by t_0. If $c(0) = bH$ and $c(t_0) = aH$, then $\varphi_{t_0} = L_{ab^{-1}}$.*

Proof. $\sigma_{c(s)}(c(t)) = c(2s - t)$, hence $\sigma_{c(t_0/2)}\sigma_{c(0)}(c(t)) = c(t + t_0)$, thus φ_{t_0} acts on c as a translation by t_0.

Since $\sigma_{c(s)}$ is an isometry, $T\sigma_{c(s)}$ maps every parallel vector field X along c to a parallel vector field along $\sigma_{c(s)}(c) = c$. Therefore, because $T_{c(s)}\sigma_{c(s)}(X) = -X$, $T\sigma_{c(s)}(X) = -X$ and $T\varphi_{t_0}(X) = X$ hold everywhere.

Now (after a left translation by b) let $c(t) = e^{tX}H$ with $X \in \mathfrak{m}$ and σ as in Theorem 7.4.3. Then

$$\varphi_{t_0} = \sigma_{e^{t_0 X/2}H}\sigma_{eH} = L_{e^{t_0 X/2}} \circ \sigma \circ L_{e^{-t_0 X/2}} \circ \sigma = L_{e^{t_0 X}} \circ \sigma \circ \sigma = L_{e^{t_0 X}}. \qquad \square$$

Corollary 7.4.7 *The parallel transport along geodesics on a symmetric space is induced by the point reflections. The Levi-Civita connection (and thus Ω) depends only on the point reflections σ_p, it is independent of the choice of a suitable symmetric metric on G/H.*

Proof. The parallel transport along geodesics determines ∇ uniquely: Given $X \in T_pM$, let c be the geodesic with $\dot{c}(0) = X$ and let $(e_j)_j$ be a basis of parallel vector fields along c. Then every vector field Y around p on c is a linear combination $Y_{|c(t)} = \sum_j f_j(t)e_j(t)$ and $\nabla_X Y = X.f_j \cdot e_j$. $\qquad \square$

This implies once more in a different way that self-intersecting geodesics are periodic (Lemma 6.8.8).

Lemma 7.4.8 *A homogeneous manifold $M = G/H$ is locally symmetric if and only if every G-invariant tensor ω is parallel (i.e. if $\nabla\omega = 0$).*

In particular, in this case every G-invariant differential form is closed, since $d\omega$ can be expressed in terms of $\nabla\omega$. Thus, by Exercise 6.8.26, for G compact the G-invariant forms represent the cohomology classes.

Proof. "$\Leftarrow$" follows from Theorem 7.2.5 with $\omega = R$.
"$\Rightarrow$" Let $c(t) = e^{tX}H$. Because $(L_{c(t)})^*\omega = \omega$, ω is parallel along c according to Lemma 7.4.6, hence $\nabla_{\dot{c}(0)}\omega = 0$. $\qquad \square$

Lemma 7.4.9 *Let c be a geodesic through $eH \in M$ parametrized by arc length, let $-\lambda_j \in \mathbb{R}$ be the eigenvalues of the symmetric endomorphism $\Omega_{eH}(\dot{c}, \cdot)\dot{c} = \mathrm{ad}_{\dot{c}}\mathrm{ad}_{\dot{c}} \in \mathrm{End}(\mathfrak{m})$ and let $(V_j)_j$ be an orthonormal basis of eigenvectors (extended in parallel along c), thus $K(\dot{c} \wedge V_j) = \lambda_j$. Then the Jacobi fields Y along c are linear combinations of Jacobi fields of the form*

$$Y_t = V_j \cdot \begin{cases} \cos\sqrt{\lambda_j}t, \ \sin\sqrt{\lambda_j}t & \lambda_j > 0 \\ 1, \ t & \text{if } \lambda_j = 0 \ . \\ \cosh\sqrt{-\lambda_j}t, \ \sinh\sqrt{-\lambda_j}t & \lambda_j < 0 \end{cases} \qquad (7.2)$$

Proof. Since Ω is parallel, each V_j is an eigenvector along all of c. Therefore the Jacobi ODE for $f \in C^\infty(\mathbb{R})$, $Y = fV_j$ says that

$$(\nabla_{\partial/\partial t})^2(f(t)V_j) = \Omega_{c(t)}(\dot{c}, Y_t)\dot{c} = -\lambda_j f(t)V_j$$

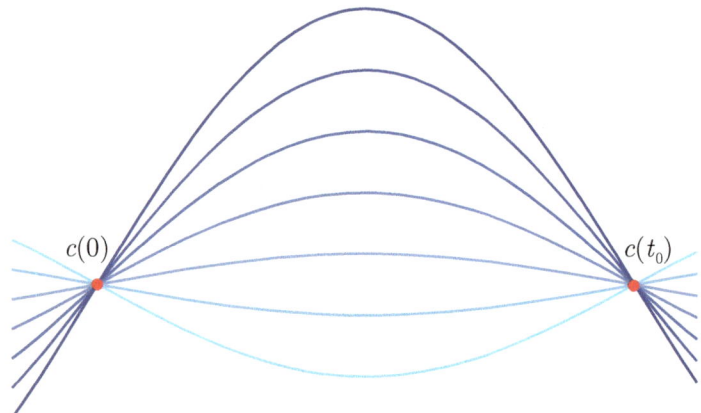

Fig. 7.7 Geodesic variation on a symmetric space.

or $f''(t) = -\lambda_j f(t)$. Since the vector fields of this form span a $2n$-dimensional space, each Jacobi field has this form. □

Lemma 7.4.10 *Let c be a geodesic parameterized by arc length on a symmetric space and let Y be a Jacobi field as in equation (7.2) with zeros at $0, t_0$. Then there is a variation c^s of geodesics parameterized by arc length such that $c^s(0) = c(0), c^s(t_0) = c(t_0)$ (Fig. 7.7).*

On general spaces, there are only variations where $c^s(t_0)$ and the intersections with c are arbitrarily close to the point $c(t_0)$.

Proof. Without restriction, if t_0 is the first zero after 0 and $c(t_0/2) = eH$, then $Y = \sin\frac{\pi t}{t_0} \cdot V$ is parallel to V. Thus $\nabla Y_{t_0/2} = 0$ holds. Let $\widetilde{X}, \widetilde{Y}$ be the Killing fields associated to $X, Y_{t_0/2} \in \mathfrak{m}$. Then $(\nabla_{\widetilde{X}}\widetilde{Y})_{|eH} = T\pi(-\frac{1}{2}[X, Y_{t_0/2}]) = 0$ holds, hence $\nabla\widetilde{Y}_{|eH} = 0$. By Exercise 5.3.14, $\widetilde{Y}_{|c}$ is a Jacobi field, and since Jacobi fields are uniquely determined by value and first derivative at a point, it follows that $Y_t = \widetilde{Y}_{c(t)}$. So $\widetilde{Y}$ vanishes at $c(0), c(t_0)$, and correspondingly the isometries $L_{e^{sY_{t_0/2}}}$ have fixed points there. Thus $c^s := L_{e^{sY_{t_0/2}}} c$ is the variation we are looking for. □

The **diameter** of a Riemannian manifold M is defined as

$$\operatorname{diam} M := \sup_{p,q \in M} \operatorname{dist}(p,q) \in \mathbb{R}^+ \cup \{\infty\}.$$

Theorem 7.4.11 *Consider a symmetric space* $M = G/H$ *with* $\mathrm{Ric} > 0$ *(i.e. if* $B < 0$*). Then the diameter satisfies*

$$
\mathrm{diam}\, M \quad \le \quad \frac{\pi}{\min\limits_{\substack{X \in \mathfrak{m} \\ \|X\|=1}} \max\limits_{\substack{Y \in X^\perp \subset \mathfrak{m} \\ \|Y\|=1}} \sqrt{K(X \wedge Y)}}
$$

$$
\underset{B<0,\ \text{normal}}{=} \quad \frac{\pi}{\min\limits_{\substack{X \in \mathfrak{m} \\ \|X\|=1}} \max\limits_{\substack{Y \in X^\perp \subset \mathfrak{m} \\ \|Y\|=1}} \|[X, Y]\|},
$$

M is compact and $\pi_1(M)$ *is finite.*

Proof. Consider $X \in \mathfrak{m}$ with $\|X\| = 1$. By Lemma 3.4.7 one gets $\mathrm{Ric}(X, X) = \sum_j K(X \wedge e_j) \cdot \|X \wedge e_j\|^2$. Hence it follows that $\lambda := \sup_Y K(X \wedge Y) > 0$. According to Lemmas 7.4.9 and 7.4.10 the geodesic $\pi(e^{tX})$ intersects another of the same length starting at eH at $\frac{\pi}{\sqrt{\lambda}}$ at the latest. So it is no longer a shortest path after that point according to Corollary 5.2.12. Because of the completeness it follows that $\mathrm{diam}\, M \le \max_X \frac{\pi}{\sqrt{\lambda}}$. The same is true for the lifts of the geodesic variation to the universal covering $\widetilde{M}$, which is thus also compact and $\#\pi_1(M) = \frac{\mathrm{vol}\, \widetilde{M}}{\mathrm{vol}\, M} < \infty$. □

Remark Similarly, for any ball B_r on which $\exp_p$ is injective, $r \le \frac{\pi}{\max \sqrt{K}}$ holds. On the other hand, one can show that $\exp_p$ is injective on $\overline{B}_{\frac{\pi}{2\max \sqrt{K}}}$ ([Kl2, Th. 2.2.26]).

Remark For general complete Riemannian manifolds, Myers's Theorem gives similar results under the condition that there exists an $a \in \mathbb{R}^+$ such that $\mathrm{Ric} - ag \ge 0$ with similar proof.

Example 1) Sphere S^n for the standard round metric: $K \equiv 1, \mathrm{diam}\, S^n = \pi = \frac{\pi}{\sqrt{1}}$, hence $\pi_1(S^n) = 0$,

2) Real projective space $\mathbb{P}^n\mathbb{R}$ with the metric induced by S^n: $\mathrm{diam}\, \mathbb{P}^n\mathbb{R} = \pi/2 < \frac{\pi}{\sqrt{1}}$ and by example (1) $\pi_1(\mathbb{P}^n\mathbb{R}) = \mathbb{Z}/2\mathbb{Z}$,

3) Complex projective space $\mathbb{P}^n\mathbb{C}$ with the metric induced by S^n: For any $X \ne 0$ one obtains $K(X \wedge T_pM) = [1, 4]$, and from Exercise 6.3.13 we know that $\mathrm{diam}\, \mathbb{P}^n\mathbb{C} = \pi/2 = \frac{\pi}{\sqrt{4}}$, hence $\pi_1(\mathbb{P}^n\mathbb{C}) = 0$.

As an application, the Lie groups with biinvariant metric can be described in more detail.

Corollary 7.4.12 (Weyl) *A Lie group* G *is compact and semisimple if and only if* $B < 0$.

Proof. "$\Rightarrow$": If G is compact, it carries a biinvariant metric by Corollary 7.1.2. So because $\mathfrak{z} = 0$, by Theorem 7.1.9 one finds $B < 0$.

"$\Leftarrow$": By Theorem 7.4.4, $\mathrm{Ric} > 0$ for the biinvariant metric $-B$ on G. Thus the assertion follows from Theorem 7.4.11. □

Theorem 7.4.13 *If* G *is a Lie group with* $n := \dim G$, *the following are equivalent:*

1) $\mathfrak{g}$ has a Lie subalgebra $\mathfrak{h}$ such that $\mathfrak{g} = \mathfrak{z} \oplus \mathfrak{h}$ and $B_{|\mathfrak{h} \times \mathfrak{h}} = B^{\mathfrak{h}} < 0$,

2) there exists a biinvariant metric on G,

3) there is a short exact sequence $0 \to Z(G) \overset{\text{can.}}{\to} G \overset{\text{Ad}}{\to} H \to 0$ with the center $Z(G)$ and a compact subgroup $H \subset SO(n)$ satisfying $\text{Ric}^H > 0$.

Proof. (1)$\Rightarrow$(2): Choose any Euclidian scalar product $\langle \cdot, \cdot \rangle$ on $\mathfrak{z}$. Then for $X, Y \in \mathfrak{z}, W \in \mathfrak{g}$ one finds

$$\underbrace{\langle \text{ad}_W X, Y \rangle}_{=0} = -\langle X, \text{ad}_W Y \rangle.$$

Therefore $g := \langle \cdot, \cdot \rangle \oplus (-B)$ is an ad-invariant scalar product and induces a biinvariant metric on G.

(2)$\Rightarrow$(3): Set $H := \text{im Ad} \subset \text{Aut}(\mathfrak{g})$. If g is biinvariant, then $H \subset O(\mathfrak{g})$. As an image of the connected set G under a continuous map, H is connected, so $H \subset SO(\mathfrak{g}) \cong SO(n)$. As an image of a Lie group homomorphism, H is a subgroup of $SO(n)$. Consider $h \in \ker \text{Ad}$. and $a \in G$. By Corollary 7.1.4 there exists an $A \in \mathfrak{g}$ such that $a = e^A$, hence

$$hah^{-1} = C_h e^A \overset{\text{Th. 1.6.17}}{=} e^{\text{Ad}_h A} = e^A = a.$$

Therefore $h \in Z(G)$ and thus $\ker \text{Ad} \subset Z(G)$. On the other hand, for any $h \in Z(G)$ differentiating $C_h a = a$ yields $\text{Ad}_h = \text{id}$, hence in combination one gets $\ker \text{Ad} = Z(G)$ and $0 \to Z(G) \to G \to H \to 0$ is exact.

Consequently, the center of H is trivial. Since H carries a biinvariant metric, it is compact by Theorem 7.1.9 and Corollary 7.4.12, hence it is a Lie subgroup.

Let $\tilde{\mathfrak{h}} \subset \mathfrak{g}$ be the subspace on which $B < 0$ holds. Because $B(X, X) < 0 \Leftrightarrow \text{ad}_X \neq 0$ (proof of Lemma 6.8.11), one obtains $\tilde{\mathfrak{h}} \cong \text{im ad} = \mathfrak{h}$, thus $\mathfrak{g} = \mathfrak{z} \oplus \mathfrak{h}$ and $\mathfrak{h}$ has a negative definite Killing form. Therefore H has positive Ricci curvature by 7.1.9.

(3)$\Rightarrow$(1): Differentiating the sequence implies $0 \to \mathfrak{z} \to \mathfrak{g} \overset{\text{ad}}{\to} \mathfrak{h} \to 0$. Set $\tilde{\mathfrak{h}} := B_{<0} = \{X \in \mathfrak{g} \mid B(X, X) < 0\} \cup \{0\}$. Then $\tilde{\mathfrak{h}} \cap \mathfrak{z} = \{0\}$ holds and hence $\dim \tilde{\mathfrak{h}} \leq \dim \mathfrak{h}$. On the other hand, it follows from $\text{ad}_X \in \mathfrak{h} \setminus \{0\}$ and $\mathfrak{h} \subset \mathfrak{so}(n)$ that the eigenvalues of ad_X are in $i\mathbb{R}$ and at least one is $\neq 0$, so $B(X, X) < 0$ as before. Thus $\text{ad}_{|\tilde{\mathfrak{h}}}$ is surjective and in combination one finds $\tilde{\mathfrak{h}} \cong \mathfrak{h}$, since ad is a Lie algebra homomorphism. $\square$

Remark The condition $B \leq 0$ is not equivalent to the statements above, as the Heisenberg group is a counterexample (see Exercise 7.4.22). But (1) in the theorem above shows that a Lie group can carry a biinvariant metric if and only if $B \leq 0$ and $\mathfrak{g}$ is reductive (in the sense of Lie algebras).

Corollary 7.4.14 *The statements in Theorem 7.1.9 are equivalent to the existence of a short exact sequence $0 \to Z(G) \overset{\text{can.}}{\to} G \overset{\text{Ad}}{\to} H \to 0$ with $H \subset SO(n)$, $\text{Ric}^H > 0$ and finite (abelian) center $Z(G)$.*

Proof. $T_e Z(G) = \mathfrak{z}$, thus here $\dim Z(G) = \dim \mathfrak{z} = 0$ holds. According to Corollary 7.4.12 G is compact, hence $Z(G)$ is finite. $\square$

Exercises

Exercise 7.4.15 Let M, N be Riemannian manifolds. Prove that $M \times N$ is symmetric if and only if both M and N are symmetric.

Exercise 7.4.16 Let $\pi : \widetilde{SL}(2, \mathbb{R}) \to SL(2, \mathbb{R})$ be the universal covering of $SL(2, \mathbb{R})$ and let $\rho : \widetilde{SL}(2, \mathbb{R}) \to GL(n, \mathbb{R})$ be an embedding as a matrix group.

1) Using Exercise 6.8.25(4), show that the fundamental group of $SL(2, \mathbb{R})$ is isomorphic to $\mathbb{Z}$.
2) Prove analogously that $SL(2, \mathbb{C})$ is simply connected (remember that S^3 is simply connected).
3) Show that $T_e\rho : \mathfrak{sl}(2, \mathbb{R}) \to \mathfrak{gl}(n, \mathbb{R})$ via $T_e\rho_{\mathbb{C}}(X + iY) := T_e\rho(X) + iT_e\rho(Y)$ induces a Lie algebra homomorphism $T_e\rho_{\mathbb{C}} : \mathfrak{sl}(2, \mathbb{C}) \to \mathfrak{gl}(n, \mathbb{C})$.
4) Deduce the existence of a Lie group homomorphism $\varphi : SL(2, \mathbb{C}) \to GL(n, \mathbb{C})$ for which the following diagram commutes:

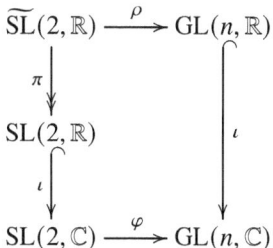

5) Deduce a contradiction from (4).

Exercise 7.4.17 1) Find a transitive operation of $GL^+(n, \mathbb{R})$ on the Euclidean scalar products of an n-dimensional vector space V. Identify the space of these scalar products with a symmetric space $M = GL^+(n, \mathbb{R})/H$.
2) A **real structure** on a complex-n-dimensional complex vector space V is an involution $J \in \mathrm{End}_{\mathbb{R}}(V)$ satisfying $J(\lambda v) = \bar{\lambda}J(v)$ for all $\lambda \in \mathbb{C}, v \in V$. Show that $U(n)$ acts transitively on the set N of these real structures via $J \mapsto A \circ J \circ \overline{A}^t$. Identify N with a symmetric space.
3) Compare the curvatures of M and N (for the standard L^2-metrics on the matrices) using the orthogonal complements of the Lie algebras $\mathfrak{h}$ of the isotropy groups.

Exercise 7.4.18 Find a metric on $\mathfrak{m} \subset \mathfrak{so}(n, 1)$ for which the homogeneous space $M := SO_0(1, n)/SO(n)$ becomes isometric to the hyperbolic space. Determine the geodesics through $[e]$ in the case $n = 2$.

Exercise 7.4.19 Let G/H be a symmetric space with involution $\sigma : G \to G$, let $G' \subset G$ be a σ-invariant closed subgroup, and $H' := G' \cap H$. Show that $G'/H' \subset G/H$ is a symmetric space and a totally geodesic submanifold. *Remark: In fact, every totally geodesic submanifold of G/H is of this type [KoN, XI Th. 4.2].*

Exercise* 7.4.20 Consider the **Graßmann manifolds**

$$SO(p + q)/SO(p) \times SO(q)$$

for $p, q \in \mathbb{N}^+$ equipped with the metric induced by the canonical metric on $SO(p+q)$. Show that these spaces are symmetric (cf. Exercise 6.8.21).

Exercise 7.4.21 Analogously to Exercise 7.4.20, show that $SU(p + q)/SU(p) \times SU(q), Sp(p + q)/Sp(p) \times Sp(q)$ are symmetric spaces for the canonical metric.

Exercise 7.4.22 Prove that the Killing form of the Heisenberg group H (Exercise 1.6.28) vanishes, and show that H cannot carry a biinvariant metric.

7.5 Compact Type and Non-Compact Type

compact type	dual symmetric spaces
non-compact type	complexification
compact Lie algebra	

According to Lemma 6.8.12, $B_{|\mathfrak{h} \times \mathfrak{h}} < 0$ holds for any almost faithful operation of G on G/H. Under the additional assumption that Ric (i.e. $-B_{|\mathfrak{m}}/2$) is positive or negative definite, two types of symmetric spaces arise which will turn out to be dual to each other.

Definition 7.5.1 *A symmetric space $M = G/H$ as in Theorem 7.4.3 with G compact semisimple (i.e. $B < 0$) is called **of compact type**. A symmetric space with $B_{|\mathfrak{h}} < 0, B_{|\mathfrak{m}} > 0$ is called **of non-compact type**.*

Such spaces exist for given Lie algebras with the corresponding properties:

Theorem 7.5.2 *Let $\mathfrak{g}$ be a semisimple Lie algebra with a decomposition $\mathfrak{g} = \mathfrak{h} \oplus \mathfrak{m}$ as in Theorem 7.4.3 such that $B < 0$. Then there is a Lie group G with Lie algebra $\mathfrak{g}$ having a compact Lie subgroup H with Lie algebra $\mathfrak{h}$. In this case G/H is compact and symmetric for the metric induced by $-B$.*

Proof. For ad $: \mathfrak{g} \to \text{End}(\mathfrak{g}) = \mathfrak{gl}(\mathfrak{g})$, ker ad $= \mathfrak{z}$ is equal to 0, since $\mathfrak{g}$ is semisimple. Lemma 7.4.1 applied to ad yields a subgroup $\widetilde{G} \subset GL(\mathfrak{g})$. Let G be the universal covering of $\widetilde{G}$. By Corollary 7.4.12, G is compact, hence so is the connected component $H := G_0^\sigma$ of the fixed point set. Thus, the compact space G/H is symmetric by Theorem 7.4.3. $\qquad\square$

Remark The subgroup H is not necessarily semisimple and may also have non-compact coverings with the same Lie algebra $\mathfrak{h}$, for instance in the case $SO(3)/SO(2) \cong S^2$.

Theorem 7.5.3 *Let* $\mathfrak{g}^*$ *be a semisimple Lie algebra with a decomposition* $\mathfrak{g}^* = \mathfrak{h} \oplus \mathfrak{m}$ *as in Theorem 7.4.3 such that* $B^*_{|\mathfrak{h} \times \mathfrak{h}} < 0$, $B^*_{|\mathfrak{m} \times \mathfrak{m}} > 0$ *for the Killing form* B^* *of* $\mathfrak{g}^*$. *Then there is a Lie group* G^* *with Lie algebra* $\mathfrak{g}^*$ *having a Lie subgroup* H^* *with Lie algebra* $\mathfrak{h}$. *In this case* G^*/H^* *is non-compact and symmetric for the metric induced by* B^*.

Proof. Construct G^* simply connected, H^* connected as in Theorem 7.5.2. As in the step "(1)⇒(2)" of the proof of Theorem 7.1.9, ad_X is skew-symmetric with respect to the Killing form for any $X \in \mathfrak{g}$, and as in the proof of Theorem 7.4.3, $B^*(\mathfrak{h}, \mathfrak{m}) = 0$ holds. By Theorem 7.4.3, G^*/H^* is symmetric. □

Lemma 7.5.4 *Consider a symmetric space* G/H *of compact type. On* $\mathfrak{g} = \mathfrak{h} \oplus \mathfrak{m}$, *define*

$$[\![X, Y]\!] := -[X, Y], \quad [\![X, V]\!] := [X, V], \quad [\![V, W]\!] := [V, W]$$

for any $V, W \in \mathfrak{h}, X, Y \in \mathfrak{m}$. *Then* $\mathfrak{g}^* := (\mathfrak{g}, [\![\cdot, \cdot]\!])$ *is a Lie algebra satisfying* $B^*_{|\mathfrak{h} \times \mathfrak{h}} < 0$, $B^*_{|\mathfrak{m} \times \mathfrak{m}} > 0$. *Conversely, a Lie algebra of non-compact type becomes one with* $B < 0$ *by this transformation.*

Remark Alternatively, one can define $\mathfrak{g}^* := \mathfrak{h} \oplus i\mathfrak{m} \subset \mathfrak{g} \otimes \mathbb{C}$.

Proof. To check the Lie algebra structure one has to verify the Jacobi identity. If $X, Y \in \mathfrak{m}, V \in \mathfrak{h}$, then

$$[\![[\![X, Y]\!], V]\!] + [\![[\![Y, V]\!], X]\!] + [\![[\![V, X]\!], Y]\!]$$
$$= [-[X, Y], V] - [[Y, V], X] - [[V, X], Y] = 0.$$

If at least two of the three vectors lie in $\mathfrak{h}$, no sign differences occur. If all three vectors are in $\mathfrak{m}$, each summand gets a sign change. Furthermore $[\![X, [\![X, Y]\!]]\!] = -[X, [X, Y]]$ and $[\![X, [\![X, V]\!]]\!] = -[X, [X, V]]$ hold. Therefore $B^*_{|\mathfrak{m}} > 0$, while $B^*_{|\mathfrak{h}}$ does not change. □

Remark Thus, given a symmetric space G^*/H^* of noncompact type, there always exists a compact Lie group H for the Lie algebra $\mathfrak{h}$ of H^*. Such Lie algebras are called **compact**.

Definition 7.5.5 *Spaces* G/H, G^*/H^* *of compact and non-compact type, respectively, whose Lie algebra decompositions arise from each other by the transformation in Lemma 7.5.4 are called* **dual to each other**.

Example Consider $M = (G \times G)/G$ with G compact semisimple. Its dual Lie algebra corresponds to $\mathfrak{g} \oplus i\mathfrak{g} = \mathfrak{g} \otimes \mathbb{C}$ with the complex conjugation as involution. The associated simply connected Lie group $G_{\mathbb{C}}$ is called the **complexification** of G, and the symmetric space dual to G is $G_{\mathbb{C}}/G$.

Lemma 7.5.6 *Given dual symmetric spaces* $M = G/H, M^* = G^*/H^*$ *with metric on* $\mathfrak{m}$ *proportional to* $B_{|\mathfrak{m} \times \mathfrak{m}} = -B^*_{|\mathfrak{m} \times \mathfrak{m}}$ *and* $X, Y \in \mathfrak{m}$ *orthonormal one obtains*

$$K(X \wedge Y) = \|[X, Y]\|^2 \quad and \quad K^*(X \wedge Y) = -\|[X, Y]\|^2.$$

Example This makes S^n dual to H^n.

Remark A symmetric space need not be normal homogeneous, the metric above is just particularly convenient to handle. In general, however, there is a decomposition of the scalar product on $\mathfrak{m}$ into components proportional to B which is compatible with the Lie algebra structure, cf. [KoN, ch. XI p. 257], [Kl2, Lemma 2.2.23, 2.2.24]. This way it can be shown that $K \geq 0$ and $K \leq 0$ always hold for the compact and non-compact types, respectively.

Proof. According to Theorem 7.4.4 one finds

$$K(X \wedge Y) = g([[X,Y],X],Y) \overset{\underset{\text{symmetric}}{\text{ad}_X \text{ skew-}}}{=} g([X,Y],[X,Y]).$$

The same way one gets

$$K^*(X \wedge Y) = g([[[X,Y]],X]],Y) = g([-[X,Y],X],Y) = -g([X,Y],[X,Y]).$$
$\square$

A symmetric space of compact type need not be simply connected, say $\mathbb{P}^n\mathbb{R} = SO(n)/S(O(1) \times O(n-1))$. However, by Theorem 7.4.11, the fundamental group is always finite. If G is simply connected with $B < 0$, then one can show that H must be connected ([ChEb, Th. 5.13]) and hence M is simply connected. For the non-compact type the situation is simpler:

Lemma 7.5.7 *Every symmetric space G/H of noncompact type is diffeomorphic to an $\mathbb{R}^n$, and exactly one geodesic passes through two points. Moreover, G is diffeomorphic to $H \times \mathbb{R}^n$ and thus H is connected.*

Proof. This follows by Theorem 6.8.9 from the Hadamard–Cartan Theorem. The last part follows by Theorem 6.8.18.
$\square$

Exercises

Exercise* 7.5.8 (cf. Exercise 6.6.6) Let $\mathfrak{g}$ be semisimple and $\mathfrak{h} \subset \mathfrak{g}$ be an ideal. Prove **Cartan's Criterion for Semisimplicity**:

1) In this setting $\mathfrak{h}^\perp$ (with respect to the Killing form) is also an ideal, $\mathfrak{h}$ and $\mathfrak{h}^\perp$ are semisimple and $\mathfrak{g} = \mathfrak{h} \oplus \mathfrak{h}^\perp$.
2) Every semisimple Lie algebra is a direct sum of simple Lie algebras.
3) Conversely, deduce that the Killing form of a direct sum of simple Lie algebras is non-degenerate.

Exercise* 7.5.9 Prove that $G := SL(n)$ together with the involution $\sigma : G \to G, A \mapsto (A^{-1})^t$ yields a symmetric space $SL(n)/SO(n)$ of non-compact type. Find a symmetric space of compact type dual to it.

Exercise 7.5.10 Determine a non-compact dual to the Graßmann manifolds from Exercise 7.4.20.

Exercise 7.5.11 Consider a Lie algebra $\mathfrak{g}$ with negative definite Killing form and a Lie group $G_{\mathbb{C}}$ with Lie algebra $\mathfrak{g} \otimes \mathbb{C}$. Show that $G_{\mathbb{C}}$ is diffeomorphic to $G \times \mathbb{R}^{\dim G}$.

7.6 Isotropy-Irreducible Spaces

isotropy-irreducible | Einstein manifold

In this section, homogeneous and symmetric spaces are further analyzed using another concept. For more extensive investigations in this direction see [Wolf1], [Wolf2], [Besse, Ch. 7], [WZ].

Lemma 7.6.1 *Let G be a Lie group, M a Riemannian manifold with isometry group G and $H_p \subset G$ the isotropy group of $p \in M$. If the operation of H_p on $T_p M$ is irreducible for every p, then M is a homogeneous space.*

Proof. Choose a $p \in M$ having an orbit of maximal dimension. If the orbit is $G \cdot p \cong G/H_p$, then H_p acts on $T_p(G \cdot p)$, so $T_p(G \cdot p) = T_p M$ or $T_p(G \cdot p) = 0$. In the first case, G/H_p is a closed submanifold of M of the same dimension, hence $M = G/H_p$. In the second case, G is discrete since $G \cdot p$ has maximal dimension, therefore a $q \in M$ exists with $H_q = e$, and $T_q M$ is reducible for $\dim M > 1$. If $\dim M = 1$, then M is Lie group. $\qquad\square$

Definition 7.6.2 *A reductive space G/H is called **isotropy-irreducible** if 0 and $\mathfrak{m}$ are the only H-invariant subspaces of $\mathfrak{m}$. A Riemannian manifold is called an **Einstein manifold** if there exists a $\lambda \in \mathbb{R}$ such that $\mathrm{Ric} = \lambda g$.*

The latter corresponds to a vacuum solution with cosmological constant λ (resp. dark energy) of the field equation (Theorem 8.4.4) of the gravitational field in General Relativity, except that there one uses Lorentzian metrics.

Theorem 7.6.3 *The metric of an isotropy-irreducible space G/H is uniquely determined (up to multiples) and Einstein.*

Proof. Consider a homogeneous metric g on G/H. For any $h \in H$ the Ricci curvature Ad_h is invariant since $h \cdot eH = eH$ and hence $\mathrm{Ric}_{eH}(T_e L_h X, T_e L_h Y) = \mathrm{Ric}_{eH}(X, Y)$. Moreover, by Theorem 6.7.9, $T_e L_h X = \mathrm{Ad}_h X$ holds. In particular, ad_X is skew-symmetric with respect to Ric for any $X \in \mathfrak{h}$.

Now let g' be a bilinear form on $\mathfrak{m}$ with respect to which ad_X is skew-symmetric for all $X \in \mathfrak{h}$. As in the proof of Proposition 7.1.10, irreducibility implies $g' = \lambda' g$. So all homogeneous metrics are proportional on G/H, and G/H is Einstein. $\qquad\square$

In particular one finds $B_{|\mathfrak{m} \times \mathfrak{m}} = \lambda g_{\mathfrak{m}}$ for some $\lambda \in \mathbb{R}$.

Lemma 7.6.4 *Any symmetric space G/H (as in Theorem 7.4.3) with $\mathfrak{g}$ simple is isotropy-irreducible, and $\mathfrak{h} = [\mathfrak{m}, \mathfrak{m}]$ holds.*

Proof. Let $\mathfrak{m}' \subset \mathfrak{m}$ be $\mathrm{ad}_{\mathfrak{h}}$-invariant and $\mathfrak{m}'' := \mathfrak{m}'^{\perp B}$. Then

$$B([\mathfrak{m}', \mathfrak{m}''], [\mathfrak{m}', \mathfrak{m}'']) \subset B([\mathfrak{m}', \mathfrak{m}''], \mathfrak{h}) \subset B(\mathfrak{m}'', [\mathfrak{m}', \mathfrak{h}]) \subset B(\mathfrak{m}'', \mathfrak{m}') = 0.$$

Because $\mathfrak{g}$ is simple, B is non-degenerate, so $[\mathfrak{m}', \mathfrak{m}''] = 0$. Set $\mathfrak{a} := \mathfrak{m}' + [\mathfrak{m}', \mathfrak{m}']$. Then by the Jacobi identity $[\mathfrak{a}, \mathfrak{m}''] = 0$ and because of $\mathrm{ad}_{\mathfrak{h}}$-invariance, $[\mathfrak{a}, \mathfrak{h}] \subset \mathfrak{a}$ and $[\mathfrak{a}, \mathfrak{m}'] = [\mathfrak{m}', \mathfrak{m}'] \subset \mathfrak{a}$ hold. Therefore $\mathfrak{a}$ is an ideal. As $\mathfrak{g}$ is simple one obtains $\mathfrak{m}' = \mathfrak{m}$ because $[\mathfrak{m}', \mathfrak{m}'] \subset \mathfrak{h}$.   □

However, not every symmetric space can be represented by spaces of this form, for example $(SO(3) \times SO(3))/SO(3)$. To show the following lemma we need Cartan's Criterion for solvable Lie algebras (see [FH, Prop. C.4]): Given a Lie algebra $\mathfrak{m}$, its Killing form satisfies $B^{\mathfrak{m}}([\mathfrak{m}, \mathfrak{m}], \mathfrak{m}) = 0$ if and only if $\mathfrak{m}$ is solvable, i.e. if the sequence $\mathfrak{m} \supset [\mathfrak{m}, \mathfrak{m}] \supset [[\mathfrak{m}, \mathfrak{m}], [\mathfrak{m}, \mathfrak{m}]] \supset \cdots$ eventually arrives at 0.

Lemma 7.6.5 *Any isotropy-irreducible space G/H with non-semisimple Lie group G is flat.*

Proof. Without restriction assume that G acts faithfully (by Lemma 6.7.6). By Lemma 6.8.12 one gets $B_{|\mathfrak{h}} < 0$. Because G is non-semisimple, $B_{|\mathfrak{m}} = \lambda g_{\mathfrak{m}}$ implies that $\lambda = 0$.

Thus, in particular, $B([\mathfrak{m}, \mathfrak{m}], \mathfrak{m}) = 0$. On the other hand, $[\mathfrak{m}, \mathfrak{h}] \subset \mathfrak{m}$ and

$$B([\mathfrak{m}, \mathfrak{m}], \mathfrak{h}) = B(\mathfrak{m}, [\mathfrak{m}, \mathfrak{h}]) = 0,$$

hence $\mathfrak{m} \subset \mathfrak{g}$ is an ideal and therefore $B^{\mathfrak{m}} = B_{|\mathfrak{m} \times \mathfrak{m}}$. By Cartan's Criterion, $\mathfrak{m}$ is solvable. In particular, $\mathfrak{m}' := [\mathfrak{m}, \mathfrak{m}]$ is a proper subset of $\mathfrak{m}$. Because of the irreducibility of the $\mathfrak{h}$-operation, one obtains $\mathfrak{m}' = 0$. Hence $\mathfrak{m}$ is abelian, thus the curvature of G/H vanishes and G/H equals $\mathbb{R}^n \times T^m$.   □

Remark For a Lie algebra $\mathfrak{m}$, semisimplicity does not follow from $[\mathfrak{m}, \mathfrak{m}] = \mathfrak{m}$ (see Exercise 7.6.7), hence we cannot argue with it here.

One can show (essentially using de Rham's Decomposition Theorem [KoN, Ch. IV, Th. 6.2]) that simply connected symmetric spaces can be decomposed into a product of isotropy-irreducible spaces ([Wolf1, Th. 8.2.4, Th. 8.3.8]). Thus, because $B_{|\mathfrak{m}} = \lambda g_{\mathfrak{m}}$ on the factors, these are flat or of compact or non-compact type (cf. also [Hel, Prop. V.4.2, p. 244]).

Remark The isotropy group of $\mathbb{R}^n \times T^m$ is the product of the individual isotropy groups, thus the spaces in Lemma 7.6.5 have the form $M = \mathbb{R}^n$ or $M = T^n$.

Theorem 7.6.6 *Let G be semisimple and G/H be symmetric isotropy irreducible with faithful G-operation. Then either $\mathfrak{g}$ is simple or G/H is a simple compact Lie group.*

Proof. By Exercise 7.5.8, $\mathfrak{g}$ decomposes into simple ideals. By the proof of Lemma 6.7.6, $\mathfrak{h}$ and $\mathfrak{g}$ have no common ideals because of the faithful operation. Let $\mathfrak{g}_1, \mathfrak{g}_2 \subset \mathfrak{g}$ be nontrivial ideals such that $\mathfrak{g} = \mathfrak{g}_1 \oplus \mathfrak{g}_2$ and $\mathfrak{g}_1$ is simple. As a Lie algebraic automorphism, the reflection symmetry $A := T_e \sigma$ permutes the simple ideals. Suppose $A(\mathfrak{g}_1) = \mathfrak{g}_1$. Then $\mathfrak{m}_1 := \mathfrak{g}_1 \cap \mathfrak{m} \neq 0$ and it is $\mathrm{Ad}_{\mathfrak{h}}$-invariant, therefore isotropy irreducibility implies $\mathfrak{m} = \mathfrak{m}_1$ and thus $\mathfrak{g} = \mathfrak{g}_1$, hence $\mathfrak{g}$ is simple.

On the other hand, assume $A(\mathfrak{g}_1) \subset \mathfrak{g}_2$. Then $\{X - AX \mid X \in \mathfrak{g}_1\} \subset \mathfrak{m}$ is nontrivial and $\mathrm{Ad}_{\mathfrak{h}}$-invariant, so as before $\mathfrak{g} = \mathfrak{g}_1 \oplus \sigma(\mathfrak{g}_1)$. Thus

$$\mathfrak{m} = \mathrm{Eig}_A(-1) = \{X - AX \mid X \in \mathfrak{g}\} = \{X - AX \mid X \in \mathfrak{g}_1\} \cong \mathfrak{g}_1.$$

In the same way $\mathfrak{h} \cong \mathfrak{g}_1$ is embedded diagonally.   □

Exercises

Exercise 7.6.7 Show that $\mathfrak{m} := \mathfrak{sl}(2) \times \mathbb{R}^2$ with the Lie bracket

$$[(A, v), (B, w)] := ([A, B], Aw - Bv)$$

becomes a Lie algebra which satisfies $[\mathfrak{m}, \mathfrak{m}] = \mathfrak{m}$, but which is not semisimple.

Chapter 8
General Relativity

One of the most interesting applications of Riemannian geometry outside mathematics is General Relativity, in which our universe is modeled by a manifold and the gravitational field g is interpreted as a non-positive definite quadratic form. The purpose of this chapter is not so much to examine the cosmological and astronomical consequences of the theory as to make plausible the fundamentals, such as the field equation of gravity, by deriving them, following Hilbert, from some simple assumptions.

In the first section, the use of Lorentzian metrics is motivated by deducing their necessity from a single physical observation. This is not done for general manifolds, but only for the Minkowski space $(\mathbb{R}^4, \langle \cdot, \cdot \rangle_L)$, since the purpose is only to justify the model. There are analogous results for manifolds, but they would lead too far here. The Minkowski space corresponds to the vacuum case in the general model. After deriving the field equation from simpler assumptions, the non-curvature terms of the equation are determined for various typical physical model situations: For free particles, dust, isentropic fluids, and electromagnetic fields.

Large parts of the algebraic calculus for Riemannian manifolds are valid also in the Lorentzian case, in particular the connection and curvature calculus (but not the Hopf–Rinow theorem). These analogous formulas will not be specially emphasized. The term "orthonormal basis" is to be understood in an extension for Minkowski forms.

For further reading on the subject, the textbooks by O'Neill [ON2] and its sequel [ON3], as well as Besse [Besse, ch. 3,4], Sachs and Wu [SWu], and for a more physical view Misner, Thorne and Wheeler [MTW], Hawking and Ellis [HE] are suitable. The physics of Special Relativity is taught, for example, in the textbook by Taylor and Wheeler [TaWh].

© The Editor(s) (if applicable) and The Author(s), under exclusive license to Springer-Verlag GmbH, DE, part of Springer Nature 2024
K. Köhler, *Differential Geometry and Homogeneous Spaces*, Universitext,
https://doi.org/10.1007/978-3-662-69721-4_8

8.1 Constant Speed of Light

quadric	light cone
doubly ruled surface	Minkowski space
double cone	Lorentz isometry
Minkowski form	Poincaré-isometry
light-like	Lorentz group
time-like	Poincaré group
space-like	Alexandrov–Ovchinnikova Theorem

The following experimental physical observation is one of the reasons for the use of the Lorentz group in relativity: Two different observers both measure the speed of light as the same speed. This observation was already made in 1728 by James Bradley with an accuracy of 1% when he measured the speed of light of the stars γ Draconis and η Ursae Maioris. Since then, it has been confirmed in numerous experiments with ever increasing accuracy and generality. In this section it is explained why this result alone in a space-time simplified as flat already forces the use of the Lorentz transformations. It is especially remarkable that no continuity of the coordinate transformation between the observers has to be assumed (or even linearity).

Let a **quadric** $Q \subset \mathbb{R}^n$ be a solution set of a quadratic equation $q \equiv 0, q : \mathbb{R}^n \to \mathbb{R}$. A light cone in $\mathbb{R}^4$ will be described by a quadric, and in general various special quadrics will play an important role in the following deduction.

Proposition 8.1.1 *Assume ℓ to be an affine straight line which has three points in common with a quadric Q. Then $\ell \subset Q$.*

Proof. The equation of the quadric, restricted to the straight line, yields a quadratic polynomial there. If this polynomial has three zeros on the straight line, it must be zero everywhere. □

The one-sheet hyperboloid has the normal form $\{x^2 + y^2 - z^2 = 1\}$. That is, by an affine bijective mapping, any one-sheet hyperboloid can be brought to this form. The hyperbolic paraboloid has the normal form $\{xy = z\}$ (Fig. 8.1).

Proposition 8.1.2 *Let ℓ_1, ℓ_2, ℓ_3 be three affine pairwise skew straight lines in $\mathbb{R}^3$. Then there is a quadric Q containing these three straight lines, and Q is either a hyperbolic paraboloid or a one-sheet hyperboloid.*

Proof. A quadric in $\mathbb{R}^3$ is characterized as the zero set of a linear combination of $1, x, y, z, xy, xz, yz, x^2, y^2, z^2$. Three points on each of the three straight lines give a total of nine equations for the ten coefficients. So there is a nontrivial solution for the coefficients, and by Proposition 8.1.1 the associated Q contains the straight lines.

Of the (up to affine transformations) 15 quadrics in $\mathbb{R}^3$, only the two mentioned contain three skew straight lines. □

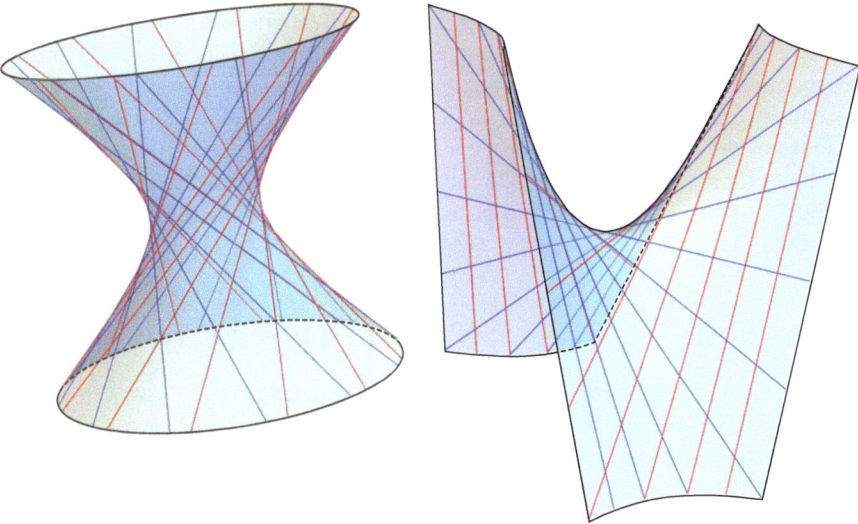

Fig. 8.1 One-sheet hyperboloid and hyperbolic paraboloid.

In the following, "almost all" shall stand for "all except for at most a finite number of exceptions".

Definition 8.1.3 *A **doubly ruled surface** Q is defined as a subset of $\mathbb{R}^n$ which is the union of each of two sets of straight lines, such that each straight line from the 1st family intersects almost all straight lines from the 2nd family, but two lines from the same family never intersect.*

Thus, each point $p \in Q$ shall lie on a straight line from the 1st and one from the 2nd family. The condition implies that exactly one such straight line from each family passes through p and straight lines from the 1st and 2nd family cannot lie on each other.

The hyperbolic paraboloid and the one-sheet hyperboloid carry two different sets of straight lines. Considering the normal form of the hyperbolic paraboloid, the approach $(x_0 + tx_1)(y_0 + ty_1) = z_0 + tz_1$ leads to the unique (up to parameterization of the straight lines) solutions $t \mapsto (t, \alpha, \alpha t)$ and $t \mapsto (\alpha, t, \alpha t)$. Analogously, in the case of the normal form of the one-sheet hyperboloid, one obtains the unique solution $t \mapsto (\sin \alpha + t \cos \alpha, \cos \alpha - t \sin \alpha, \pm t)$. For $n \geq 3$, let a hyperbolic paraboloid or a one-sheet hyperboloid in $\mathbb{R}^n$ be a corresponding quadric in a three-dimensional subspace.

Theorem 8.1.4 (Hilbert, Cohn-Vossen 1932 [HCV, p. 14]) *A doubly ruled surface Q in $\mathbb{R}^n$ is either a plane, a hyperbolic paraboloid, or a one-sheet hyperboloid.*

Proof. Let ℓ_1, ℓ_2, ℓ_3 be three straight lines from the 1st family, $\ell_j(t_j) =: a_j + b_j t_j$.

1st case: Two of the straight lines are parallel: Then each straight line intersecting both lies in a plane, and Q is this plane.

2nd case: All three straight lines are skew to each other: Then through almost all points p of ℓ_1 passes exactly one straight line, which intersects ℓ_2 and ℓ_3. This is because the corresponding equation is $\ell_2(t_2) - p = s \cdot (\ell_3(t_3) - p)$, i.e. the linear system of equations $b_2 t_2 + (p - a_3)s - b_3 st_3 = p - a_2$ in the variables t_2, s, st_3. Because of the skewness of the straight lines, this system of equations has rank 3 for almost all p and $s \neq 0$ holds. On the other hand, it has a solution for almost all p since p lies on a straight line from the 2nd family. Consequently, almost all the straight lines from the 2nd family are uniquely determined by ℓ_1, ℓ_2, ℓ_3 and have the form

$$s \mapsto \ell_1(t_1) + s(\ell_3(t_3) - \ell_1(t_1)) = a_1 + t_1 b_1 + s(a_3 - a_1) + b_3 st_3 - b_1 st_1,$$

thus they lie in the 3-dimensional affine subspace spanned by $a_3 - a_1, b_1, b_3$. Therefore the same follows for all straight lines from the 1st family, because each of them intersects three of these straight lines from the 2nd family, and thus for Q, which is therefore uniquely determined by ℓ_1, ℓ_2, ℓ_3. On the other hand, by Proposition 8.1.2 applied to ℓ_1, ℓ_2, ℓ_3, there exists a hyperbolic paraboloid or a one-sheet hyperboloid Q' in this 3-dimensional subspace. Hence $Q = Q'$. $\qquad\square$

A more elementary argument in the 2nd case is as follows: The straight lines through p and ℓ_2 form a plane P minus a straight line parallel to ℓ_2. Because of the skewness ℓ_3 does not lie in the plane and generically has an intersection with P.

Let the quadric C_0 in an n-dimensional $\mathbb{R}$-vector space V be a **double cone**, i.e. determined up to a linear transformation by the normal form $x_0^2 - x_1^2 - \cdots - x_{n-1}^2 = 0$. It will model the points that can be reached by a light ray (or other objects moving at the speed of light) from the zero point, as well as the points from which a light ray can pass through the zero point. Because in a period of time $|x_0|$ a path of length $c|x_0| = \sqrt{x_1^2 + \cdots + x_{n-1}^2}$ is covered by the light ray according to the definition of the physical term "speed". The following proposition states that an experimentally observed light cone essentially fixes this quadratic equation and thereby uniquely determines a Minkowski form (up to a factor).

Proposition 8.1.5 *The double cone C_0 determines its quadratic equation $q = 0$ uniquely up to a constant, thus it determines up to a positive real factor the associated* **Minkowski form** $g(v, w) = \frac{1}{4}(q(v + w) - q(v - w))$.

This statement applies analogously to other quadrics. We set $\|v\|^2 := q(v)$. A vector v is called **light-like** if $\|v\|^2 = 0$ (i.e. $v \in C_0$), **time-like** if $\|v\|^2 > 0$ (i.e. v lies inside the cone), and otherwise **space-like**. Let the **light cone** to a point $x \in V$ be $C_x := x + C_0$; thus, this models the points of the spacetime that can be connected to x by a light ray.

Proof. Let $e_0 \in V$ be any fixed time-like vector and g be a Minkowski form such that $C_0 = \{v \in V \mid g(v, v) = 0\}$. Let $v \in V \setminus \mathbb{R} \cdot e_0$. The straight line $\ell : \mathbb{R} \to V$, $t \mapsto tv + e_0$ intersects C_0 at least once, since the intersection of C_0 with the plane spanned by v, e_0 consists of two straight lines and at least one of them has different slope than ℓ.

If ℓ intersects only one of these light lines, v is parallel to the other one and $\|v\|^2 = 0$. Otherwise, the equation

$$0 = \|tv + e_0\|^2 = t^2\|v\|^2 + 2tg(v, e_0) + \|e_0\|^2$$

has two (different) real solutions t_1, t_2 describing the intersection of ℓ and C_0. Because $e_0 \notin C_0$, one finds $t_1, t_2 \neq 0$. The product of the solutions satisfies

$$\|v\|^2 = \frac{\|e_0\|^2}{t_1 t_2}$$

and hence t_1, t_2 and $\|e_0\|^2$ uniquely determine the value $\|v\|^2$.   □

The pair (V, g) is called **Minkowski space**. Let a **Lorentzian isometry** be an isometry of (V, g), let a **Poincaré isometry** be an affine Lorentzian isometry, i.e. a composition of a Lorentzian isometry and a translation. The groups consisting of these isometries are the **Lorentz group** $\cong SO(1, n - 1)$ and the **Poincaré group**, respectively.

The signature of the Minkowski form (in terms of the diagonal elements of the Gram matrix in normal form) is $(1, -1, -1, \ldots, -1)$. Let e_0 be a time-like vector. Then by Sylvester's Law of Inertia g is negative definite on $e_0^\perp$. In particular, $\langle e_0, v \rangle \neq 0$ holds for any light-like vector $v \neq 0$, and $v^\perp$ is a space spanned by v and space-like vectors.

Proposition 8.1.6 *If $y \in C_0 \setminus \{0\}$, then $C_0 \cap C_y = \mathbb{R} \cdot y$.*

Proof. Consider $z \in C_0 \cap C_y$. Then $\|z\|^2 = 0 = \|z - y\|^2$ and thus also $g(z, y) = 0$. Thus g vanishes on the subspace spanned by y, z, which must therefore be 1-dimensional.   □

Alexandrov–Ovchinnikova Theorem 8.1.7 (1953 [AO],[A]) *Let $(\mathbb{R}^n, g)$ be a Minkowski space with $n \geq 3$ and $\varphi : \mathbb{R}^n \to \mathbb{R}^n$ a bijective map which maps exactly the light-like vectors to light-like vectors; more precisely $y \in C_x \Leftrightarrow \varphi(y) \in C_{\varphi(x)}$. Then φ is multiple of a Poincaré isometry.*

The physical interpretation is the following: Two observers A, B equip the 4-dimensional spacetime around them with coordinates. Since for both the space-time has the same points, the conversion of the coordinates of A into those of B is a bijective map φ. A motion of a third object is now measured by each observer as light-like, if the movement takes place in a light cone. Thus, under the condition that A measures a motion as light-like exactly when B does so as well, their coordinates must differ by a multiple of a Poincaré transformation. Nothing about continuity or even linearity of φ needs to be artificially assumed. By the way, for $n = 2$ (Fig. 8.2) this statement is wrong:

Example In the two-dimensional case a double cone has a normal form $xy = 0$, with which one can easily find counterexamples. They are formulated for the double cone

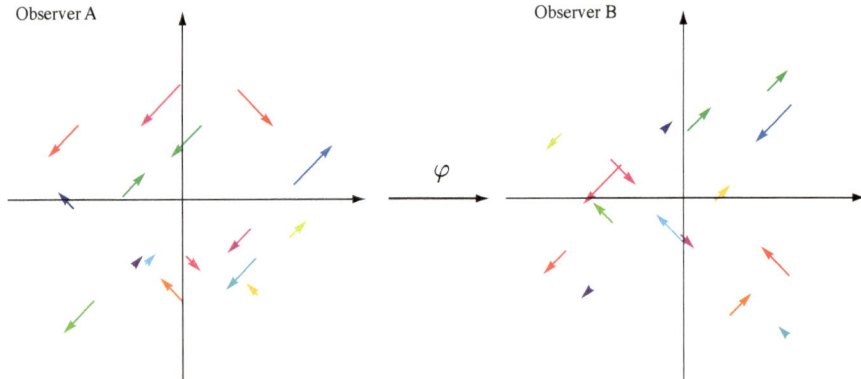

Fig. 8.2 Assumption in the Alexandrov–Ovchinnikova Theorem

$\{x^2 - y^2 = 0\}$ here by rotating about $45°$. With a nonlinear bijective map $f : \mathbb{R} \to \mathbb{R}$ and

$$\varphi : \mathbb{R}^2 \to \mathbb{R}^2, \begin{pmatrix} a \\ b \end{pmatrix} \mapsto \begin{pmatrix} f(a+b) + f(a-b) \\ f(a+b) - f(a-b) \end{pmatrix},$$

one obtains

$$\varphi \begin{pmatrix} t+a \\ t+b \end{pmatrix} - \varphi \begin{pmatrix} a \\ b \end{pmatrix} = \begin{pmatrix} f(2t+a+b) + f(a-b) - f(a+b) - f(a-b) \\ f(2t+a+b) - f(a-b) - f(a+b) + f(a-b) \end{pmatrix}$$

with constant difference of the components, analogously for $\varphi \begin{pmatrix} t+a \\ -t+b \end{pmatrix} - \varphi \begin{pmatrix} a \\ b \end{pmatrix}$. Thus φ maps light cones to light cones, but is not linear. In particular for $f(x) = x^3$ we get the counterexample

$$\varphi : \mathbb{R}^2 \to \mathbb{R}^2, \begin{pmatrix} a \\ b \end{pmatrix} \mapsto 2 \begin{pmatrix} a(a^2 + 3b^2) \\ b(3a^2 + b^2) \end{pmatrix}.$$

It is remarkable that the statement of the theorem cannot be extended to just this case $n = 2$, which is often used in Special Relativity for motivations and derivations.

Proof. 1) If $x \neq y$, $y \in C_x$, then $\varphi(y) \in C_{\varphi(x)}$, hence because of the bijectivity $\varphi(C_x) = C_{\varphi(x)}$. Now let $\ell_x \subset C_x$ be a straight line through $y \in C_x$. Then $\ell_x \subset C_y$ also holds and by Proposition 8.1.6 one gets $\ell_x = C_x \cap C_y$. Thus $\varphi(\ell_x) = C_{\varphi(x)} \cap C_{\varphi(y)}$ is also a light-like straight line according to Proposition 8.1.6 (Fig. 8.3).

2) Let P be a 2-dimensional affine plane intersecting C_x in two distinct straight lines (Fig. 8.4). Assertion: $\varphi(P)$ is either a plane, a one-sheet hyperboloid or a hyperbolic paraboloid with two families of light-like straight lines.

P carries two sets of light-like straight lines, namely the parallels to ℓ_x, ℓ'_x. So $\varphi(P)$ is also a surface with two flocks of light-like straight lines, because through every point of a chosen straight line from the 1st family passes exactly one straight line from the 2nd family. The assertion follows by Theorem 8.1.4.

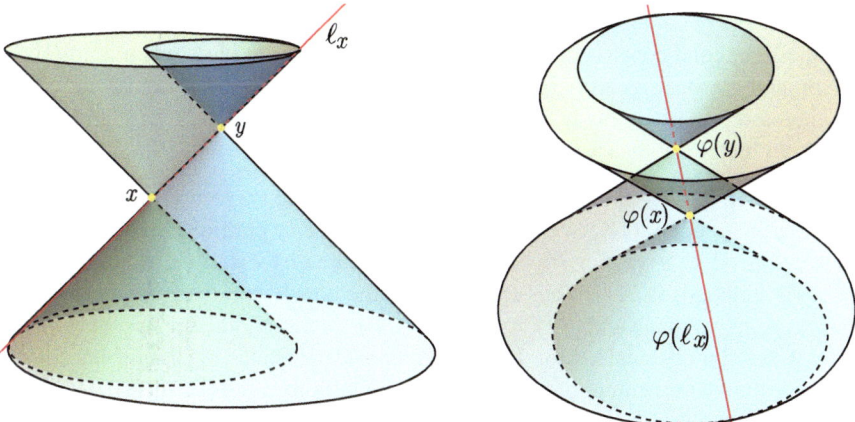

Fig. 8.3 Step (1) in the proof of the Alexandrov–Ovchinnikova Theorem.

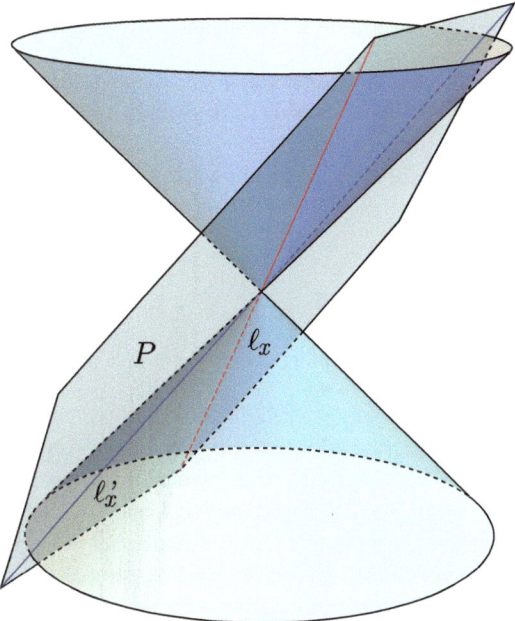

Fig. 8.4 Step (2) in the proof of the Alexandrov–Ovchinnikova Theorem

3) Assertion: $\varphi(P)$ is a plane. In the one-sheet hyperboloid there is a straight line from the 1st family and one from the 2nd family which do not intersect (namely the parallel opposite ones). This is not the case in P, so $\varphi(P)$ cannot be a hyperboloid.

In the hyperbolic paraboloid all straight lines from the 1st family are parallel to a plane Q. But light-like straight lines parallel to a 2-dimensional plane Q containing a light-like vector v are all parallel to one or to one of two straight lines $Q \cap C_0$. This is because either Q contains a time-like vector, so the signature of $g_{|Q}$ becomes $(1,-1)$ and the light cone consists of two straight lines. Or Q does not contain a time-like vector, and $g_{|Q}$ has signature $(0,1)$. The light cone in Q then becomes $\mathbb{R} \cdot v$.

Therefore, infinitely many straight lines from the 1st family are parallel to each other, but this is not the case in the hyperbolic paraboloid.

4) Let $\ell \subset \mathbb{R}^n$ be any straight line, $x \in \ell$ and let ℓ_1, ℓ_2 be two more straight lines in the open inner cone to C_x such that the slopes of ℓ, ℓ_1, ℓ_2 are linearly independent (here we need $n \geq 3$). Let P_j be the plane spanned by ℓ and ℓ_j. As in step (3), $g_{|P_j}$ has signature $(1,-1)$. Thus P_j intersects the cone C_x in two straight lines, so $\varphi(P_j)$ is again a plane and $\varphi(\ell) = \varphi(P_1) \cap \varphi(P_2)$ is a straight line. As a map on an at least two-dimensional vector space mapping straight lines to straight lines, φ is affine by the main theorem of affine geometry (Darboux 1880 [Dar],[Fi]), i.e. there are $A \in \mathbb{R}^{n \times n}, v_0 \in \mathbb{R}^n$ such that $\varphi(v) = v_0 + Av$.

5) The linear map A maps C_0 to itself according to the condition in the theorem. By Proposition 8.1.5, C_0 uniquely determines a Minkowski form g up to a constant, so φ preserves this Minkowski form up to a constant.   □

Alternatively, instead of (2),(3), one can argue that, for the plane P_{ℓ_x} tangent to C_x along ℓ_x, the union of all light cones on ℓ_x is given by $\bigcup_{y \in \ell_x} C_y = (\mathbb{R}^n \setminus P_{\ell_x}) \cup \ell_x$. So P_{ℓ_x} is mapped to just such a plane.

By choosing a fixed g, (5) can also be shown as follows: Choosing a fixed time-like e_0 such that $\|e_0\|^2 = 1$ and $v \perp e_0$, one obtains $\|v + \sqrt{-\|v\|^2} e_0\|^2 = 0 = \|v - \sqrt{-\|v\|^2} e_0\|^2$, hence

$$\|Av\|^2 + 2\sqrt{-\|v\|^2} \langle Av, Ae_0 \rangle - \|v\|^2 \|Ae_0\|^2 = 0$$
$$= \|Av\|^2 - 2\sqrt{-\|v\|^2} \langle Av, Ae_0 \rangle - \|v\|^2 \|Ae_0\|^2.$$

Thus $Av \perp Ae_0$ and $\|Av\|^2 = -\|v\|^2 \cdot \|Ae_0\|^2$ hold.

There are numerous generalizations of the Alexandrov–Ovchinnikova Theorem for weakened conditions (no bijectivity, no equivalence, for geodesics on manifolds, more general bilinear forms, local versions) which we do not discuss here (cf. [Giu]). In the Euclidean case there is, among others, the analogous theorem of Beckman and Quarles ([BQ], 1953), which also holds for $n = 2$. A comparable result is Wigner's Theorem, which is fundamental to quantum mechanics. [Wi, p. 233–236].

Exercise* 8.1.8 Verify the formula for the straight line on the one-sheet hyperboloid in its normal form.

Exercise 8.1.9 Find explicitly the intersections of two straight lines $t \mapsto (t, \alpha, \alpha t)$, $s \mapsto (\beta, s, \beta s)$ on the hyperbolic paraboloid and analogously those of two straight lines in the one-sheet hyperboloid.

Exercise 8.1.10 1) Show for $e_0 \in \mathbb{R}^n$ time-like (e.g. by Proposition 6.1.4) that for all $v \in \mathbb{R}^n \setminus \mathbb{R} \cdot e_0$ we have $g(e_0, v)^2 > \|e_0\|^2 \|v\|^2$.
2) Deduce for v, w time-like that $\|v + w\| \geq \|v\| + \|w\|$.

8.2 The Lorentz Group

Pauli matrices	irreducible representation
symmetric power	

In this section we study the Lorentz group $SO(1,3)$ of $\mathbb{R}^4$ and its Lie algebra to make its representation theory more accessible. Using the quadratic form $\det$, the space $(\mathbb{C}^{2\times 2}_{\mathrm{herm}}, \det)$ is isometric to the 4-dimensional Cartesian Minkowski space, since the **Pauli matrices**

$$\begin{pmatrix} 1 & 0 \\ 0 & 1 \end{pmatrix}, \quad \begin{pmatrix} 0 & 1 \\ 1 & 0 \end{pmatrix}, \quad \begin{pmatrix} 0 & -i \\ i & 0 \end{pmatrix}, \quad \begin{pmatrix} 1 & 0 \\ 0 & -1 \end{pmatrix}$$

form an orthonormal basis with respect to the bilinear form

$$\langle A, B \rangle = \frac{1}{4}\left(\det(A + B) - \det(A - B) \right)$$

induced via the polarization identity.

Lemma 8.2.1 *The map* $\varphi : SL(2, \mathbb{C}) \to SO_0(1,3), A \mapsto (X \mapsto AX\overline{A}^t)$ *for* $X \in \mathbb{C}^{2\times 2}_{\mathrm{herm}}$ *is a 2-fold covering.*

Proof. The kernel of φ is $\{A \in SL(2, \mathbb{C}) \mid AX\overline{A}^t = X \; \forall X \in \mathbb{C}^{2\times 2}_{\mathrm{herm}}\}$. If $X =$id, then $\overline{A}^t = A^{-1}$ follows, so the condition for $X \in \mathbb{C}^{2\times 2}_{\mathrm{Herm}}$ becomes $AX = XA$. Thus, $A(iX) = (iX)A$ is also true for the skew-Hermitian matrices iX. Thus it follows that $AX = XA$ for any $X \in \mathbb{C}^{2\times 2}$, and hence $A = \lambda$id. Because $1 = \det A = \lambda^2$, one obtains $A = \pm$id.

As $\ker \varphi$ is discrete, φ is a submersion. Since $SL(2, \mathbb{C})$ and $SO_0(1,3)$ are connected, φ is surjective according to Theorem 1.6.19. $\square$

Analogous to the exterior power, the q**th symmetric power** of a vector space V^* is the vector space

$$\mathrm{Sym}^q V^*$$
$$:= \{\omega \in V^{*\otimes q} \mid \forall \sigma \in \mathfrak{S}_q, v_1, \ldots, v_q \in V : \omega(v_1, \ldots, v_q) = \omega(v_{\sigma(1)}, \ldots, v_{\sigma(q)})\}.$$

This space is nothing else than the familiar vector space of homogeneous polynomials of degree q on V. A $\omega \in \mathrm{Sym}^q V^*$ thereby becomes the polynomial $V \to \mathbb{R}, X \mapsto \omega(\underbrace{X, \ldots, X}_{q-\mathrm{mal}})$. Correspondingly, $\bigoplus_{q \geq 0} \mathrm{Sym}^q V^*$ equipped with the product of polynomials is an algebra, simply the polynomial algebra of polynomials in $\dim V$-many variables, and $\dim \mathrm{Sym}^q V^* = \binom{\dim V^* + q - 1}{q}$.

The irreducible $\mathbb{C}$-representations of $\mathfrak{sl}(2, \mathbb{C})$ are, up to isomorphism, the symmetric powers of the standard representation, i.e. the spaces of homogeneous polynomials in two variables (Exercise 1.6.32). In this context, the complex Lie algebra

$\mathfrak{so}(1,3) \otimes \mathbb{C} \cong \mathfrak{so}(4) \otimes \mathbb{C}$ as a complexification of the Lie algebra of the connected compact group SO(4) is somewhat easier to handle than the real one for several reasons, not least because of the following lemma.

Lemma 8.2.2 *There is a canonical isomorphism of Lie algebras* $\mathfrak{so}(1,3) \otimes_{\mathbb{R}} \mathbb{C} \overset{\mathrm{can.}}{\cong}$ $\mathfrak{sl}(2,\mathbb{C}) \times \mathfrak{sl}(2,\mathbb{C})$. *Under this identification, the complexified standard representation* ρ *on* $\mathbb{R}^4 \otimes_{\mathbb{R}} \mathbb{C}$ *becomes* $\rho_1 \otimes_{\mathbb{C}} \rho_2$ *with*

$$(\rho_1 \otimes_{\mathbb{C}} \rho_2)(X,Y)v \otimes w := \rho_1(X)v \otimes w + v \otimes \rho_2(Y)w,$$

for $v, w \in \mathbb{C}^2, X, Y \in \mathfrak{sl}(2,\mathbb{C})$.

Proof. Using J :=multiplication by i in $\mathfrak{sl}(2,\mathbb{C})$ define the projections

$$\pi_{1,2} : \mathfrak{sl}(2,\mathbb{C}) \to \mathfrak{sl}(2,\mathbb{C}) \otimes_{\mathbb{R}} \mathbb{C}$$
$$A \mapsto \frac{1}{2}(A \mp iJA).$$

Because $[JA,B] = [A,JB] = J[A,B]$, $[\pi_1 A, \pi_2 B] = 0$ and $[\pi_1 A, \pi_1 B] = \pi_1[A,B]$ hold. Thus π_1, π_2 are (real) Lie algebra isomorphisms on their image. By Lemma 8.2.1 it follows that

$$\mathfrak{so}(1,3) \otimes \mathbb{C} \cong \mathfrak{sl}(2,\mathbb{C}) \otimes_{\mathbb{R}} \mathbb{C} \cong \mathfrak{sl}(2,\mathbb{C}) \times \mathfrak{sl}(2,\mathbb{C}).$$

The analogous decomposition $(\pi_1, \pi_2) : \mathbb{C}^2 \otimes_{\mathbb{R}} \mathbb{C} \overset{\mathrm{can.}}{\cong} \mathbb{C}^2 \times \mathbb{C}^2, v \mapsto (\frac{1}{2}(1-iJ)v, \frac{1}{2}(1+iJ)v)$ implies $\rho(\pi_1 A + \pi_2 B)(\pi_1 v \otimes \pi_2 w) = \pi_1(\rho_1(A)v) \otimes \pi_2 w + \pi_1 v \otimes \pi_2(\rho_2(B)w)$. □

This works in general for any $\mathfrak{g}_{\mathbb{C}}$ with $\rho_1 \otimes \bar{\rho}_2$ (cf. [FH, p. 439 Ex. 26.14]), but here additionally $\bar{\rho} \cong \rho$ is real.

Remark Accordingly, there is a twofold covering $\varphi : \mathrm{SL}(2,\mathbb{C}) \times \mathrm{SL}(2,\mathbb{C}) \to \mathrm{SO}(1,3,\mathbb{C})$. Conjugation by $\begin{pmatrix} i & & \\ & 1 & \\ & & 1 \\ & & & 1 \end{pmatrix} \in \mathrm{GL}(4,\mathbb{C})$ is a group isomorphism from $\mathrm{SO}(4,\mathbb{C})$ to $\mathrm{SO}(1,3,\mathbb{C})$.

Exercises

Exercise* 8.2.3 Compare on the quaternions $\mathbb{H} = \left\{ \begin{pmatrix} z & w \\ -\bar{w} & \bar{z} \end{pmatrix} \middle| z, w \in \mathbb{C} \right\}$ the quaternionic scalar product and the bilinear form det and find an orthonormal basis on $\mathbb{H}$ as a four-dimensional Euclidean vector space.

8.3 Adjoints of Connections

Hodge *(star)-operator	musical isomorphism
L^2 scalar product	formal adjoint
divergence	gauge invariance

In the next section we need some general facts about the formal adjoint of the Levi-Civita connection to differentiate the Hilbert action. Instead of Riemannian metrics, we consider Lorentzian metrics. In this section, let g be non-degenerate with arbitrary signature $(1, \ldots, 1, -1, \ldots, -1)$. Let $(e_k)_k$ be an oriented local orthonormal basis of TM and let $e_k^\vee := \pm e_k$ with $(e_k^\vee)^\flat = g(\cdot, e_k^\vee) = e^k$. On $T_q^p M$, $\Lambda^\bullet T^* M$ and $\mathrm{Sym}^\bullet T^* M$, g induces metrics, for which tensors of the form $e^{j_1} \otimes \cdots \otimes e^{j_q} \otimes e_{\ell_1} \otimes \cdots \otimes e_{\ell_m}$, $e^{j_1} \wedge \cdots \wedge e^{j_q}$ and $e^{j_1} \cdots e^{j_q}$, respectively, can be taken as elements of an orthonormal basis. These metrics are independent of the choice of the orthonormal basis.

Definition 8.3.1 *The **Hodge *(star) operator** $* \in \Gamma(M, \mathrm{End}(\Lambda T^* M))$ is defined as the linear operator satisfying $\eta \wedge *\omega = g(\eta, \omega)_\Lambda \, d\mathrm{vol}$ with $d\mathrm{vol} = e^1 \wedge \cdots \wedge e^n$.*

This implies for any oriented orthonormal basis that $*e^1 \wedge \cdots \wedge e^q = \prod_{j=1}^q \|e^j\|^2 \cdot e^{q+1} \wedge \cdots \wedge e^n$ (cf. Exercise 3.1.30). By changing the order of the basis vectors to another oriented basis it follows in general that

$$*e^{k_1} \wedge \cdots \wedge e^{k_q} = \prod_{j=1}^q \|e^{k_j}\|^2 \cdot e^{k_{q+1}} \wedge \cdots \wedge e^{k_n} \cdot \mathrm{sign} \begin{pmatrix} 1 & \cdots & n \\ k_1 & \cdots & k_n \end{pmatrix}.$$

Lemma 8.3.2 *Assume that the signature of g contains (-1) m-times. Then the Hodge $*$ operator has the following properties:*

1) *$*^2 = (-1)^{q(n-q)+m} = (-1)^{q(n+1)+m}$ on $\Lambda^q T^* M$,*
2) *for all $\eta, \omega \in \Lambda^q T^* M$ we have $g(*\eta, *\omega) \, d\mathrm{vol} = (-1)^m g(\eta, \omega) \, d\mathrm{vol}$,*
3) *for all $\omega \in \Lambda^q T^* M$, $\vartheta \in \Lambda^{n-q} T^* M$ we have*

$$g(\vartheta, *\omega) d\mathrm{vol} = (-1)^{q(n-q)} g(*\vartheta, \omega) d\mathrm{vol},$$

4) *for all $X \in TM$, $\omega \in \Lambda^q T^* M$ we have $*\iota_X \omega = (-1)^{q-1} X^\flat \wedge *\omega$.*

So in the 4-dimensional Lorentzian case $*^2 = (-1)^{q+1}$. If m is even, by (2) $*$ is an isometry. In the case when n is odd or on the subalgebra $\bigoplus_q \Lambda^{2q} T^* M$, by (3) $*$ is symmetric.

Proof. 1) By switching to a basis with differently arranged basis vectors, the only remaining case to be tested is

$$*^2(e^1 \wedge \cdots \wedge e^q) = * \prod_{j=1}^{q} \|e^j\|^2 \cdot e^{q+1} \wedge \cdots \wedge e^n$$

$$= \prod_{j=1}^{n} \|e^j\|^2 \cdot e^1 \wedge \cdots \wedge e^q \cdot \mathrm{sign} \begin{pmatrix} 1 & \cdots & n-q & n-q+1 & \cdots & n \\ q+1 & \cdots & n & 1 & \cdots & q \end{pmatrix}.$$

2) By (1) one finds

$$g(*\eta, *\omega)\, d\mathrm{vol} = *\eta \wedge *^2\omega = (-1)^{q(n-q)+m} *\eta \wedge \omega$$
$$= (-1)^m \omega \wedge *\eta = (-1)^m g(\eta, \omega)\, d\mathrm{vol}.$$

3) $g(\vartheta, *\omega)d\mathrm{vol} = *\omega \wedge *\vartheta = (-1)^{q(n-q)} *\vartheta \wedge *\omega = (-1)^{q(n-q)} g(*\vartheta, \omega)d\mathrm{vol}.$

4) 1st case: $\|X\|^2 \neq 0$. Extend $e_1 = \dfrac{X}{\sqrt{|\|X\|^2|}}$ to an orthonormal basis. Let without restriction ω be a product of these basis vectors. If e^1 does not occur in the product representation of ω, both sides vanish. So, up to multiples and permutations of the basis vectors, let $\omega = e^1 \wedge \cdots \wedge e^q$. Then one obtains

$$*(\iota_{e_1}\omega) = *e^2 \wedge \cdots \wedge e^q = \prod_{k=2}^{q} \|e^k\|^2 \cdot e^1 \wedge e^{q+1} \wedge \cdots \wedge e^n$$

$$\cdot \mathrm{sign} \begin{pmatrix} 1 & \cdots & q-1 & q & q+1 & \cdots & n \\ 2 & \cdots & q & 1 & q+1 & \cdots & n \end{pmatrix}$$

$$= (-1)^{q-1}\|e^1\|^2 e^1 \wedge *\omega.$$

Multiplication by $\sqrt{|\|X\|^2|}$ yields the assertion.

2nd case: $\|X\|^2 = 0$. The equation is linear in X. Decompose $X = X_1 + X_2$ such that $\|X_1\|^2, \|X_2\|^2 \neq 0$. $\qquad\square$

Thus one gets $*\vartheta \wedge *\omega = (-1)^m \vartheta \wedge \omega$, $*\eta \wedge \omega = (-1)^{q(n-q)+m}\eta \wedge *\omega$.

Remark 8.3.3 In this chapter it will be more convenient to define a contraction $\widetilde{\mathrm{Tr}}_{jk}$ of the jth and kth factors without distinguishing between co- and contravariant factors. That is, for $j, k \in \{1, \dots, n\}$ we set

$$\widetilde{\mathrm{Tr}}_{jk}\left(\sum X_1 \otimes \cdots \otimes X_p \otimes \alpha_{p+1} \otimes \cdots \otimes \alpha_{p+q}\right)$$

$$:= \sum \alpha_k(X_j) X_1 \otimes \cdots \otimes \widehat{X_j} \otimes \cdots \otimes X_p \otimes \alpha_{p+1} \otimes \cdots \otimes \widehat{\alpha_k} \otimes \cdots \otimes \alpha_{p+q}.$$

Contractions $\widetilde{\mathrm{Tr}}_{g,jk}$ with an index g are meant to be such that a musical isomorphism is first applied to a component before contracting with it, and the jth, kth factors are contracted with sequential numbering of co- and contravariant components.

Definition 8.3.4 *Given tensors* $\alpha, \beta \in T_q^p M$ *and* $\gamma \in T_p^q M$, *their* L^2-***scalar products*** *are defined as* $\langle \alpha, \gamma \rangle_{L^2} := \int_M \alpha(\gamma) \, d\mathrm{vol}_g$, $(\alpha, \beta)_{L^2} := \int_M g(\alpha, \beta) d\mathrm{vol}_g$. *The* ***divergence*** $\mathrm{div} : \Gamma(M, T_q^p M) \to \Gamma(M, T_q^{p-1} M)$ *is defined as*

$$\mathrm{div}\,\alpha := -\widetilde{\mathrm{Tr}}_{12} \nabla^{T_q^p M} \alpha.$$

Considering the ***musical isomorphism*** $T_q^p M \to T_p^q M, \alpha \mapsto \alpha^\natural$, *the formal adjoint* $\nabla^* : \Gamma(M, T_q^p M) \to \Gamma(M, T_{q-1}^p M)$ *of* $\nabla^{T_q^p M}$ *is given as* $\nabla^* \alpha := \left(\mathrm{div}\,\alpha^\natural \right)^\natural$,

$$\nabla^* \alpha = -\widetilde{\mathrm{Tr}}_{g,12} \nabla^{T_q^p M} \alpha = -\sum_{j=1}^n \iota_{e_j^\vee} \nabla_{e_j}^{T_q^p M} \alpha.$$

Given such a vector field X, one thus finds

$$\mathrm{div}\,X = -\mathrm{Tr}\,\nabla X = -\sum_{j=1}^n g(e_j^\vee, \nabla_{e_j} X) \in C^\infty(M, \mathbb{R})$$

for any local orthonormal basis $(e_j)_j$. For other tensors this definition is not a uniform standard in the literature. The dual ∇^* is often also called the divergence, especially when operating on symmetric forms. Obviously, this can lead to misunderstandings, which is why it is omitted here.

Proposition 8.3.5 *Assume that the signature of* g *contains* (-1) m-*times. If* $\omega \in \mathfrak{A}^q(M)$, *then* $d * \omega = (-1)^q * \nabla^* \omega$ *holds, thus* $*d * \omega = (-1)^{(q+1)n+1+m} \nabla^* \omega$. *Given any vector field* X, *it follows that* $d * X^\flat = - * \mathrm{div}\,X$.

Proof. One gets

$$d * \omega \overset{\text{Theorem 3.3.4}}{=} \sum_{k=1}^n e^k \wedge \nabla_{e_k}(*\omega) = \sum_k e^k \wedge *\nabla_{e_k}\omega$$

$$= (-1)^{q-1} \sum_k *\|e^k\|^2 \iota_{e_k} \nabla_{e_k} \omega = (-1)^q * \nabla^* \omega. \qquad \square$$

The Stokes–Cartan Theorem implies

Corollary 8.3.6 *If* $X \in \Gamma_c(M, TM)$, *then* $\int_M \mathrm{div}\,X \cdot d\mathrm{vol} = 0$ *holds.*

∇^* is a **formal adjoint** of $\nabla^{T_q^p M}$ in the sense of the following lemma.

Lemma 8.3.7 *If* $\alpha \in \Gamma(M, T_{q+1}^p M), \beta \in \Gamma_c(M, T_q^p M)$, *then*

$$(\alpha, \nabla^{T_q^p M} \beta)_{L^2} = (\nabla^* \alpha, \beta)_{L^2} = \langle \mathrm{div}\,\alpha^\natural, \beta \rangle_{L^2}.$$

Proof. Consider $\gamma \in \Gamma_c(M, T^*M), \gamma(X) := g(\iota_X\alpha, \beta)$, then

$$\operatorname{div}\gamma^{\#} = -\sum(\nabla_{e_j}\gamma)(e_j^{\vee}) = -\sum e_j \cdot \left(g(\iota_{e_j^{\vee}}\alpha, \beta)\right) + g(\iota_{\nabla_{e_j}e_j^{\vee}}\alpha, \beta).$$

Thus one obtains

$$g(\alpha, \nabla\beta) = \sum_j g(\alpha, e^j \otimes \nabla_{e_j}\beta) = \sum_j \left(e_j \cdot \left(g(\iota_{e_j^{\vee}}\alpha, \beta)\right) - g(\nabla_{e_j}(\iota_{e_j^{\vee}}\alpha), \beta)\right)$$

$$= \sum_j \left(e_j \cdot \left(g(\iota_{e_j^{\vee}}\alpha, \beta)\right) - g(\iota_{\nabla_{e_j}e_j^{\vee}}\alpha, \beta) - g(\iota_{e_j^{\vee}}\nabla_{e_j}\alpha, \beta)\right)$$

$$= -\operatorname{div}\gamma^{\#} + g(\nabla^*\alpha, \beta).$$

By Corollary 8.3.6, the integral over the divergence term vanishes.   □

This is also true with literally the same proof in the algebras $\Lambda^{\bullet}T^*M$ and $\operatorname{Sym}^{\bullet}T^*M$ with the same formula for ∇^*. While in these cases the metrics differ by a factor $q!$ from that on $T_q^0 M$, in the projection of $e_j \otimes \nabla\beta$ a factor $1/q$ arises which cancels this effect.

Now let $\mathbf{F}$ be a functional from the space L of Lorentzian or Riemannian metrics on a manifold to $\mathbb{R}$. Assume that for any diffeomorphism $\varphi : M \to N, \mathbf{F}(\varphi^*g) = \mathbf{F}(g)$ holds. If $\mathbf{F}$ is sufficiently differentiable on the subset L of the vector space of all bilinear forms on M, then $0 = \frac{\partial}{\partial t}\big|_{t=0}\mathbf{F}(\Phi_t^{X*}g) = T_g\mathbf{F}(L_Xg)$ follows for any vector field as an infinitesimal diffeomorphism (cf. [MeMi], [FrKr] for a corresponding calculus; here we assume differentiability of $\mathbf{F}$ and the chain rule as a condition on the considered functionals $\mathbf{F}$). $T_g\mathbf{F}$ is a linear form on $\Gamma(M, \operatorname{Sym}^2T^*M)$; thus, if it were continuous on $\Gamma_{L^2}(M, \operatorname{Sym}^2T^*M)$, it would have the form $(\cdot, \omega)_{L^2}$ for an L^2-form ω according to the Riesz–Fréchet Representation Theorem. The assumption in the next lemma requires this with a differentiable ω.

Lemma 8.3.8 *Choose a fixed Lorentzian or Riemannian metric g. If $T_g\mathbf{F}(h) = \langle h, \omega\rangle_{L^2}$ for some $\omega \in \Gamma(M, \operatorname{Sym}^2TM)$, then* $\operatorname{div}\omega = 0$ *holds.*

Proof. Given any 1-form α, Lemma 3.3.5 implies that

$$0 = \frac{1}{2}\langle L_{\alpha^{\#}}g, \omega\rangle_{L^2} = \langle\nabla\alpha, \omega\rangle_{L^2} = \langle\alpha, \operatorname{div}\omega\rangle_{L^2}.$$ □

In particular, ω is divergence free if and only if the **gauge invariance**

$$\langle L_Xg, \omega\rangle_{L^2} = 0 \qquad \text{for all } X$$

holds.

Exercises

Exercise* 8.3.9 Given a vector field X and $f \in C_0^\infty(M, \mathbb{R})$, prove that

$$\int_M X.f \, d\text{vol} = \int_M f \, \text{div} \, X \cdot d\text{vol}.$$

Exercise* 8.3.10 Let E be a vector bundle, ∇^E a connection on E and ∇^{E^*} the induced connection on E^*. Using Exercise 8.3.9, show that with respect to the L^2-product associated to the canonical pairing $(T^*M \otimes E) \times (TM \otimes E^*) \to \mathbb{R}$ the operator

$$\text{div} : \Gamma(M, TM \otimes E^*) \to \Gamma(M, E^*)$$
$$X \otimes \mu \mapsto -\nabla_X^{E^*}\mu + \mu \cdot \text{div} \, X$$

satisfies the equation

$$(\text{div} \, (X \otimes \mu), s)_{L^2} = (X \otimes \mu, \nabla^E s)_{L^2} = (\mu, \nabla_X^E s)_{L^2}$$

for any $X \in \Gamma(M, TM), \mu \in \Gamma(M, E^*), s \in \Gamma_c(M, E)$.

Exercise 8.3.11 Deduce from Exercise 8.3.10 for a given vector bundle E equipped with a metric h and a metric connection ∇^E that for $s \in \Gamma(M, E)$ the formal adjoints of the operators ∇^E, ∇_X^E with respect to the metric are given by

$$\nabla^*(X^\flat \otimes s) = (\nabla_X^E)^* s = -\nabla_X^E s + (\text{div} \, X) \cdot s.$$

Describe ∇^* analogously to Definition 8.3.4.

Exercise 8.3.12 Prove that the divergence of Killing fields vanishes by showing in general that $\text{div} \, X = -\frac{1}{2}\text{Tr}_g L_X g$.

Exercise 8.3.13 Verify Lemma 8.3.7 in a shorter way in the case $\beta \in \mathfrak{A}^q(M)$, $\alpha \in \mathfrak{A}^{q+1}(M)$, by using Proposition 8.3.5 to compute $d(\beta \wedge *\alpha)$.

8.4 Derivation of the Hilbert Action

gravitational field

Lagrangian measure

Lagrangian function

action

Lagrangian functional

Bianchi map

gravitational waves

Hilbert action

Ricci curvature

cosmological constant

field equation of gravity

stress-energy tensor

In Hilbert's original article [H], submitted on November 20, 1915, the field equation arises as a solution of a variational problem for a Lagrangian functional. He states three axioms as a model assumption, which in modern terms (with different numbering) have more or less the following form:

I) The spacetime is a 4-dimensional manifold M which can carry Lorentzian metrics.

II) On M there is a (signed) measure $\mathcal{L}$ with density function L which depends pointwise on the 0th, 1st and 2nd derivatives of a Lorentzian metric g as well as the 0th and 1st derivatives of other fields q_j (Hilbert refers to the electro-magnetic field). Physically realized g, q_j correspond to points of vanishing variation of $\int_M \mathcal{L}$.

III) $\mathcal{L} = \mathcal{L}_1 + \mathcal{L}_2$, where $\mathcal{L}_1$ does not depend on the q_j, is linear in the 2nd derivatives of g, and $\mathcal{L}_2$ does not depend on the 1st and 2nd derivatives of g.

The Lorentzian metric g is the **gravitational field**. The measures $\mathcal{L}$ are called **Lagrangian measures**. If $\mathcal{L} = L\,d\mathrm{vol}_g$ (in the sense of Remark 2.5.9), then L is called a **Lagrangian function**. $\int_M \mathcal{L}$ is called an **action functional** or **Lagrangian functional**. $\mathcal{L}_2$ contains information about other components of the universe such as particles, electromagnetic field, etc., typically modeled as sections q_j in bundles over M.

Since the above axioms do not contain any reference to quantum mechanics, it goes without saying that they can only approximately describe a part of physics. Even in the framework of General Relativity combined with electromagnetism they would be used in a more complicated form, cf. [MTW, §21.2] and the end of section 8.7. On the other hand, these axioms in [H] had led at that time to the Lagrangian function of gravity (and thus by simple differentiation to the field equation), for which there is so far no practicable proposal for improvement. So, in particular, they provide a very elegant motivation for this equation as well as a basis for further developments.

The derivative of $-\mathcal{L}_2$ will be identified with the stress-energy tensor in later sections for various model situations. In this section we will see that these axioms enforce that $\mathcal{L}_1$ is a constant multiple of the scalar curvature s. By Corollary 4.2.16, the very requirement of the existence of a Lorentzian metric is a nontrivial topological condition on M.

Axiom II is justified by analogies to the classical field theory of mechanics and to electrodynamics, in which in the Lagrangian functions zeroth and first derivatives of the fields with respect to the space-time coordinates appear. But invariants of the metric which depend pointwise only on 1st derivatives of the metric do not exist. This was shown in the section on geodesics by Theorem 5.3.6: In a canonical chart $\exp_p^{-1}$ around each $p \in M$ the metric has the Taylor expansion

$$(\exp_p^* g)_X = g_p + \frac{1}{3} \sum_{k,\ell=1}^{n} g_p(\Omega_p(\cdot, e_k)\cdot, e_\ell) \cdot x_k x_\ell + O(\|X\|^3).$$

Consequently, any invariant that depends on at most the 2nd derivatives of g is a function of g_p and R_p. However, the identification of this chart with a subset of $\mathbb{R}^4$ depended not only on the point p, but additionally on the choice of an isometry $T_pM \to \mathbb{R}^4$, i.e., the choice of an orthonormal basis in T_pM. Thus, by Axiom II, the invariants we are looking for are real-valued functions of g_p and R_p which are invariant, in particular, under the operation of the oriented isometry group $\cong SO(1,3)$ on g_p and R_p. This group acts trivially on g_p. So it remains to find $SO(1,3)$-invariant maps from the space of possible values of curvature tensors to $\mathbb{R}$, which are then in addition supposed to be linear by Axiom III. Axiom III is motivated by the approach of truncating a Taylor expansion from $\mathcal{L}$ to g after the linear term: A constant term in the Lagrangian measure can be taken to be part of $\mathcal{L}_2$, and the next possible term is the linear one. Hilbert postulates it in his article separately from Axiom I and II (which he numbers in a different order) and does not call it an "axiom".

For any Lorentzian or Riemannian metric, the curvature tensor at $p \in M$ lies by Theorem 3.4.1 in the subspace A of those elements of $\mathrm{Sym}^2 \Lambda^2 T_p^* M$ which satisfy the 1st Bianchi identity. The linear $\mathbb{C}$-valued invariants are thus found to be the trivial $SO(1,3,\mathbb{C})$-subrepresentations.

Lemma 8.4.1 *In the $SO(1,3)$ representation A the only one-dimensional subrepresentation consists of multiples of the scalar curvature. In particular, $\mathcal{L}_1 \in \mathbb{R} \cdot s \cdot d\mathrm{vol}$ holds.*

Proof. As a representation of $\mathfrak{so}(4,\mathbb{C}) \cong \mathfrak{sl}(2,\mathbb{C}) \times \mathfrak{sl}(2,\mathbb{C})$, in terms of the respective standard complex representations $E = V_1 \otimes V_2$ one obtains

$$\Lambda^2 E = \mathrm{Sym}^2 V_1 \otimes_\mathbb{C} \underbrace{\Lambda^2 V_2}_{\cong \mathbb{C}} \oplus \underbrace{\Lambda^2 V_1}_{\cong \mathbb{C}} \otimes_\mathbb{C} \mathrm{Sym}^2 V_2,$$

hence

$$\mathrm{Sym}^2 \Lambda^2 E \cong \mathrm{Sym}^2(\mathrm{Sym}^2 V_1) \otimes \mathbb{C} \oplus \mathbb{C} \otimes \mathrm{Sym}^2(\mathrm{Sym}^2 V_2) \oplus \mathrm{Sym}^2 V_1 \otimes \mathrm{Sym}^2 V_2.$$

There is a canonical embedding

$$\mathrm{Sym}^4 V_\ell \hookrightarrow \mathrm{Sym}^2(\mathrm{Sym}^2 V_\ell),$$
$$\alpha_1 \cdot \alpha_2 \cdot \alpha_3 \cdot \alpha_4 \mapsto (\alpha_1 \cdot \alpha_2) \cdot (\alpha_3 \cdot \alpha_4) + (\alpha_1 \cdot \alpha_3) \cdot (\alpha_2 \cdot \alpha_4).$$

Because $\dim \mathrm{Sym}^4 V_\ell = 5 = \dim \mathrm{Sym}^2(\mathrm{Sym}^2 V_\ell) - 1$, it follows that $\mathrm{Sym}^2(\mathrm{Sym}^2 V_\ell) \cong \mathrm{Sym}^4 V_\ell \oplus \mathrm{Sym}^0 V_\ell$ and

$$\mathrm{Sym}^2 \Lambda^2 E$$
$$= \mathrm{Sym}^4 V_1 \otimes \mathbb{C} \oplus \mathbb{C} \otimes \mathbb{C} \oplus \mathbb{C} \otimes \mathrm{Sym}^4 V_2 \oplus \mathbb{C} \otimes \mathbb{C} \oplus \mathrm{Sym}^2 V_1 \otimes \mathrm{Sym}^2 V_2.$$

Using the **Bianchi map** $b : \mathrm{Sym}^2 \Lambda^2 E \to \Lambda^4 E$,

$$b(R)(X,Y,Z,W) := R(X,Y,Z,W) + R(Y,Z,X,W) + R(Z,X,Y,W)$$

one of the trivial components corresponds to im $b = \Lambda^4 E$. The other one necessarily yields the scalar curvature. □

Remark One gets $A \otimes_{\mathbb{R}} \mathbb{C} = \ker b = \mathbb{C} \oplus \text{Sym}_0^2 E \oplus W$ and $\text{Sym}^2 E = \text{Sym}^2 V_1 \otimes \text{Sym}^2 V_2 \oplus \mathbb{C}$, so $W = \text{Sym}^4 V_1 \oplus \text{Sym}^4 V_2$ becomes the space of Weyl curvature tensors in the above decomposition (cf. [Besse, ch. 1.G]). Since the Ricci curvature component of R is determined via T by pointwise fields or matter, the component in W determines gravitational waves in vacuum.

Definition 8.4.2 *The **Hilbert action** is the Lagrangian functional*

$$\mathbf{F}_1(g) := \int_M s \, d\text{vol}.$$

Next, we differentiate the Hilbert action by g to obtain the field equation of gravity. In the formulation of the following theorems we take a fixed variation g_t of the metric $g = g_0$ and $'$ means $\frac{\partial}{\partial t}|_{t=0}$.

Proposition 8.4.3 *There is a vector field X such that $\text{Tr}_g \text{Ric}' = \text{div } X$.*

Proof. The connections $\nabla : \mathfrak{A}^1(M, \text{End}(TM)) \to \mathfrak{A}^2(M, \text{End}(TM))$ and

$$\nabla^{T_2^1 M} : \Gamma(M, T_2^1 M) \to \Gamma(M, T_3^1 M)$$

satisfy (see Corollary 3.3.6)

$$(\nabla \alpha)(X, Y, Z) = (\nabla_X^{T_2^1 M} \alpha)(Y, Z) - (\nabla_Y^{T_2^1 M} \alpha)(X, Z).$$

Now $\text{Ric}' = \widetilde{\text{Tr}}_{14}(\nabla^2)' = \widetilde{\text{Tr}}_{14}(\nabla' \circ \nabla + \nabla \circ \nabla') = \widetilde{\text{Tr}}_{14}(\nabla(\nabla'))$ holds, thus

$$
\begin{aligned}
\text{Tr}_g \text{Ric}' \quad &= \quad \text{Tr}_g \widetilde{\text{Tr}}_{14}(\nabla^{T_2^1 M}(\nabla')) - \text{Tr}_g \widetilde{\text{Tr}}_{24}(\nabla^{T_2^1 M}(\nabla')) \\
&\overset{\text{interchange the Tr}}{=} \quad \text{Tr } \widetilde{\text{Tr}}_{g,23}(\nabla^{T_2^1 M}(\nabla')) - \text{Tr}_g \widetilde{\text{Tr}}_{24}(\nabla^{T_2^1 M}(\nabla')) \\
&= \quad \text{Tr } \nabla\left(\widetilde{\text{Tr}}_{g,12}\nabla' - (\widetilde{\text{Tr}}_{13}\nabla')^{\#}\right) \\
&= \quad -\text{div }\left(\widetilde{\text{Tr}}_{g,12}\nabla' - (\widetilde{\text{Tr}}_{13}\nabla')^{\#}\right).
\end{aligned}
$$

The idea here is to make the second trace independent of the additional covariant component in ∇ to interchange ∇ and Tr. □

Theorem 8.4.4 *Consider $\mathbf{F}_1(g) := \int_M s \, d\text{vol}$ and a variation of $g = g_0$ with $\frac{\partial}{\partial t} g_t$ having compact support. Then*

$$T_g \mathbf{F}_1(h) = (h, \frac{1}{2} s g - \text{Ric})_{L^2}.$$

Even if the integral $\mathbf{F}_1(g)$ does not converge, this gives a finite value for $T_g \mathbf{F}_1$ for variations with compact support.

Proof. If ω is 2-fold covariant and independent of g, then (with appropriate Gram matrices) $(\mathrm{Tr}_g \omega)' = (\mathrm{Tr}\, g^{-1}\omega)' = -\langle h, \omega \rangle_g$. Furthermore one gets

$$\sqrt{-\det g_{jk}}\,' = \frac{-(\det g_{jk})'}{2\sqrt{-\det g_{jk}}} = \frac{-\det g_{jk}}{2\sqrt{-\det g_{jk}}} \cdot \mathrm{Tr}\, g^{-1}h$$

$$= \frac{1}{2}\sqrt{-\det g_{jk}}\,\langle g, h \rangle_g.$$

Thus with the vector field X from Proposition 8.4.3 one finds

$$(s\,d\mathrm{vol})' = (\mathrm{Tr}_g \mathrm{Ric} \cdot d\mathrm{vol})' = (-\langle h, \mathrm{Ric} \rangle_g + \mathrm{div}\, X + \langle h, \tfrac{1}{2}sg \rangle_g)d\mathrm{vol}. \qquad \square$$

Remark More generally, if $\mathbf{F}(g) := \int_M (s + \Lambda)\, d\mathrm{vol}$ with a constant $\Lambda \in \mathbb{R}$ (the **cosmological constant** or "dark energy"), one obtains

$$T_g\mathbf{F}(h) = (h, \frac{1}{2}(s + \Lambda)g - \mathrm{Ric})_{L^2}.$$

In particular, by Lemma 8.3.8 it follows that:

Corollary 8.4.5 *The identity* $\nabla^*(\mathrm{Ric} - \frac{1}{2}sg) = 0$ *holds, i.e.* $(\mathrm{Ric} - \frac{1}{2}sg)^{\natural}$ *is divergence free.*

According to Axiom (III), $\mathbf{F}(g) = \int_M (s + L_2)\, d\mathrm{vol}$, where L_2 is to depend pointwise on g, other fields and their 1st derivatives, but not on derivatives of g. Then, according to Theorem 8.4.4 one obtains

$$T_g\mathbf{F}(h) = \int_M \left(\langle h, \frac{1}{2}sg - \mathrm{Ric} \rangle_g + \frac{\partial L_2}{\partial g}(h) + \frac{1}{2}L_2 \cdot \langle g, h \rangle_g \right) d\mathrm{vol},$$

which with a constant κ and $\kappa T := -\frac{\partial L_2}{\partial g}^{\#} - \frac{1}{2}L_2 \cdot g \in \Gamma(M, \mathrm{Sym}^2 T^*M)$ leads to the **field equation of gravity** (for the stationary points of $\mathbf{F}$)

$$\frac{1}{2}sg - \mathrm{Ric} = \kappa T \tag{8.1}$$

([H, p. 404, last line]). The symmetric tensor T is called the **stress-energy tensor**.

Exercises

Exercise* 8.4.6 Show Corollary 8.4.5 by using the 2nd Bianchi identity in place of Lemma 8.3.8.

8.5 Free Particles

particle	energy-momentum vector
rest mass	velocity
energy	force
momentum	

In the remaining sections we investigate and motivate what the term $T_g F_2$ associated to the Lagrangian measure $\mathcal{L}_2$ can look like and what physical interpretation it has. To do this, we slowly work our way up to more complicated matter models, in which each model motivates the next. In this section, we first work out the form of $T_g F_2$ for a particle, followed by dust (a flow of particles) and fluids (dust equipped with a density function).

A **particle** is a curve $\gamma : I \to M$, $I =]a, b[$. Let $\gamma : I \to \mathbb{R}^4$, $t \mapsto (t, vt, 0, 0)^t$ be a curve with speed $v \in \mathbb{R}$ in the spacetime. The conservation of the speed of light c presented in section 8.1 requires that $\|\dot\gamma\|^2 = 0$ if and only if $v = c$, and $\|\cdot\|^2$ is assumed to be invariant under the operation of $SO(3)$ on the spatial part. So $\|(x_0, x_1, x_2, x_3)^t\|^2$. We choose the latter as the Minkowski form.

Let γ be a particle that is at rest relative to an observer A, i.e., $\gamma(t) = (m_0 t, 0, 0, 0)^t$. Let m_0 be the **rest mass** of the particle. According to section 8.1, a second observer B sees the particle as $\gamma(t) = mt \cdot (1, v_1, v_2, v_3)^t$ for $m, v_1, v_2, v_3 \in \mathbb{R}$ with $\|\tilde\gamma\|^2 = \|\dot\gamma\|^2 = c^2 m_0^2$, i.e. the rest mass is invariant under Lorentz transformations.

Given an (observer) orthonormal basis (e_0, e_1, e_2, e_3) with e_0 time-like at a point $p \in M$, let $e^R := (e^1, e^2, e^3) : TM \to \mathbb{R}^3$. Then for an observer moving on a world curve with derivative e_0, $E := c \cdot e^0(\gamma')$ is the **energy** and $p := e^R(\gamma')$ is the **momentum** of the particle. Therefore γ' is called the **energy-momentum vector**.

In general, given an (observer) chart $\varphi = (\varphi_Z, \varphi_R) : U \to \mathbb{R} \times \mathbb{R}^3$, the **velocity** of γ equals

$$\frac{\partial \varphi_R(\gamma)}{\partial \varphi_Z(\gamma)} = \frac{d\varphi_R(\gamma)/ds}{d\varphi_Z(\gamma)/ds} = \frac{ce^R(\gamma')}{e^0(\gamma')}$$

by the chain rule. The **force** acting on the particle is defined as the derivative of the momentum with respect to the proper time, more precisely

$$F = \frac{e^R(\nabla_{\partial/\partial t}\gamma')}{e^0(\nabla_{\partial/\partial t}\gamma')}.$$

The Lagrangian measure of the rest of the universe (particles, fields etc.) is assumed to have support in $M \setminus U$ for an open subset U satisfying $\gamma(I) \subset U$, $\partial\gamma(I) \not\subset U$. Thus, the particle is free from external action except possibly at its starting and ending points if γ is not closed. So $\mathbf{F}(g) = \int_M s\, d\mathrm{vol} + \int_M \mathcal{H} + \sum_j \int_{I_j} \ell_j\, dt + $etc. for various particles, indexed by j, and a global Lagrangian $s\, d\mathrm{vol} + \mathcal{H}$.

Lemma 8.5.1 *For all $X \in \Gamma_c(U, TM)$ we have $T_g \mathbf{F}_2(L_X g) = 0$.*

Proof. Given the Hilbert action $\mathbf{F}_1(g) = \int_M s\,d\mathrm{vol}$, by the Remark before Lemma 8.3.8 one finds

$$0 = T_g\mathbf{F}_2(L_Xg) + \underbrace{T_g\mathbf{F}_1(L_Xg)}_{=0}. \qquad\qquad \square$$

Theorem 8.5.2 *Let* $\gamma : I =]a, b[\to M$ *be an embedding as a submanifold and let* $\mathbf{F}_2$ *have at a metric* g *a derivative of the form*

$$T_g\mathbf{F}_2(h) = \int_I h(\widehat{T}(t))\,dt$$

with $\widehat{T} \in \Gamma(I, \gamma^*\mathrm{Sym}^2TM \setminus \{0\})$. *Then the path of* γ *is that of a geodesic with respect to* g *and* $\widehat{T} = \mathrm{const.} \cdot \frac{\dot\gamma \otimes \dot\gamma}{\|\dot\gamma\|_g}$.

The choice $\mathbf{F}_2(g) := 2\int_I \|\dot\gamma\|\,dt$ has this derivative.

Proof. Write $\widehat{T}$ as $\widehat{T} = \sum_{k=1}^n f_k n_k \otimes n_k$ in terms of an orthogonal basis $(n_k)_{k=1}^4$ along γ with $n_1 = \dot\gamma$, $f_k \in C^\infty(I)$. For any $j \in \{2, \dots, n\}$ let $N_j \in \Gamma(M, TM)$ be such that $N_j|_{\gamma(I)} = n_j$ and $X = r \cdot N_j$ for a function $r \in C_c^\infty(U)$, $r|_{\gamma(I)} = 0$. Then one obtains

$$(\tfrac{1}{2}L_Xg)(Y, Y) = \tfrac{1}{2}X.\|Y\|^2 - g([X, Y], Y) = g(\nabla_Y X, Y) \qquad (8.2)$$
$$= Y.(g(X, Y)) - g(X, \nabla_Y Y) \qquad (8.3)$$

and therefore

$$\int_I (\tfrac{1}{2}L_Xg)(\widehat{T})\,dt \overset{(8.2)}{=} \sum_k \int_I f_k g(\nabla_{n_k} X, n_k)\,dt$$

$$= \sum_k \int_I f_k g(n_k.r \cdot N_j, n_k)\,dt = \int_I f_j n_j.r \cdot \|n_j\|^2\,dt.$$

Given a suitable choice of r, $n_j.r$ does not vanish, so it follows that $f_j \equiv 0$ and $\widehat{T} = f_1\dot\gamma \otimes \dot\gamma$.

Via the reparametrization $\widetilde{\gamma}(u) := \gamma(t(u))$ with $u(t) := \int_a^t \frac{ds}{f_1(s)}$, i.e. $f_1(t(u)) = t'(u)$ and $t(J) = I$, one thus finds that

$$\int_I h(\widehat{T})\,dt = \int_J h(\widetilde{\gamma}', \widetilde{\gamma}')\,du.$$

Given any X, because $X_{\gamma(a)} = 0$, $X_{\gamma(b)} = 0$, one gets

$$\int_J (\tfrac{1}{2}L_Xg)(\widetilde{\gamma}', \widetilde{\gamma}')\,du \overset{(8.3)}{=} \int_J (\frac{d}{du}(g(X, \widetilde{\gamma}')) - g(X, \nabla_{\partial/\partial u}^{\gamma^*TM}\widetilde{\gamma}'))\,du$$

$$= \underbrace{g(X, \widetilde{\gamma})|_{t^{-1}(a)}^{t^{-1}(b)}}_{=0\text{ by assumption}} - \int_J g(X, \nabla_{\partial/\partial u}^{\gamma^*TM}\widetilde{\gamma}')\,du.$$

Therefore $\widetilde{\gamma}$ is a geodesic and

$$\text{const.} \equiv \|\widetilde{\gamma}'(u)\|_g^2 = \|\dot{\gamma}(t(u))\|_g^2 (t'(u))^2 = (\|\dot{\gamma}\|_g^2 f_1^2)_{|t(u)}. \qquad \square$$

In particular, this $\mathbf{F}_2$ depends only on the 0th and 1st derivatives of γ. Because $\|\widetilde{\gamma}'\|^2 \equiv$ const., the velocity of γ is everywhere light-like or time-like if it is so at at least one point.

Exercises

Exercise 8.5.3 Show that any diffeomorphism-invariant functional $\mathbf{F}$ with pointwise support vanishes. More precisely, show that if $p \in M$ and $\widehat{T} \in \mathrm{Sym}^2 T_p M$ such that for all X the tensor $(L_X g)(\widehat{T})$ vanishes, then $\widehat{T} = 0$.

8.6 The Stress-Energy Tensor

dust	internal energy
energy density	energy density
pressure	spatially isotropic
energy density	Euler equations
isentropic perfect fluid	Navier–Stokes
elastic potential	

Physical observation (more exactly the Special Theory of Relativity) suggests that gravity is generated by E/c^2 (the so-called "gravitational mass") or better its spatial density. This does not fit exactly; granted, the definition of force in the last paragraph specifies an "inertial mass" at least in the case of particles. But the gravitational field g is generated by the whole 4×4 tensor T and depends via the non-linear field equation in a very complicated way on the values of T further away from the particle, so there is no such thing as "gravitational mass" in this generality at all. The term E contributes the lion's share to gravity only because of the magnitude of c^2.

Now the Special Theory of Relativity says further that, although energy is not a coordinate-independent term, the energy-momentum vector in TM is. Likewise, "spatial density" is well-defined only with respect to a chosen time direction in TM. So T should associate to any time-like vector v_0 the density of the energy-momentum vector with respect to the hyperplane $v_0^{\perp}$. More precisely, given a flow with energy-momentum vector field X and a vector v_0, this value is the density of the flow through the 3-dimensional hyperplane $v_0^{\perp}$ on a subset of volume $\|v_0\|$ multiplied by the vector X_p. By dualizing with the metric one thus obtains $T \in T^* M \otimes T^* M$. As an example, we now use a convenient model of matter, dust, to check which tensor T is obtained

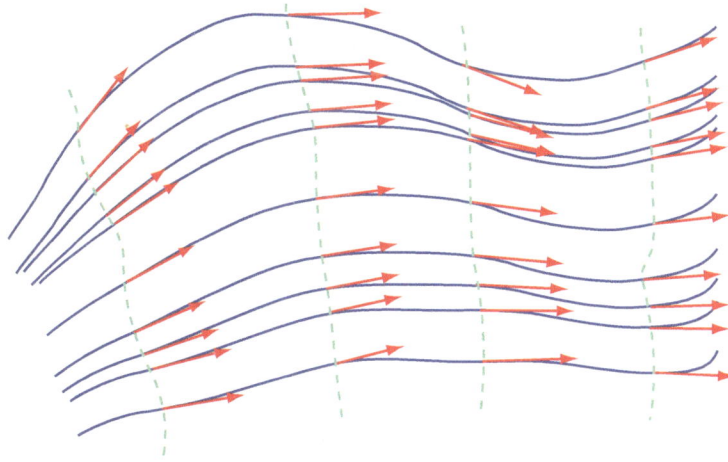

Fig. 8.5 Variable-density dust flow through space-like hyperplanes.

by this motivation, and compare the result with $T_g F_2$ or $\widehat{T}$ for free particles. At the end of the section, this model is generalized to that of a fluid.

Dust is taken here to denote a dense collection of particles which should exert no influence on each other except gravity. As a model for the dust, we choose a flow Φ whose integral curves are to be the particles, and a measure $\omega = \rho\, d\mathrm{vol} \in \Gamma(M, \Lambda^4 T^* M)$ with a density function ρ (Fig. 8.5). The number of particles remains constant during the flow, i.e., the density is supposed to distribute with Φ from a 3-dimensional space-like initial hyperplane over time. Thus, we want $\Phi^* \omega = \omega$ to hold or $L_X \omega = 0$ with the vector field X to Φ. According to Proposition 8.3.5 this condition means

$$0 = L_X \omega = d(\iota_X \rho\, d\mathrm{vol}) = d(\iota_{\rho X} d\mathrm{vol}) = \pm \mathrm{div}\,(\rho X) \cdot d\mathrm{vol},$$

thus it is equivalent to $\mathrm{div}\,\rho X = 0$. In contrast to the approach here, the literature additionally requires $\|X\| \equiv m_0 c =$ const. Moreover, some books assume that ρ should be independent of g and not ω.

1st approach (interpretation using the axioms via Theorem 8.5.2): We construct a map $\widetilde{T} : T_p M \to T_p M$ which associates to a time-like direction v_0 the energy-momentum vector X of a particle flow weighted by the spatial density of the flow. To the flow and a time-like direction v_0 we assign the density of the flow through the hyperplane $v_0^\perp$ on a subset of volume $\|v_0\|$ multiplied by the vector X_p (Fig. 8.6). The density (=number of particles traversing $v_0^\perp$) is the volume of the 4-dimensional parallelepiped generated by an appropriately sized 3-dimensional parallelepiped in

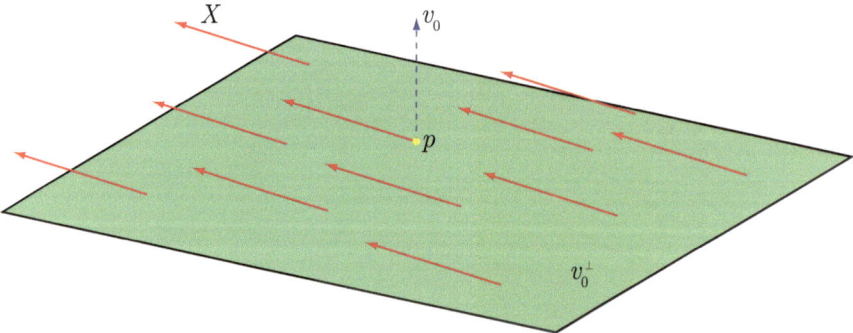

Fig. 8.6 Flow through hyperplane $v_0^\perp$ (the latter drawn here two-dimensional).

$v_0^\perp$ and a unit vector in the direction X times the density of the flow, i.e.

$$\left(\left(\frac{X}{\|X\|}\right)^\flat \wedge \iota_{v_0}\omega\right)(d\text{vol}^\natural) = \rho \cdot \left(g\left(\frac{X}{\|X\|}, v_0\right) - \iota_{v_0}(\frac{X}{\|X\|})^\flat \wedge\right) d\text{vol}(d\text{vol}^\natural)$$

$$= \rho g\left(\frac{X}{\|X\|}, v_0\right).$$

Simplified, this can be measured physically by choosing a box with one side perpendicular to v_0 and measuring how much goes in on that side and out on the other. Thus, the flow density through $v_0^\perp$ is $\widetilde{T}(v_0) = \rho g(\frac{X}{\|X\|}, v_0) \cdot X$. Thus one obtains $\widetilde{T} = \rho \frac{g(X,\cdot)X}{\|X\|}$, which is a symmetric endomorphism. The analogous form of the stress-energy tensor for single particles in Theorem 8.5.2 motivates that $\widetilde{T}^\# = \rho \frac{X \otimes X}{\|X\|}$ is identified with the (dualized) stress-energy tensor $T^\natural \in \text{Sym}^2 TM$ of dust.

This way the components of the matrix of T with respect to $(e_j)_j$ can be interpreted: $\iota_{e_0}T \in T_p^*M$ is the energy-momentum density in space with respect to an observer with time-like vector e_0, in particular $T(e_0, e_0)$ is the **energy density** in space. Given a space-like vector v_0, the space-like component of $\iota_{v_0}T$ thus becomes momentum per time and surface area. With force=momentum per time, this becomes a tensile stress exerted on matter in or behind the surface $e^R(v_0^\perp)$; the component perpendicular to $e^R(v_0^\perp)$ is more accurately called **pressure** (or "tension" depending on the sign). The time-like component of $\iota_{v_0}T$ represents energy per time and area and is called the **energy flux density**; because of the symmetry of T, this form is dual to the momentum density. Thus the symmetric Gram matrix of T has the form

$$T = \begin{pmatrix} \text{energy density} & \text{energy flux density} \\ \hline \text{momentum density} & \begin{matrix} \text{pressure} & & \text{tensile} \\ \text{tensile} & \text{pressure} & \text{stress} \\ \text{stress} & & \text{pressure} \end{matrix} \end{pmatrix}$$

The 3×3 stress tensor of the space-like components on the lower right was already introduced by Cauchy as a symmetric tensor. Its present name is a little bit unfortunate, because "tensor" originates from the Latin **tensio**, meaning **tension**.

2nd approach (very heuristic): Let the stress-energy tensor be an element of $\text{Sym}^2 T^* M$ which grows proportionally with the mass of matter. Thus, if only X is given as the direction, there is also only X to choose from for the 2nd vector, i.e. $T^\natural \sim X \otimes X$. Since T should grow linearly with mass and density, it becomes $T^\natural = \frac{\rho}{\|X\|} X \otimes X$.

Lemma 8.6.1 *The condition $\nabla^* T = 0$ is equivalent to the flow lines of X being paths of geodesics.*

Proof. $0 = \text{div} (\rho X)$ implies that

$$\text{div} \, T^\natural = \text{div} \, (\rho X) \frac{X}{\|X\|} - X.(\frac{1}{\|X\|})\rho X - \frac{\rho}{\|X\|} \nabla_X X$$

$$= \frac{\rho g(\nabla_X X, X)}{\|X\|^3} X - \frac{\rho}{\|X\|} \nabla_X X = -\frac{\rho}{\|X\|} (\nabla_X X)^{\perp X}.$$

This vanishes if and only if $\nabla_X X = aX$ for some $a \in C^\infty(M)$. If $r \in C^\infty(M)$ is any solution of the linear differential equation of 1st order $-X.r = ra$, it follows that $\nabla_{rX}(rX) \equiv 0$, so the integral curves to X are then the paths of geodesics. □

The Lagrangian measure $\mathcal{L}(g) := 2\omega\|X\|$, where X and ω are considered to be independent of the metric, satisfies

$$T_g(\mathcal{L}(g))(h) = \frac{\rho h(X, X)}{\|X\|} d\text{vol} = g(h, T) d\text{vol}.$$

This measure thus provides the assumed stress-energy tensor. It depends not only on g but also on X, therefore $\nabla^* T$ does not vanish automatically.

An important generalization of the dust model for cosmological models is that of the **isentropic perfect fluid**. This refers to a fluid in thermodynamic equilibrium (see Hawking–Ellis [HE, p. 69f]). As before, let $L_X \omega = 0$ and $\omega = \rho \, d\text{vol}$. Let the Lagrangian measure be $\mathcal{L}(g) := 2\omega\|X\|(1 + \varepsilon(\rho))$ with the **elastic potential** (or **internal energy**) $\varepsilon : \mathbb{R} \to \mathbb{R}$.

Lemma 8.6.2 *The stress-energy tensor associated to the Lagrangian measure $2\omega\|X\|(1 + \varepsilon(\rho))$ is given by $T = (\mu + p)\frac{X^\flat \otimes X^\flat}{\|X\|^2} - pg$ with the **pressure***

$$p := \rho^2 \varepsilon'(\rho)\|X\|$$

*and the **energy density***

$$\mu := ((1 + \varepsilon(\rho))\rho)\|X\| - p = (1 + \varepsilon(\rho) - \rho\varepsilon'(\rho))\rho\|X\|.$$

Proof. One gets

$$0 = (T_g\omega)(h) = (T_g\rho)(h) \cdot d\text{vol} + \rho \cdot (T_g d\text{vol})(h) = (T_g\rho)(h) + \langle h, \tfrac{g}{2}\rangle_g \rho) \, d\text{vol},$$

hence $T_g\rho(h) = -\langle h, \tfrac{g}{2}\rangle_g \rho$ and

$$T_g(\mathcal{L}(g))(h) = \frac{\rho h(X, X)}{\|X\|}(1 + \varepsilon(\rho)) \, d\text{vol} - \rho^2\|X\|\varepsilon'(\rho)\langle h, g\rangle_g \, d\text{vol}.$$

Substituting the definitions of p and μ yields the lemma. $\qquad\square$

Remark Unlike here, Hawking–Ellis [HE, p. 69f] set $\|X\|\omega := \rho \, d\text{vol}$ and also $L_X\omega = 0$ (hence div $\rho\frac{X}{\|X\|}) = 0$) and $\mathcal{L}(g) := 2\omega\|X\|(1 + \varepsilon(\rho))$. Then one obtains

$$T_g\rho = T_g\frac{\|X\|\omega}{d\text{vol}} = T_g\frac{\|X\|}{\sqrt{-\det g_{jk}}}\frac{\omega}{dx_1 \wedge \cdots \wedge dx_n}$$

$$= \left(\frac{h(X, X)}{2\|X\|} - \langle h, \tfrac{g}{2}\rangle_g\|X\|\right) \cdot \underbrace{\frac{\omega}{d\text{vol}}}_{\rho/\|X\|} \cdot$$

Hence one finds

$$T_g(\mathcal{L}(g))(h) = \frac{h(X, X)}{\|X\|^2}\rho(1 + \varepsilon(\rho)) \, d\text{vol} + \rho^2\varepsilon'(\rho)\langle h, \frac{X^\flat \otimes X^\flat}{\|X\|^2} - g\rangle_g \, d\text{vol}.$$

Thus $T = (\mu + p)\frac{X^\flat \otimes X^\flat}{\|X\|^2} - pg$ with $p := \rho^2\varepsilon'(\rho)$ and $\mu := (1 + \varepsilon(\rho))\rho$ in this case.

With respect to the time direction $V := X/\|X\|$ the Gram matrix of T equals

$$T = \begin{pmatrix} \mu & 0 & 0 & 0 \\ 0 & p & 0 & 0 \\ 0 & 0 & p & 0 \\ 0 & 0 & 0 & p \end{pmatrix}.$$

Thus p is the pressure and μ is the energy density. In particular, T is **spatially isotropic**, i.e., there exists a time-like vector e_0 with respect to which $T_{|e_0^\perp}$ is pointwise a multiple of $g_{|e_0^\perp}$. The divergence equation implies that

$$0 = \text{div}\left(V \otimes V(\mu + p) - p \cdot g^\sharp\right)$$

$$= (\mu + p)\text{div}(V) \cdot V - V.(\mu + p) \cdot V - (\mu + p)\nabla_V V + dp^\#.$$

Because $\nabla_V V \perp V$, this splits into temporal and spatial components with respect to the time-like vector V as follows

$$V.\mu - (\mu + p)\text{div } V = 0, \qquad (\mu + p)\nabla_V V - (\text{grad } p)^{\perp V} = 0.$$

From these equations one can obtain the classical **Euler equations** (or **Navier–Stokes** without friction and heat loss) by passing to suitable limits. So in the case of dust ($p = 0$) one finds again $\nabla_V V = 0$. An interesting interpretation of pressure using thermal motion is described in [SWu, Ex. 3.15.6].

8.7 Electromagnetism

electromagnetic field strength tensor	inhomogeneous Maxwell equation
Faraday tensor	vacuum magnetic permeability
electric field strength	four-current density
magnetic field strength	charge density
quadrupole potential	current density
homogeneous Maxwell equation	charge
electric potential	Lorentz force
vector potential	Cartan 1-form

The **electromagnetic field strength tensor** (or **Faraday tensor**) is a 2-form $F \in \Gamma(M, \Lambda^2 T^* M)$. Given an observer vector field e_0 (i.e., time-like), F can be decomposed into a summand F_1 containing multiples of e^0 and a summand F_2 not containing e^0. Thus, F_1 has the form $e^0 \wedge \iota_{e_0} F$ with the space-like 1-form $\iota_{e_0} F =: E/c$. Because $*F_2$ contains the factor e^0, it follows in the same way that $*F_2 = e^0 \wedge B$, hence

$$F_{|\ker e^0} = *(B_1 e^1 \wedge e^0 + B_2 e^2 \wedge e^0 + B_3 e^3 \wedge e^0)$$
$$= B_1 e^2 \wedge e^3 - B_2 e^1 \wedge e^3 + B_3 e^1 \wedge e^2.$$

Thus $F = e^0 \wedge E/c + *(B \wedge e^0)$ or as a Gram matrix for an orthonormal basis extending e_0 it is given by

$$F = \begin{pmatrix} 0 & -E_1/c & -E_2/c & -E_3/c \\ E_1/c & 0 & -B_3 & B_2 \\ E_2/c & B_3 & 0 & -B_1 \\ E_3/c & -B_2 & B_1 & 0 \end{pmatrix}.$$

The 1-forms E, B are called the **electric** and **magnetic field strength**, respectively. Because $*F = e^0 \wedge B + *(-E/c \wedge e^0)$, $*$ interchanges the roles of E/c and B and

$$*F = \begin{pmatrix} 0 & -B_1 & -B_2 & -B_3 \\ B_1 & 0 & E_3/c & -E_2/c \\ B_2 & -E_3/c & 0 & E_1/c \\ B_3 & E_2/c & -E_1/c & 0 \end{pmatrix}.$$

The Pfaffian $\mathrm{Pf}(F)$ (Definition 4.2.1) is given observer-independently by $\mathrm{Pf}(F)d\mathrm{vol} = \frac{1}{2!}F \wedge F = \frac{1}{2!}(2\langle e^0 \wedge E, B \wedge e^0\rangle_L d\mathrm{vol}) = -\langle E/c, B\rangle_L d\mathrm{vol}$. Therefore the determinant of the Gram matrix is $(\mathrm{Pf}(F))^2 = \frac{g(E,B)_L^2}{c^2}$.

For exact F, any $\alpha \in \Gamma(M, T^*M)$ with $F = d\alpha$ is called a **quadrupole potential**. In particular, in this case $dF = 0$ holds (the **homogeneous Maxwell equation**). Then $\varphi := \alpha(e_0) \in C^\infty(M)$ is the **electric potential** and $\alpha^{\perp e_0} \in \Gamma(M, T^*M^{\perp e_0})$ is the **vector potential**. The inhomogeneous Maxwell equation is $\nabla^* F = \mu_0 \mathbf{j}$ with the **vacuum magnetic permeability** $\mu_0 \in \mathbb{R}$ and the **four-current density** $\mathbf{j} \in \Gamma(M, T^*M)$. Here $\mathbf{j}(e_0)/c$ is the **charge density** and $\mathbf{j}^{\perp e_0} \in \Gamma(M, T^*M^{\perp e_0})$ is the **current density**.

Now, given any 2-form F, consider the Lagrangian function

$$\mathcal{L}(g)(d\mathrm{vol}^\natural) := \kappa_e \|F\|_{g,\wedge^2 T^*M}^2 = \frac{\kappa_e}{2}\|F\|_{g,T^*M^{\otimes 2}}^2$$
$$= \frac{\kappa_e}{2}\mathrm{Tr}_g\widetilde{\mathrm{Tr}}_{g,24}(F \otimes F) = \frac{\kappa_e}{2}\| * F\|_g^2$$

(in the literature $\kappa_e := -\frac{1}{4\pi}$ is often chosen) or $\mathcal{L}(g) = \kappa_e F \wedge *F$. The metric induced by the embedding into $\bigotimes T^*M$ (with $\|e^1 \otimes \cdots \otimes e^q\|^2 = \|e^1\|^2 \ldots \|e^q\|^2$) is on q-forms $q!$ times as large as the metric on $\wedge T^*M$ (with $\|e^1 \wedge \cdots \wedge e^q\|^2 = \|e^1\|^2 \ldots \|e^q\|^2$). This would be different for the metric induced by the quotient. In terms of E, B one gets $\|F\|_{g,\wedge^2 T^*M}^2 = BB^t - EE^t = -\|B\|_g^2 + \|E/c\|_g^2$ (by our convention, g is negative definite on space-like vectors).

Lemma 8.7.1 *The stress-energy tensor $\kappa_e T$ associated to this Lagrangian measure is given by*

$$T = -\widetilde{\mathrm{Tr}}_{g,24}(F \otimes F) + \frac{1}{4}\|F\|_{g,T^*M^{\otimes 2}}^2 \cdot g.$$

The corresponding Gram matrix with respect to $(e_j)_j$ thus equals

$$T = \begin{pmatrix} \frac{\|E/c\|_g^2 + \|B\|_g^2}{2} & E/c \times B \\ (E/c \times B)^t & \frac{\|E/c\|_g^2 + \|B\|_g^2}{2}\mathrm{id}_{\mathbb{R}^3} - E^t E/c^2 - B^t B \end{pmatrix}.$$

Proof. As in the calculation of s', it follows with $(\mathrm{Tr}_g\omega)' = -\langle h, \omega\rangle_g$ and $d\mathrm{vol}' = \langle h, \frac{g}{2}\rangle d\mathrm{vol}$ that

$$\left(\frac{1}{2}\|F\|_g^2 \, d\mathrm{vol}\right)'$$
$$= \left(\frac{1}{2}\langle h, -\widetilde{\mathrm{Tr}}_{g,13}F \otimes F\rangle_g + \frac{1}{2}\langle h, -\widetilde{\mathrm{Tr}}_{g,24}F \otimes F\rangle_g + \langle h, \frac{1}{4}\|F\|_g^2 g\rangle_g\right) d\mathrm{vol}$$
$$= \left(\langle h, -\widetilde{\mathrm{Tr}}_{g,24}F \otimes F\rangle_g + \langle h, \frac{1}{4}\|F\|_g^2 g\rangle_g\right) d\mathrm{vol}. \qquad \square$$

Theorem 8.7.2 *The stress-energy tensor satisfies*

$$\nabla^* T = F((*d * F)^\sharp, \cdot) - (*F)((*dF)^\sharp, \cdot).$$

Vanishing divergence with linear independence of the summands and generic F thus implies Maxwell's equations in a vacuum $\nabla^* F = 0$, $dF = 0$.

Proof. Consider a vector field $X \in \Gamma_c(M, TM)$ with flow Φ. Then the Hodge $*$-operator $\Phi_t^* *$ associated to the metric $\Phi_t^* g$ satisfies

$$(F \wedge *F)' = F \wedge (L_X *)F = F \wedge L_X(*F) - F \wedge *L_X F$$
$$= F \wedge d\iota_X * F + F \wedge \iota_X d * F - F \wedge *d\iota_X F - F \wedge *\iota_X dF.$$

Now one finds

$$-F \wedge \iota_X d * F = *F \wedge *\iota_X d * F = *F \wedge X^\flat \wedge (*d * F)$$
$$= -(*d * F) \wedge X^\flat \wedge *F = (*d * F) \wedge *\iota_X F$$
$$= -(d * F) \wedge \iota_X F = (*F) \wedge d\iota_X F - d((*F) \wedge \iota_X F)$$
$$= F \wedge *d\iota_X F - d((*F) \wedge \iota_X F),$$

and thus with the substitution $F \mapsto *F$ the integrals over the other two terms are also equal. Hence it follows that

$$2 \int \nabla^* T(X) \, d\mathrm{vol} = \int \langle L_X g, T \rangle \, d\mathrm{vol} = \int (F \wedge *F)'$$
$$= -2(F, \iota_X dF)_{L^2} - 2(*F, \iota_X d * F)_{L^2}$$
$$= -2(X^\flat \wedge F, dF)_{L^2} - 2(X^\flat \wedge *F, d * F)_{L^2}$$
$$= -2(X^\flat, \iota_{(*dF)^\sharp} * F)_{L^2} + 2(X^\flat, \iota_{(*d*F)^\sharp} F)_{L^2}. \qquad \square$$

The homogeneous Maxwell equation states $dF = 0$. In the literature this is often deduced from the assumption $F = d\alpha$; but as below, according to Dirac, F can also be understood as the curvature of a complex line bundle. So the term $F((\mathrm{div}\, F^\sharp), \cdot)$ has to cancel with the divergence of the stress-energy tensor of another matter field. For example, when considering a dust (X, ρ) in which particles have constant **charge** e, with $X = m_0 V$ (hence $\nabla_X X \perp X$) and $\mathrm{div}\, F^\sharp := e\rho V$ (motivated as a four-current density) as well as $\kappa = \kappa_e$ (the latter is a choice for the effect of "charge") one gets

$$\rho \nabla_X X = e\rho F(X, \cdot)^\flat \qquad \text{or} \qquad m_0 \nabla_V V = e F(V, \cdot)^\flat.$$

This is the law of action of the **Lorentz force**.

Assuming $F = d\alpha$, when varying α in place of g the action $\int \|d\alpha\|^2 \, d\mathrm{vol}$ causes $\int \langle \alpha', \nabla^* d\alpha \rangle \, d\mathrm{vol} = 0$ for all α', thus $\nabla^* F \equiv 0$ holds in a vacuum.

The Aharonov–Bohm effect [AhBo] shows that the potential α also acts where F vanishes, i.e., it is actually physically measurable. On the other hand, F need not be exact. As a solution of the dilemma (in the direction of a unification with Quantum Theory) a Hermitian line bundle $L \to M$ is postulated (Kaluza–Klein 1924[1][Kal]). The potential is then represented by a Hermitian connection ∇^L on L. Locally, according to Lemmas 3.2.15, 3.2.16, $\nabla^L = d + \alpha$ with $\alpha \in \Gamma(M, T^*M \otimes \text{End}_{\text{schiefherm}}(L)) = i\mathfrak{A}^1(M)$ and curvature given by

$$\Omega^L = (d + \alpha)^2 = d \circ \alpha - \alpha \circ d + \alpha \wedge \alpha = d\alpha =: iF \in i\mathfrak{A}^2(M).$$

Kaluza and Klein further proposed to consider the circle bundle $\widetilde{M} := \{s \in L \mid \|s\| = 1\}$ associated to L as a 5-dimensional spacetime $\widetilde{M}$, with the induced $(+, -, -, -, -)$-metric. Theorem 8.7.3 shows that the scalar curvature $\widetilde{s}$ of this space is exactly (the lift of) the Lagrangian function $s_M - \frac{1}{4}\|F\|^2$ with $F := \Omega^L$. A choice of the (fixed) radius of the circle conveniently yields the 2π factor in front of F.

The connection on the line bundle induces a horizontal structure on the circle bundle via a right-invariant **Maurer–Cartan 1-form** $\vartheta : T\widetilde{M} \to \mathbb{R}$ (cf. the definition given after Theorem), by setting

$$\nabla^L_X s =: ds(\widetilde{X}) + \vartheta(\widetilde{X}) \cdot s$$

for any right-invariant lifts $\widetilde{X}$. If $T^H\widetilde{M} := \ker \vartheta$, then any horizontal X, Y satisfy

$$d\vartheta(X, Y) = X.\vartheta(Y) - Y.\vartheta(X) - \vartheta([X, Y]) = -\vartheta(2A_X Y) \tag{8.4}$$

by O'Neill's Theorem 6.3.5. Lifting the metric on M to this horizontal structure, together with the metric of constant length 2π on the circles, yields a metric on $\widetilde{M}$. Choosing a different length yields a corresponding factor in front of the term $\|\Omega_L\|^2$.

Theorem 8.7.3 *On the 5-dimensional manifold $\widetilde{M}$, the Lagrangian functionals to gravity and electromagnetic field can be combined into*

$$\mathbf{F}(g, \nabla^L) = \int_{\widetilde{M}} \widetilde{s} \, d\widetilde{\text{vol}}.$$

Proof. By Exercise 5.2.19, the fibers are geodesics, so their 2nd fundamental form satisfies $T \equiv 0$. If U is vertical such that $\|U\|^2 = -1$ (hence $U^\vee = -U$) (Fig. 8.7) and e_j^* are horizontal lifts of a local orthonormal basis $e_j \in \Gamma(M, TM)$ then Parseval's Theorem implies that

$$\|A_X U\|^2 = \sum_k \langle e_k^*, A_X U \rangle^2 \cdot \|e_k\|^2 = \sum_k \langle A_X e_k^*, U \rangle^2 \cdot \|e_k\|^2 = -\sum_k \|A_X e_k^*\|^2 \cdot \|e_k\|^2. \tag{8.5}$$

---

[1] 1921, Theodor Franz Eduard Kaluza, 1885–1954; Oskar Benjamin Klein, 1894–1977.

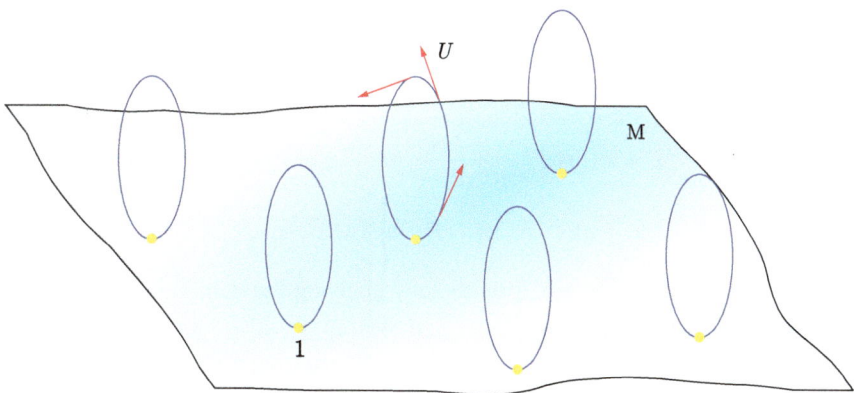

Fig. 8.7 Kaluza–Klein Theory.

By O'Neill's Formulas 6.3.11 the scalar curvature equals

$$\widetilde{s} = \sum_{j,k} \widetilde{R}(e_j^{*\vee}, e_k^{*\vee}, e_j^*, e_k^*) + \sum_j \widetilde{R}(e_j^{*\vee}, U^\vee, e_j^*, U) + \sum_j \widetilde{R}(U^\vee, e_j^{*\vee}, U, e_j^*)$$

$$= \sum_{j,k} \widetilde{R}(e_j^{*\vee}, e_k^{*\vee}, e_j^*, e_k^*) - 2 \sum_j \widetilde{R}(e_j^{*\vee}, U, e_j^*, U)$$

$$= \sum_{j,k} \left(R(e_j^\vee, e_k^\vee, e_j, e_k) - 3\|A_{e_j^*} e_k^*\|^2 \|e_j\|^2 \|e_k\|^2 \right)$$

$$+ 2 \sum_j g((\nabla_U A)_{e_j^*} U, e_j^*)\|e_j\|^2 - 2 \sum_j \|A_{e_j^*} U\|^2 \|e_j\|^2.$$

Here the symmetries of A show that $g((\nabla_U A)_{e_j^*} U, e_j^*) = -g(U, (\nabla_U A)_{e_j^*} e_j^*) = 0$ holds (Lemma 6.3.6). Hence one obtains

$$\widetilde{s} = s - 3 \sum_{j,k} \|A_{e_j^*} e_k^*\|^2 \|e_j\|^2 \|e_k\|^2 - 2 \sum_j \|A_{e_j^*} U\|^2 \|e_j\|^2$$

$$\overset{(8.5)}{=} s - \sum_{j,k} \|A_{e_j^*} e_k^*\|^2 \|e_j\|^2 \|e_k\|^2$$

$$\overset{(8.4)}{=} s - \frac{1}{4} \sum_{j,k} \|d\vartheta(e_j^*, e_k^*)\|^2 \|e_j\|^2 \|e_k\|^2$$

$$= s - \frac{1}{4}\|d\vartheta\|^2 = s - \frac{1}{4}\|\Omega^L\|^2. \qquad \square$$

Exercises

Exercise* 8.7.4 Show that

$$\nabla^* T = F(\text{div } F^\sharp, \cdot) + \frac{1}{2} \sum_{j,k} F(e_j^\vee, e_k^\vee) dF(e_j, e_k, \cdot)$$

$$= F(\text{div } F^\sharp, \cdot) + (*F)(\text{div } (*F)^\sharp, \cdot),$$

independently of Theorem 8.7.2, by directly applying the definition of div to the formula for T in Lemma 8.7.1.

Appendix A
Solutions to Selected Exercises

Exercise 1.1.8. Consider $f : \mathbb{R}^3 \to \mathbb{R}$, $(x, y, z) \mapsto (\sqrt{x^2 + y^2} - R)^2 + z^2 - r^2$, thus $M = f^{-1}(0)$. On M, $x^2 + y^2 \neq 0$ holds, as $f(0, 0, z) = R^2 + z^2 - r^2 > 0$. Hence we get

$$f'_{|(x,y,z)} = \left(2x \left(1 - \frac{R}{\sqrt{x^2 + y^2}} \right), 2y \left(1 - \frac{R}{\sqrt{x^2 + y^2}} \right), 2z \right).$$

If $z \neq 0$, then the third component of f' is not equal to 0 and thus rank $f' = 1$. In the case $z = 0$ we find $1 - \frac{R}{\sqrt{x^2+y^2}} = \frac{r^2}{\sqrt{x^2+y^2}} \neq 0$ on M, hence rank $f' = 1$ holds again because $(x, y) \neq (0, 0)$. □

Exercise 1.2.17. Choose an enumeration $\mathbb{N}^+ \to \mathbb{Q}, n \mapsto r_n$, and set $U_{r_n} := \,]r_n - 2^{-n}, r_n + 2^{-n}[$. Then the U_r have total length 2 and thus they cannot cover $\mathbb{R}$. □

Exercise 1.2.18. Given a cover $(U_j)_j$ of N and a point $p \in \overline{N}$, choose a neighborhood V_p in M with $V_p \cap N \subset U_{j_p}$ for a j_p (via local identification with a submanifold of $\mathbb{R}^n$). Reduce the cover of M by V_p and $M \setminus \overline{N}$ to a countable one and replace in it each V_p by U_{j_p}. □

Exercise 1.2.20. Let

$$g : \mathbb{R}^2 \to M, \quad (t, s) \mapsto \begin{pmatrix} (R + r \cos 2\pi s) \cos 2\pi t \\ (R + r \cos 2\pi s) \sin 2\pi t \\ r \sin 2\pi s \end{pmatrix},$$

let $I \subset \mathbb{R}$ be an open interval of length < 1, $x \in \mathbb{R}^2$ and let $\psi_{I,x}$ be the inverse of the local parametrization $g_{|(x+I^2)}$. Then $\psi_{I,x}$ is a chart of M. Set $f : \mathbb{R}^2/\mathbb{Z}^2 \to M$, $[(t, s)] \mapsto g(t, s)$. Because of the periodicity of $\cos, \sin$, f is well-defined. Then using the charts $\varphi_{I,x}$ of $\mathbb{R}^2/\mathbb{Z}^2$, $\psi_{I,x} \circ f \circ \varphi_{I,x}^{-1} : (x + I^2) \to (x + I^2)$ is the identity, so in particular it is a diffeomorphism from $x + I^2$ to $x + I^2$. The map f is bijective and therefore globally a diffeomorphism from $\mathbb{R}^2/\mathbb{Z}^2$ to M. □

© The Editor(s) (if applicable) and The Author(s), under exclusive license to Springer-Verlag GmbH, DE, part of Springer Nature 2024
K. Köhler, *Differential Geometry and Homogeneous Spaces*, Universitext,
https://doi.org/10.1007/978-3-662-69721-4_9

Exercise 1.3.10. Consider the transition map

$$(T\psi \circ (T\varphi)^{-1})((y, u)) = ((\psi \circ \varphi^{-1})(y), (\psi \circ \varphi^{-1})'_{|y}(u))$$

from Lemma 1.3.4. The derivative of $(\psi \circ \varphi^{-1})'_{|y}(u)$ with respect to u is given by the linear map $(\psi \circ \varphi^{-1})'_{|y}$. Let g be the derivative of $(\psi \circ \varphi^{-1})'_{|y}(u)$ with respect to y. The term $(\psi \circ \varphi^{-1})(y)$ is constant in u and its derivative with respect to y equals again $(\psi \circ \varphi^{-1})'_{|y}$. Altogether, the determinant of the Jacobi matrix is therefore given by

$$\det(T\psi \circ (T\varphi)^{-1})'_{|(y,u)} = \det \begin{pmatrix} (\psi \circ \varphi^{-1})'_{|y} & g \\ 0 & (\psi \circ \varphi^{-1})'_{|y} \end{pmatrix}$$

$$= \det((\psi \circ \varphi^{-1})'_{|y})^2 > 0. \qquad \square$$

Exercise 1.4.13. 1) $\mathcal{J}_p$ is well-defined, because if $f(p) = 0$ holds for a representative f of $[f] \in \mathcal{J}_p$, then this is also true for any other representative $\tilde{f}$ because of the equality of f and $\tilde{f}$ on an open neighborhood of p. Multiplication by $[g] \in \mathcal{F}_p$ yields $[gf]$ with $(gf)(p) = 0$, thus $[gf] \in \mathcal{J}_p$.

2) If $a : \mathcal{J}_p \to T_p^*M$, then $\ker a$ consists of the functions with a double zero at p. Hence $\ker a = (\mathcal{J}_p)^2$. On the other hand, locally (via a test function around p) $f(x_1, \ldots, x_n) := \sum a_j x_j$ is a function such that $a(f) = \sum a_j \frac{\partial}{\partial x_j}$. Therefore a is surjective and induces the desired isomorphism. $\qquad \square$

Exercise 1.5.12. Insert the definition and delete identical terms:

$$(L_X L_Y - L_Y L_X) L_Z - L_Z (L_X L_Y - L_Y L_X)$$
$$+ (L_Y L_Z - L_Z L_Y) L_X - L_X (L_Y L_Z - L_Z L_Y)$$
$$+ (L_Z L_X - L_X L_Z) L_Y - L_Y (L_Z L_X - L_X L_Z) = 0.$$

This calculation also works in any other associative algebra. $\qquad \square$

Exercise 1.6.29. 1) The diffeomorphism $\varphi : G \to G$, $g \mapsto g^{-1}$ satisfies $X^R = -\varphi_* X$ for any $X \in \mathfrak{g}$. Thus we get

$$[X^R, Y^R] = [-\varphi_* X, -\varphi_* Y] = \varphi_* [X, Y] = -[X, Y]^R.$$

2) The map $A : (\mathfrak{g}, [\cdot, \cdot]) \to (\mathfrak{g}^R, [\cdot, \cdot])$, $X \mapsto \varphi_* X = -X^R$ is a Lie algebra isomorphism, since φ_* commutes with the Lie bracket. $\qquad \square$

Exercise 1.6.30. 1) Setting $x := a^2 + bc$ one finds $\exp X = \mathrm{Id} \cdot \cosh \sqrt{x} + X \cdot \frac{\sinh \sqrt{x}}{\sqrt{x}}$ and $\exp X = \mathrm{Id} \cdot \cos \sqrt{-x} + X \cdot \frac{\sin \sqrt{-x}}{\sqrt{-x}}$, respectively.

2) According to the formulas above, $\mathrm{Tr}\, \exp X \geq -2$.

3) Use $\exp \begin{pmatrix} 0 & 2\pi \\ -2\pi & 0 \end{pmatrix} = \exp \begin{pmatrix} 0 & 0 \\ 0 & 0 \end{pmatrix}$. $\qquad \square$

Exercise 1.6.32. 1) $\begin{pmatrix} 1 & 0 \\ 0 & -1 \end{pmatrix} \cdot \begin{pmatrix} 0 & 1 \\ 0 & 0 \end{pmatrix} - \begin{pmatrix} 0 & 1 \\ 0 & 0 \end{pmatrix} \cdot \begin{pmatrix} 1 & 0 \\ 0 & -1 \end{pmatrix} = \begin{pmatrix} 0 & 2 \\ 0 & 0 \end{pmatrix}$ etc.

2) If $v \in V_\lambda$, then (1) implies $HXv = XHv + 2Xv = (\lambda + 2)Xv$ and analogously for Y.

3) q exists since H on V has at least one eigenvalue like any other endomorphism over $\mathbb{C}$. Furthermore $Y \cdot Y^k v = Y^{k+1} v \in W$, and (2) shows $H \cdot Y^k v = (q - 2k)Y^k v \in W$ and

$$
X \cdot Y^k v = \sum_{j=0}^{k-1} Y^j \underbrace{[X, Y]}_{=H} Y^{k-1-j} v + Y^k \underbrace{Xv}_{=0}
$$

$$
= \sum_{j=0}^{k-1} (q - 2(k - 1 - j))Y^{k-1}v
$$

$$
= k(q - k + 1)Y^{k-1}v \in W. \tag{A.1}
$$

Hence W is a $\mathfrak{sl}(2)$-representation and because of the irreducibility of V it is equal to V.

4) Because $Y^k v \in V_{q-2k}$, the nonvanishing $Y^k v$ are linearly independent. On the other hand, $\dim V < \infty$, so there is a smallest $k \in \mathbb{N}$ such that $Y^k v = 0$. Hence by (A.1) one gets $0 = X \cdot Y^k = k(q-k+1)Y^{k-1}v$, i.e. $q = k-1 \in \mathbb{N}_0$. By the formulas in (3), the action of Y, H, X on V is uniquely determined by q independent of the choice of v.

5) By Theorem 1.6.19, the representation of the Lie algebra uniquely determines that of the Lie group.

6) The basis $(s^k t^{q-k})_{0 \le k \le q}$ of V^q satisfies

$$
H \cdot s^k t^{q-k} = \frac{\partial}{\partial \varepsilon}\Big|_{\varepsilon=0} e^{\varepsilon H} \cdot s^k t^{q-k} = \frac{\partial}{\partial \varepsilon}\Big|_{\varepsilon=0} (e^\varepsilon s)^k (e^{-\varepsilon} t)^{q-k} = (2k - q)s^k t^{q-k}.
$$

$\square$

Exercise 2.1.15. Let O be the trivial K-line bundle and let

$$
f : \mathrm{Hom}_K(L, L) \to O, A \mapsto \mathrm{Tr}\, A
$$

be pointwise the trace. Then f is pointwise a vector space isomorphism for dimensional reasons, and f depends smoothly on the base point. $\sqcup$

Exercise 2.2.11. 1) The map is

$$
\mathrm{Hom}(V, W) \otimes \mathrm{Hom}(W, Z) \cong V^* \otimes W \otimes W^* \otimes Z \to V^* \otimes Z \cong \mathrm{Hom}(V, Z)
$$

$$
\alpha \otimes w \otimes \beta \otimes z \mapsto \alpha \otimes \beta(w) \cdot z,
$$

so it is equal to the contraction of the second term with the third term.

2) In the same way

$$\text{Hom}(V, W) \otimes V \cong V^* \otimes W \otimes V \to W$$

$$\alpha \otimes w \otimes v \mapsto \alpha(v) \cdot w$$

is the contraction of the first term with the third term.   □

Exercise 2.3.11. For any $J \subset \{1, \ldots, n\}$ set $J^c := \{1, \ldots, n\} \setminus J$. Then given a basis $(v_1, \ldots, v_n)$, one finds

$$
\begin{aligned}
(f - X \cdot \text{id})^* v^1 \wedge \cdots \wedge v^n &= \sum_{J \subset \{1, \ldots, n\}} \bigwedge_j \begin{cases} v^j \circ f & \text{if } j \in J \\ (-X) v^j & j \notin J \end{cases} \\
&= \sum_{J \subset \{1, \ldots, n\}} \text{sign} \begin{pmatrix} 1 \ldots n \\ J \ J^c \end{pmatrix} (f^* v^J) \cdot (-X)^{|J^c|} v^{J^c} \\
&= \sum_{J \subset \{1, \ldots, n\}} (-X)^{|J^c|} \text{sign} \begin{pmatrix} 1 \ldots n \\ J \ J^c \end{pmatrix} (f^* v^J)(v_J) v^J \wedge v^{J^c} \\
&= \sum_{q=0}^{n} (-X)^{n-q} \underbrace{\sum_{\substack{J \subset \{1, \ldots, n\} \\ |J| = q}} (f^* v^J)(v_J) \cdot v^1 \wedge \cdots \wedge v^n}_{=\text{Tr} f^*_{|\wedge^q V^*}}.
\end{aligned}
$$
                                                                              □

Exercise 2.3.12. Consider bases $(v_1, \ldots, v_j), (v_{j+1}, \ldots, v_n)$ of U, W. Then

$$f : \bigoplus_{q=0}^{k} \Lambda^q U^* \otimes \Lambda^{k-q} W^* \to \Lambda^k V^*$$

$$v^J \otimes v^L \mapsto v^J \wedge v^L$$

(where $J \subset \{1, \ldots, j\}, L \subset \{j+1, \ldots, n\}$) is linear and bijective.   □

Exercise 2.3.15. For any local basis $(s_1, \ldots, s_k)$ of E, $s_1 \wedge \cdots \wedge s_k$ is a generator of $\det E$. Thus if g_{jk} is a transition function, then by Lemma 2.3.4 $\det g_{jk}$ is the transition function of $\det E$.   □

Exercise 2.3.17. Given $f \in C^\infty(M, \mathbb{R})$, one gets

$$
\begin{aligned}
&(fX).\alpha(Y) - Y.\alpha(fX) - \alpha([fX, Y]) \\
&= f \cdot X.\alpha(Y) - Y.(f \cdot \alpha(X)) - \alpha(f \cdot [X, Y] - Y.f \cdot X) \\
&= f \cdot X.\alpha(Y) - (Y.f) \cdot \alpha(X) - f \cdot Y.(\alpha(X)) - f \cdot \alpha([X, Y]) + \alpha(Y.f \cdot X) \\
&= f \cdot (X.\alpha(Y) - Y.\alpha(X) - \alpha([X, Y])).
\end{aligned}
$$

Since $d\alpha(X, Y)$ is skew-symmetric in X and Y, this implies tensoriality in both variables.   □

Exercise 2.4.11. If M is compact, $H_c^\bullet(M) = H^\bullet(M)$. Given a 0-form $f \in C_c^\infty(M)$, it follows from $df = 0$ that f is constant. Hence $f \equiv 0$ holds, if M is not compact. □

Exercise 2.5.18. Let $(U_j)_j$ be a cover of N with trivializations of E. A section $s \in \Gamma(M, f^*E)$ locally has the form $s_{|f^{-1}(U_j)} = \sum g_{\ell j} f^* s_{\ell j}$ in terms of local bases $(s_{\ell j})_\ell$. Thus given a partition of unity $(\tau_k)_k$ subordinate to $(U_j)_j$, it equals

$$s = \sum_{\ell,k} g_{\ell j(k)} \tau_k \circ f \cdot f^* s_{\ell j(k)} = \sum_{\ell,k} g_{\ell j(k)} f^* (\tau_k s_{\ell j(k)}).$$

If N is compact, the cover can be reduced to a finite one. □

Exercise 3.1.24. The formula $(g_{jk})_{j,k} = \begin{pmatrix} 1 & 0 \\ 0 & r(u)^2 \end{pmatrix}$ implies

$$\mathrm{vol}(M) = \int_0^{2\pi} \int_I r(u) \, du \, d\vartheta = 2\pi \int_I r(u) \, du.$$ □

Exercise 3.1.27. 1) Let β be given in the form $\beta(t) = \alpha(t) + u(t)w(t)$ with $u : I \to \mathbb{R}$. Then $\beta' \perp w'$ implies

$$0 = \langle \alpha' + u'w + uw', w' \rangle = \langle \alpha', w' \rangle + u\|w'\|^2,$$

hence $u = -\frac{\langle \alpha', w' \rangle}{\|w'\|^2}$. Therefore β is uniquely given as

$$\beta := \alpha - \frac{\langle \alpha', w' \rangle}{\|w'\|^2} w.$$

2) Now consider another directrix $\overline{\alpha} = \alpha + \overline{u}w$ of u. Then $\overline{\beta}$ for this directrix is equal to

$$\overline{\beta} = \overline{\alpha} - \frac{\langle \overline{\alpha}', w' \rangle}{\|w'\|^2} w = \alpha + \overline{u}w - \frac{\langle \alpha' + \overline{u}'w + \overline{u}w', w' \rangle}{\|w'\|^2} w$$

$$= \alpha - \frac{\langle \alpha', w' \rangle}{\|w'\|^2} w = \beta.$$

3) Since $w \perp w'$ one gets $\beta' \times w = \lambda w'$. In particular,

$$\det(g_{jk}) = \|\partial_t u \times \partial_s u\|^2 = \|(\beta' + sw') \times w\|^2$$
$$= \lambda^2 \|w'\|^2 + s^2 \|w' \times w\|^2 = (\lambda^2 + s^2)\|w'\|^2.$$

Thus u is singular at the point (t, s) if and only if $\lambda(t) = 0$ and $s = 0$. □

Exercise 3.2.22. Given any $f \in C^\infty(M, \mathbb{R})$, one finds

$$T(fX, Y) = \nabla_{fX}Y - \nabla_Y fX - [fX, Y]$$
$$= f\nabla_X Y - f\nabla_Y X - df(Y) \cdot X - f[X, Y] + df(Y) \cdot X$$
$$= f \cdot T(X, Y).$$

So T is tensorial in the first variable, and because $T(Y, X) = -T(X, Y)$, in the second variable as well.   □

Exercise 3.2.23. Because of the skew symmetry of T one obtains

$$\nabla'_X Y - \nabla'_Y X - [X, Y] = T(X, Y) - \frac{1}{2}T(X, Y) + \frac{1}{2}T(Y, X) = 0.$$ □

Exercise 3.3.14. As in the proof of the Koszul formula, it follows that

$$
\begin{aligned}
&X.g(Y, Z) + Y.g(X, Z) - Z.g(X, Y) \\
&= g(\nabla'_X Y, Z) + g(Y, \nabla'_X Z) + g(\nabla'_Y X, Z) \\
&\quad + g(X, \nabla'_Y Z) - g(\nabla'_Z X, Y) - g(X, \nabla'_Z Y) \\
&= 2g(\nabla'_X Y, Z) + g(X, [Y, Z] + T(Y, Z)) + g(Y, [X, Z] + T(X, Z)) \\
&\quad - g(Z, [X, Y] + T(X, Y)).
\end{aligned}
$$

Subtraction of the Koszul formula provides the desired result. By Lemma 3.2.15 S is skew-symmetric in the last two components. In particular, the connection $\nabla + S$ associated to a given $T \in \Lambda^2 T^* M \otimes TM$ has torsion T.   □

Exercise 3.4.9. Consider $X \in T_p M \setminus \{0\}$ and extend $\frac{X}{\|X\|}$ to an orthonormal basis $(e_j)_j$. Then

$$\mathrm{Ric}(X, X) = \sum_{j=1}^n R(X, e_j, X, e_j) = \sum_{j=2}^n \|X\|^2 K(X \wedge e_j) \geq (n-1)\|X\|^2 K_0.$$

Taking the trace provides the inequality for s.   □

Exercise 4.1.10. $\int_{E_p} : H_c^k(E_p) \to \mathbb{R}$ is well-defined since the forms $\mathfrak{A}_c(E_p)$ with compact support have finite integral and for any $\alpha \in \mathfrak{A}_c(E_p)$, $\int_{E_p} d\alpha = 0$ holds by the Stokes–Cartan Theorem. By Lemma 4.1.8 one gets $\int_{E_p} \varphi^* U = \int_{E_p} U = 1$.   □

Exercise 4.2.17. Choose a basis $\mathcal{B}$ which diagonalizes A over $\mathbb{R}$ with 2×2-blocks $\begin{pmatrix} 0 & -\lambda \\ \lambda & 0 \end{pmatrix}$ as in the proof of Theorem 4.2.2. Then $-A^{-1}$ has blocks of the form $\begin{pmatrix} 0 & -\lambda^{-1} \\ \lambda^{-1} & 0 \end{pmatrix}$ on the diagonal. Therefore one obtains $\mathrm{Pf}(-A^{-1}) = (\lambda_1 \cdots \lambda_{k/2})^{-1}$ and $\mathrm{Pf}(A)\mathrm{Pf}(-A^{-1}) = 1$.   □

Exercise 4.2.18. As a 2-form,

$$
\begin{aligned}
\langle A \cdot, \cdot \rangle = a \cdot e^1 \wedge e^2 + b \cdot e^1 \wedge e^3 + c \cdot e^1 \wedge e^4 \\
+ e \cdot e^2 \wedge e^3 + f \cdot e^2 \wedge e^4 + g \cdot e^3 \wedge e^4.
\end{aligned}
$$

Hence $\mathrm{Pf}(A) = ag - bf + ce$ and $\det A = (ag - bf + ce)^2$.   □

Exercise 4.2.20. Given a parameterized family ∇_t of connections on E, one finds $\dot{\Omega} = [\nabla, \dot{\nabla}] = \nabla\dot{\nabla}$. Therefore

$$
\begin{aligned}
\mathcal{T}(\Omega_t^k)^{\cdot} &= \mathcal{T}\left(\sum \Omega^\ell \wedge \nabla\dot{\nabla} \wedge \Omega^{k-\ell-1}\right) \\
&\overset{\text{2nd Bianchi}}{=} \mathcal{T}\left(\nabla\sum \Omega^\ell \wedge \dot{\nabla} \wedge \Omega^{k-\ell-1}\right) \\
&= d\left(\mathcal{T}\left(\sum \Omega^\ell \wedge \dot{\nabla} \wedge \Omega^{k-\ell-1}\right)\right).
\end{aligned}
$$

Integrating over t yields the exactness of $\chi(\nabla_1^E) - \chi(\nabla_0^E)$. $\qquad\square$

Exercise 5.1.19. The Hessian of u is equal to

$$
u'' = \begin{pmatrix} \alpha'' + sw'' & w' \\ w' & 0 \end{pmatrix}.
$$

The vector $\mathbf{n} = \frac{\lambda w' + s w' \times w}{\sqrt{\lambda^2 + s^2}\,\|w'\|}$ is a normal vector. Thus

$$
\langle \mathbf{n}, w'\rangle = \lambda\|w'\|/\sqrt{\lambda^2 + s^2}.
$$

Hence

$$
\begin{aligned}
\det(II_{jk}) = \det\langle \mathbf{n}, u''\rangle &= \det\begin{pmatrix} * & \lambda\|w'\|/\sqrt{\lambda^2 + s^2} \\ \lambda\|w'\|/\sqrt{\lambda^2 + s^2} & 0 \end{pmatrix} \\
&= -\frac{\lambda^2\|w'\|^2}{\lambda^2 + s^2}.
\end{aligned}
$$

Moreover, $\det(g_{jk}) = \|\partial_t u \times \partial_s u\|^2 = (\lambda^2 + s^2)\|w'\|^2$, and together these show that

$$
K = \frac{\det(II_{jk})}{\det(g_{jk})} = -\frac{\lambda^2}{(\lambda^2 + s^2)^2}. \qquad\square
$$

Exercise 5.1.20. The second derivative in the tangential direction satisfies

$$
\begin{aligned}
\widetilde{g}(\widetilde{\nabla}_X\widetilde{\nabla}_{\widetilde{Y}}\widetilde{n}, n') &= X.\widetilde{g}(\widetilde{\nabla}_Y\widetilde{n}, n') - \widetilde{g}(\widetilde{\nabla}_Y\widetilde{n}, \widetilde{\nabla}_X n') \\
&= X.\widetilde{g}(\widetilde{\nabla}_Y\widetilde{n}, n') - g^N(\nabla_Y^N n, \nabla_X^N n') - g(T_Y n, T_X n') \\
&= g^N(\nabla_X^N\nabla_Y^N n, n') - g(T_Y n, T_X n').
\end{aligned}
$$

Together with $\widetilde{g}(\widetilde{\nabla}_{[X,Y]}\widetilde{n}, n') = g(\nabla_{[X,Y]}^N n, n')$ this proves the statement. $\qquad\square$

Exercise 5.2.19. As X is Killing, ∇X is skew-symmetric and for any Y it follows that

$$
g(\nabla_X X, Y) = -g(X, \nabla_Y X) = -\frac{1}{2}Y.\|X\|^2 = 0,
$$

hence $\nabla_X X = 0$. $\qquad\square$

Exercise 5.2.23. 4) dist$(p, q) = 0 \Leftrightarrow p = q$: Let $U = \exp_p B_r(0)$ be a ball-shaped normal neighborhood of p. In the case $q \notin U$, U is traversed by every path c, so length$(c) \geq r$ follows. In the case $q \in U$ the assertion follows from the previous theorem.

5) Let $\mathcal{T}$ be the standard topology and let $\mathcal{T}^{\text{dist}}$ be the one induced by the metric dist. Consider $U \in \mathcal{T}^{\text{dist}}$, i.e.

$$U = \bigcup_{p \in U} \bigcup_{B_r^{\text{dist}}(p) \subset U} B_r^{\text{dist}}(p).$$

Then for any p and r sufficiently small, $B_r^{\text{dist}}(p) = \exp_p B_r(0) \in \mathcal{T}$. Therefore

$$U = \bigcup_{p \in U} \bigcup_{r \text{ sufficiently small}} \exp_p B_r(0) \in \mathcal{T}.$$

Conversely, consider $U \in \mathcal{T}$, i.e. for every $p \in U$ and every chart φ_p around p there exists a sufficiently small ball $B_r(\varphi_p(p)) \subset \mathbb{R}^n$ and thus

$$U = \bigcup_{p \in U} \varphi_p^{-1}(B_r(\varphi_p(p))).$$

Choose $\varphi_p = \exp_p$ as charts. Then for r sufficiently small, $\exp_p(B_r(0)) = B_r^{\text{dist}}(p)$ and hence $U \in \mathcal{T}^{\text{dist}}$. $\qquad\square$

Exercise 5.3.16. For reasons of symmetry (reflection at the plane through c and the axis of rotation) c is a geodesic. The Jacobi fields from Example 5.3.2 span a two-dimensional vector space. Rotations about the axis of rotation are isometries. Therefore differentiating yields a Killing field $\partial/\partial\vartheta$, which is a Jacobi field along c according to Exercise 5.3.14. $\qquad\square$

Exercise 5.3.17. Using

$$\omega_t(Y, \widetilde{Y}) := g(Y_t, \nabla^{c^*TM}_{\partial/\partial t} \widetilde{Y}_t) - g(\nabla^{c^*TM}_{\partial/\partial t} Y_t, \widetilde{Y}_t)$$

one gets

$$\frac{\partial}{\partial t}\omega_t(Y, \widetilde{Y}) = g(Y_t, (\nabla^{c^*TM}_{\partial/\partial t})^2\widetilde{Y}_t) - g((\nabla^{c^*TM}_{\partial/\partial t})^2 Y_t, \widetilde{Y}_t)$$
$$= 0$$

according to the differential equation of Jacobi fields. Hence ω is independent of t. Because the Jacobi fields are uniquely determined by any $(Y, \nabla^{c^*TM}_{\partial/\partial t} Y)_{t=0} \in (T_{c(0)}M)^2$, ω is non-degenerate. $\qquad\square$

Exercise 5.3.18. The vector fields $\dot{c}(t), Y(t)$ along c can be extended as

$$\exp_*\left(\frac{\|X\|}{\|X_0\|}X\right) = \frac{\|X\|}{\|X_0\|}\mathfrak{R}, \quad Y := \exp_*\left(\frac{\|X\|}{\|X_0\|}V\right)$$

to every point $\exp_p X$ with $X \in T_p M$ of a normal neighborhood of p. Then

$$\Omega_{c(t)}(\dot{c}, Y_t)\dot{c} = \frac{1}{t^2}\Omega(\mathfrak{R}, Y)\mathfrak{R}_{|c(t)} = \|X_0\|^2 \Omega(\frac{\mathfrak{R}}{\|X\|}, Y)\frac{\mathfrak{R}}{\|X\|}$$

$$= \|X_0\|^2 \Big(\nabla_{\frac{\mathfrak{R}}{\|X\|}}\nabla_Y \frac{\mathfrak{R}}{\|X\|} - \nabla_Y \nabla_{\frac{\mathfrak{R}}{\|X\|}}\frac{\mathfrak{R}}{\|X\|} - \underbrace{\nabla_{[\frac{\mathfrak{R}}{\|X\|}, Y]}\frac{\mathfrak{R}}{\|X\|}}_{=\frac{\nabla_{\partial/\partial t}\dot{c}}{\|X_0\|^2}=0}\Big)$$

$$= \|X_0\|^2 \Big(\nabla_{\frac{\mathfrak{R}}{\|X\|}}\nabla_{\frac{\mathfrak{R}}{\|X\|}}Y + \nabla_{\frac{\mathfrak{R}}{\|X\|}}[\frac{\mathfrak{R}}{\|X\|}, Y] - \nabla_{[\frac{\mathfrak{R}}{\|X\|}, Y]}\frac{\mathfrak{R}}{\|X\|}\Big)$$

$$= (\nabla_{\partial/\partial t})^2 Y + \underbrace{\|X_0\|^2 \Big[\frac{\mathfrak{R}}{\|X\|}, [\frac{\mathfrak{R}}{\|X\|}, Y]\Big]}_{=\|X_0\|\exp_*\big[\frac{X}{\|X\|}, [\frac{X}{\|X\|}, \|X\|V]\big]} .$$

Because $[\frac{X}{\|X\|}, \|X\|V] = -\frac{X}{\|X\|}\sum_j \frac{x_j v_j}{\|X\|}$ and $X.\sum_j \frac{x_j v_j}{\|X\|} = 0$, the last summand vanishes. $\qquad\square$

Exercise 5.3.19. Using the notations from the proof of the theorem, similarly to the case of the 2nd order term one obtains

$$\frac{d^5}{dt^5}\Big|_{t=0} \|Y\|^2 = 10g((\nabla_{\partial/\partial t})^4 Y, \nabla_{\partial/\partial t}Y)$$

$$= 10g((\nabla_{\partial/\partial t})^2(\Omega(\dot{c}, Y)\dot{c}), \widetilde{V})$$

$$= 10g\big(\nabla_{\partial/\partial t}\big((\nabla_{\partial/\partial t}\Omega)(\dot{c}, Y)\dot{c} + \Omega(\dot{c}, \nabla_{\partial/\partial t}Y)\dot{c}\big), V\big)$$

$$= 20g\big((\nabla_{\partial/\partial t}\Omega_p)(\dot{c}, V)\dot{c}, V\big) . \qquad\square$$

Exercise 5.3.21. We are given the family of great circles

$$c_s(t) = p \cdot \cos t + \sin t \cdot (X \cos s + V \sin s)$$

where $p \in S^n$, $X, V \in T_p S^n$, $\|X\| = \|V\| = 1$, $X \perp V$. Then $\frac{\partial^2}{\partial t^2}c_s(t) \| c_s(t)$ in $\mathbb{R}^{n+1}$. Therefore $t \mapsto c_s(t)$ is geodesic and $Y := \frac{\partial}{\partial s}\big|_{s=0}c_s = V \sin t$ is a Jacobi field along c_0. One gets $\nabla^{S^n}_{\partial/\partial t}Y = \mathrm{proj}(\underbrace{\frac{\partial}{\partial t}Y}_{\sim V \perp T_{c_0(t)}S^n}) = \frac{\partial}{\partial t}Y$, $(\nabla^{S^n}_{\partial/\partial t})^2 Y = \mathrm{proj}(\frac{\partial}{\partial t}\nabla^{S^n}_{\partial/\partial t}Y) =$

$\frac{\partial^2}{\partial t^2}Y = -V \sin t$ and according to the Jacobi ODE $-Y = \ddot{Y} = \Omega(\dot{c}, Y)\dot{c}$. Hence

$$K_p(X \wedge V) = \lim_{t \searrow 0}\frac{-g(\Omega(\dot{c}, Y)\dot{c}, Y)}{\|\dot{c} \wedge Y\|^2} = \lim_{t \searrow 0}\frac{\|Y\|^2}{\|Y\|^2} = 1. \qquad\square$$

Exercise 6.1.12. 1) For the sphere of radius r around x_0 and the straight line with slope $X \in S^{n-1}$, the intersections at $t_{1,2}$ are determined by $\|tX - x_0\|^2 = r^2$. Thus the product of the distances equals $|t_1 \cdot t_2| = |\, \|x_0\|^2 - r^2|$.

2) Proof by cases: For circles not passing through N, the argument is as follows: Choose a sphere containing the circle that does not go through zero. Setting $\psi(x) := \varphi(x - N) + N = \frac{2x}{\|x\|^2}$, there is a constant $c \neq 0$ that depends on the sphere such that collinear points p, q on the sphere satisfy $\|p\| = \frac{c}{\|q\|} = \frac{c}{2}\|\psi(q)\|$. Because $\psi(q)$ is collinear to p, it follows that $\psi(q) = \pm\frac{2}{c}p$. Because of the continuity only one sign occurs, and the image of the sphere is the original sphere stretched by the factor $\pm 2/c$. Each circle lies in a three-dimensional subspace, thus it is mapped to an intersection of two such spheres in $\mathbb{R}^3$.

3) This follows because of the pointwise proportionality to the Euclidean metric. □

Exercise 6.1.13. Using the map φ^{-1} from Theorem 6.1.2, the distance t is determined by the point of the geodesic $\begin{pmatrix} \cosh t \\ X \sinh t \end{pmatrix} = \varphi^{-1}(u) = \frac{1}{1-\|u\|^2}\begin{pmatrix} 1+\|u\|^2 \\ 2u \end{pmatrix}$. Thus $\cosh t = \frac{1+\|u\|^2}{1-\|u\|^2}$. □

Exercise 6.1.14. Because of the transitivity of the $SO_0(1, n)$-action it is sufficient to calculate the curvature at $\mathbf{x} = N$. Let $X_s \in S^n$ be a curve and let $c_s(t) = \begin{pmatrix} \cosh t \\ X_s \sinh t \end{pmatrix}$ be the associated family of geodesics. Then

$$Y_t := \frac{\partial}{\partial s}\Big|_{s=0} c_s(t) = \begin{pmatrix} 0 \\ X_0' \sinh t \end{pmatrix}$$

is a Jacobi field along c_0. By Theorem 3.3.2 one finds

$$\nabla_{\partial/\partial t}Y = \mathrm{proj}\left(\frac{\partial}{\partial t}Y\right) = \begin{pmatrix} 0 \\ X_0' \cosh t \end{pmatrix},$$

$$(\nabla_{\partial/\partial t})^2 Y = \mathrm{proj}\left(\frac{\partial}{\partial t}\nabla_{\partial/\partial t}Y = Y\right).$$

Thus according to the Jacobi differential equation $Y = (\nabla_{\partial/\partial t})^2 Y = \Omega(\dot{c}_0, Y)\dot{c}_0$ and

$$K_N(X_0 \wedge X_0') = \lim_{t \to 0} \frac{-g(\Omega(\dot{c}_0, Y)\dot{c}_0)}{\|\dot{c}_0 \wedge Y\|^2} = \lim_{t \to 0} \frac{-\|Y\|^2}{\|Y\|^2} = -1.$$ □

Exercise 6.3.13. 1) The radial vector field $\mathfrak{R}$ on $\mathbb{C}^{n+1}$ satisfies $\mathfrak{n} = \mathfrak{R}|_{S^{2n+1}}$. Therefore, in the quotient $\mathbb{C}^{n+1} \setminus \{0\}/\mathbb{C}$ every complex line $\mathbb{C}$ is spanned by $\mathfrak{n}$, so as a real plane it is spanned by $\mathfrak{n}$ and $J\mathfrak{n}$. Thus the fibers are great circles with tangent space $J\mathfrak{n} \cdot \mathbb{R}$.

2) This follows from (1) because rotations of the sphere are isometries.

3) According to Example 5.2.3 the geodesics are the images of great circles perpendicular to $J\mathfrak{n}$.

4) The fibers are geodesics according to (1), so T vanishes. For any horizontal X, Y one gets

$$A_X(J\mathfrak{n}) = (\nabla_X^{\mathbb{C}^{n+1}}(J\mathfrak{R}))^H = (JX)^H = JX.$$

By Lemma 6.3.6 it follows that $A_X Y = \langle X, JY\rangle J\mathfrak{n}$.

5) Consider lifts X, Y of $\widetilde{X}, \widetilde{Y}$. Then (4) implies $K(\widetilde{X}\wedge\widetilde{Y}) = 1+3\frac{\langle X,JY\rangle^2}{\|X\wedge Y\|^2} \in [1,4]$.   □

Exercise 6.3.18. Let $A \subset M$ be closed and $y \subset \widetilde{M} \setminus \pi(A)$. Choose a relatively compact neighborhood B of y. Then $\pi^{-1}(\overline{B})$ is compact, hence so are $\pi^{-1}(\overline{B}) \cap A$ and $\pi(\pi^{-1}(\overline{B}) \cap A)$. Because $\pi(\pi^{-1}(\overline{B}) \cap A) \subset \overline{B} \cap \pi(A)$, $B \setminus \pi(\pi^{-1}(\overline{B}) \cap A)$ is an open neighborhood of y in $\widetilde{M} \setminus \pi(A)$.   □

Exercise 6.4.12. Consider $p \in M$ and let $f : G \to M$, $\gamma \mapsto \rho(\gamma)(p)$. Then $T_e f(X) = X'_p$ and $X'_{f(\gamma)} = \frac{\partial}{\partial t}|_{t=0}\rho(\gamma e^{tX})(p) = \frac{\partial}{\partial t}|_{t=0}f(L_\gamma e^{tX}) = T_\gamma f(T_e L_\gamma X)$ hold. By Lemma 1.4.7, the Lie bracket of the Killing fields X' thus corresponds to that of the left-invariant vector fields on G, although G acts on M on the right. Furthermore it follows that $[X'', Y''] = [-X', -Y'] = [X, Y]' = -[X, Y]''$. Thus in the case of an action on the left we get the Lie bracket of the right-invariant vector fields.   □

Exercise 6.4.13. Let $(W_j)_{j\in J}$ be a base for the topology on M. Then $\pi(W_j)_{j\in J}$ is a base for the topology on M/G: This is because, by Proposition 6.4.7, every $\pi(W_j)$ is open. And if $U \subset M/G$ is open, then $\pi^{-1}(U)$ is open. Thus there exists a subset $K \subset J$ such that $\pi^{-1}(U) = \bigcup_{j\in K} W_j$ and hence $U = \bigcup_{j\in K} \pi(W_j)$ holds.   □

Exercise 6.5.12. 2) As $|qv\overline{q}^{-1}|^2 = |q|^2 \cdot |v|^2 \cdot |\overline{q}|^{-2} = |v|^2$, one notices that ψ takes values in SO(4). The kernel $\{(q, \widetilde{q}) \in \{\pm 1\}\setminus(S^3)^2 \mid qv\widetilde{q}^{-1} = v\forall v \in \mathbb{H}\}$ is trivial. At $p = (1, 1)$, $T_1 S^3 \times T_1 S^3 = \mathbb{R}^\perp \times \mathbb{R}^\perp$ and $T_1\psi(x, y)v = xv - vy$ holds for any $x, y \in \mathbb{R}^\perp$. Hence one finds

$$\begin{aligned}
\|T_1\psi(x, y)\|^2_{\mathfrak{so}(\mathbb{H})} &= -\mathrm{Tr}_\mathbb{H} T_1\psi(x, y) \circ T_1\psi(x, y) \\
&= -\mathrm{Tr}_\mathbb{H}(v \mapsto xxv - 2xvy - vyy) \\
&= \mathrm{Tr}\,(v \mapsto |x|^2 v + v|y|^2 + 2xvy) \\
&= 4|x|^2 + 4|y|^2 + 2\mathrm{Tr}\,\underbrace{(v \mapsto xvy)}_{=:|x|\cdot|y|A}.
\end{aligned}$$

If $x, y \neq 0$, then the map $A = \psi\left(\frac{x}{|x|}, (\frac{y}{|y|})^{-1}\right) \in$ SO($\mathbb{H}$), $Av = \frac{xvy}{|x|\cdot|y|}$ is nontrivial, it has determinant equal to 1 and $A^2 = \mathrm{id}$ holds. Therefore the eigenvalues are given by $1, 1, -1, -1$ and the trace equals 0. As a local isometry (up to the factor 4) ψ is a covering by Hermann's theorem, so because of the trivial kernel it is an isomorphism.

1) The identities $q\overline{q} = 1$ and $v = -\overline{v}$ imply that $\overline{qvq^{-1}} = -qvq^{-1}$, thus $qvq^{-1} \in \mathbb{R}^\perp$. Furthermore $|qvq^{-1}|^2 = |q|^2 \cdot |v|^2 \cdot |q|^{-2} = |v|^2$, hence φ takes values in SO(3).

At $p = 1$ one finds $T_1 S^3 = \mathbb{R}^\perp$ and $T_1\varphi(x)v = xv - vx$ for any $x \in \mathbb{R}^\perp$. As in (2) one gets

$$
\begin{aligned}
\|T_1\varphi(x)\|^2_{\mathfrak{so}(\mathbb{R}^\perp)} &= -\mathrm{Tr}_{\mathbb{R}^\perp} T_1\varphi(x) \circ T_1\varphi(x) \\
&= -\mathrm{Tr}_{\mathbb{R}^\perp}(v \mapsto xxv - 2xvx - vxx) \\
&= 6|x|^2 + 2\mathrm{Tr}_{\mathbb{H}}(v \mapsto xvx) - 2\mathrm{Tr}_{\mathbb{R}}(v \mapsto xvx) \\
&\overset{(2)}{=} 6|x|^2 - xx = 8|x|^2.
\end{aligned}
$$

The rest of the argument follows as in (2).

Alternative proof: By embedding $\mathbb{H} \subset \mathrm{End}(\mathbb{C}^2)$ one obtains $\mathbb{R}^\perp \cong \mathfrak{su}(2) \subset \mathrm{End}(\mathbb{C}^2)$ and

$$
\mathrm{SU}(2) = \left\{ \begin{pmatrix} w & -\overline{z} \\ z & \overline{w} \end{pmatrix} \,\middle|\, |w|^2 + |z|^2 = 1 \right\} \cong S^3.
$$

This way one observes $\varphi = \mathrm{Ad} : \{\pm 1\}\backslash\mathrm{SU}(2) \to \mathrm{End}(\mathfrak{su}(2))$. With respect to the metric $(A, B) \mapsto \mathrm{Tr}\, A\overline{B}^t$ on $\mathfrak{su}(2)$, Ad takes values in the isometries. As $\mathrm{SU}(2)$ is connected, φ thus can be written as $\mathrm{Ad} : \{\pm 1\}\backslash\mathrm{SU}(2) \to \mathrm{SO}(\mathfrak{su}(2))$. According to Lemma 1.6.22(1) this map is a group homomorphism. From the formula for the Killing form for $\mathfrak{so}(n)$ (not shown here) it follows that ad is an isometry up to a factor.   □

Exercise 6.6.6. To show (1) compare the first part of the proof of Theorem 7.1.9, which shows for all $Z \in \mathfrak{g}$ the skew-symmetry of ad_Z with respect to B. Thus, for $X \in \mathfrak{h}^\perp, Y \in \mathfrak{h}, Z \in \mathfrak{g}$, it holds that $B(\mathrm{ad}_Z X, Y) = -B(X, \mathrm{ad}_Z Y) = 0$. Because $B_{|\mathfrak{h}^\perp} < 0$, (2) follows by induction.   □

Exercise 6.8.20. The inequality $m \leq \frac{n(n+1)}{2}$ implies $n \geq -\frac{1}{2} + \frac{1}{2}\sqrt{1 + 8m}$. Hence $\dim H = m - n \leq m + \frac{1}{2} - \frac{1}{2}\sqrt{1 + 8m}$.   □

Exercise 6.8.23. If $X \in \mathfrak{m}$, $A \in \mathfrak{g}$, then $\mathrm{ad}_X \mathrm{ad}_X A \in \mathfrak{m}$. Therefore

$$
\mathrm{Tr}\, \mathrm{ad}_X \mathrm{ad}_X = \mathrm{Tr}_{|\mathfrak{m}} \mathrm{ad}_X \mathrm{ad}_X.   \qquad \square
$$

Exercise 7.1.18. By Lemma 1.6.22 one obtains

$$
\mathrm{ad}_{\mathrm{Ad}_h X} Y = [\mathrm{Ad}_h X, Y] = \mathrm{Ad}_h[X, \mathrm{Ad}_h^{-1} Y] = (\mathrm{Ad}_h \circ \mathrm{ad}_X \circ \mathrm{Ad}_h^{-1})(Y),
$$

hence

$$
g(\mathrm{Ad}_h X, \mathrm{Ad}_h Y) = -\mathrm{Tr}\, (\mathrm{Ad}_h \circ \mathrm{ad}_X \circ \mathrm{Ad}_h^{-1} \circ \mathrm{Ad}_h \circ \mathrm{ad}_Y \circ \mathrm{Ad}_h^{-1}) = g(X, Y). \quad \square
$$

Exercise 7.4.20. Choose a p-dimensional oriented subspace V of the oriented $\mathbb{R}^{p+q}$. Let $A \in \mathrm{O}(p+q)$ be the reflection at V and $\sigma : \mathrm{SO}(p+q) \to \mathrm{SO}(p+q), B \mapsto ABA$. Then the fixed point set of σ is equal to $\mathrm{S}(\mathrm{O}(p) \times \mathrm{O}(q))$ (the isometries of V and of $V^\perp$), and $H := \mathrm{SO}(p) \times \mathrm{SO}(q)$ is the connected component of the neutral element.

The Killing form of $SO(p+q)$ is negative definite. So the metric is symmetric by Exercise 7.1.21. Alternatively, one can use that the scalar product $-\mathrm{Tr}\,AB$ on $\mathfrak{so}(p+q)$ is Ad-invariant. □

Exercise 7.5.8. 1) If $X \in \mathfrak{h}, Y \in \mathfrak{h}^\perp, Z \in \mathfrak{g}$, then $B(X,[Y,Z]) = \underbrace{B([Z,X],Y)}_{\in\mathfrak{h}} = 0$

holds. Thus $\mathfrak{h}^\perp$ is an ideal. Furthermore $\mathfrak{h} \cap \mathfrak{h}^\perp$ is abelian, because for any $X, Y \in \mathfrak{h} \cap \mathfrak{h}^\perp, Z \in \mathfrak{g}$ one finds $B([X,Y],Z) = B(\underbrace{X}_{\in\mathfrak{h}^\perp},\underbrace{[Y,Z]}_{\in\mathfrak{h}}) = 0$ and thus $[X,Y] = 0$.

For any subspace $\mathfrak{a} \subset \mathfrak{g}$ satisfying $\mathfrak{g} = \mathfrak{a} \oplus \mathfrak{h} \cap \mathfrak{h}^\perp$ it follows that

$$\mathrm{ad}_X\mathrm{ad}_Z : \begin{array}{l} \mathfrak{a} \to \mathfrak{h} \cap \mathfrak{h}^\perp, \\ \mathfrak{h} \cap \mathfrak{h}^\perp \to 0, \end{array}$$

thus $B(X,Z) = \mathrm{Tr}\,\mathrm{ad}_X\mathrm{ad}_Z = 0$. Hence, because B is non-degenerate, it follows that $\mathfrak{h} \cap \mathfrak{h}^\perp = \{0\}$, and for the same reason $\dim \mathfrak{g} = \dim \mathfrak{h} + \dim \mathfrak{h}^\perp$. Therefore $\mathfrak{g} = \mathfrak{h} \oplus \mathfrak{h}^\perp$. In particular, $B_{|\mathfrak{h}}$ is non-degenerate.

2) This follows directly as in Exercise 6.6.6 by induction.

3) Assume that $\mathfrak{g}$ is simple. The annihilator of B is an ideal, so it is either 0 or $\mathfrak{g}$. In the case $B \equiv 0$, $\mathfrak{g}$ would be abelian, so B is non-degenerate. By Exercise 6.8.23, the Killing form of a sum of simple Lie algebras is equal to the direct sum of the Killing forms. □

Exercise 7.5.9. Set $\mathfrak{h} = \mathfrak{so}(n)$, $\mathfrak{m} = \{A \in \mathbb{R}^{n\times n} \mid \mathrm{Tr}\,A = 0, A \text{ symmetric}\}$. The map $(A,B) \mapsto \mathrm{Tr}\,AB$ induces an invariant metric on $\mathfrak{m}$. One finds $\mathfrak{h} \oplus i\mathfrak{m} = \mathfrak{su}(n)$. Hence $SU(n)/SO(n)$ is a symmetric space dual to $SL(n)/SO(n)$. □

Exercise 8.1.8. Let $(x_0, y_0, z_0) + t(x_1, y_1, z_1)$ be a straight line in the one-sheet hyperboloid given by $x^2 + y^2 - z^2 = 1$, i.e.

$$1 = (x_0 + tx_1)^2 + (y_0 + ty_1)^2 - (z_0 + tz_1)^2$$

or

$$0 = 2t(x_0x_1 + y_0y_1 - z_0z_1) + t^2(x_1^2 + y_1^2 - z_1^2).$$

Thus $x_0x_1 + y_0y_1 - z_0z_1 = 0$, $x_1^2 + y_1^2 - z_1^2 = 0$ hold. This shows $z_1 \neq 0$. So let $z_1 = 1$, $z_0 = 0$ by rescaling t. According to the 2nd equation there exists an α such that $x_1 = \cos\alpha$, $y_1 = \sin\alpha$. The first equation then gives exactly 2 solutions for $(x_0, y_0, 0)$ with $x_0^2 + y_0^2 = 1$. □

Exercise 8.2.3. The basis

$$\begin{pmatrix} 1 & 0 \\ 0 & 1 \end{pmatrix}, \quad \begin{pmatrix} i & 0 \\ 0 & -i \end{pmatrix}, \quad \begin{pmatrix} 0 & i \\ i & 0 \end{pmatrix}, \quad \begin{pmatrix} 0 & -1 \\ 1 & 0 \end{pmatrix}$$

of $\mathbb{H}$ is an orthonormal basis for the signature $(1,1,1,1)$. □

Exercise 8.3.9. Apply Lemma 8.3.7 and set $\alpha^\sharp := X, \beta := f$. Or in a more direct way:
By the proof of Proposition 8.3.5 we know that $L_X d\mathrm{vol} = -\mathrm{div}\, X\, d\mathrm{vol}$. Therefore

$$
\begin{aligned}
X.f\, d\mathrm{vol} \quad &= \quad L_X(f\, d\mathrm{vol}) - f L_X d\mathrm{vol} \\
&\overset{\substack{\text{Cartan}\\\text{homotopy formula}}}{=} d\left[\iota_X f\, d\mathrm{vol}\right] + \iota_X \underbrace{d[f\, d\mathrm{vol}]}_{=0} + f \mathrm{div}\, X\, d\mathrm{vol}. \qquad \square
\end{aligned}
$$

Exercise 8.3.10. If $s \in \Gamma_c(M, E)$, then

$$
\int (\mu(\nabla_X^E s) + (\nabla_X^{E^*}\mu)s)\, d\mathrm{vol} = \int X.(\mu(s))\, d\mathrm{vol} = \int \mu(s)\mathrm{div}\, X\, d\mathrm{vol}.
$$

Furthermore, one finds

$$
(s, \nabla^*(X^\flat \otimes \mu))_{L^2} = (\nabla s, X^\flat \otimes \mu)_{L^2} = (\nabla_X s, \mu))_{L^2} = (s, \nabla_X^*\mu)_{L^2}. \qquad \square
$$

Exercise 8.4.6. One obtains

$$
\begin{aligned}
0 \quad &= \quad (\nabla\Omega)(X, Y, Z, V, W) \\
&= \quad (\nabla_X\Omega)(Y, Z, V, W) + (\nabla_Y\Omega)(Z, X, V, W) \\
&\quad + (\nabla_Z\Omega)(X, Y, V, W) \\
&\overset{\substack{Y=e_j, V=e_j^\vee,\\Z=e_k, W=e_k^\vee}}{=} (\nabla_X\Omega)(e_j, e_k, e_j^\vee, e_k^\vee) + (\nabla_{e_j}\Omega)(e_k, X, e_j^\vee, e_k^\vee) \\
&\quad + (\nabla_{e_k}\Omega)(X, e_j, e_j^\vee, e_k^\vee) \\
&= \quad (\nabla_X\Omega)(e_j, e_k, e_j^\vee, e_k^\vee) - (\nabla_{e_j}\Omega)(e_j^\vee, e_k^\vee, X, e_k) \\
&\quad - (\nabla_{e_k}\Omega)(e_k^\vee, e_j^\vee, X, e_j).
\end{aligned}
$$

Summation yields (since taking the trace commutes with ∇)

$$
0 = -\nabla_X s + (\widetilde{\mathrm{Tr}}_{g,12}\nabla\mathrm{Ric})(X) + (\widetilde{\mathrm{Tr}}_{g,12}\nabla\mathrm{Ric})(X).
$$

Hence

$$
\begin{aligned}
0 = -ds - 2\nabla^*\mathrm{Ric} &= -\sum_j \nabla_{e_j} s \cdot g(e_j^\vee, \cdot) - 2\nabla^*\mathrm{Ric} \\
&= -\sum_j \nabla_{e_j}(sg)(e_j^\vee, \cdot) - 2\nabla^*\mathrm{Ric} = \nabla^*(sg - 2\mathrm{Ric}). \qquad \square
\end{aligned}
$$

Exercise 8.7.4. One finds

$$
\begin{aligned}
-\nabla^* T &= -\widetilde{\mathrm{Tr}}_{g,12}\widetilde{\mathrm{Tr}}_{g,35}\nabla^\otimes(F \otimes F) + \frac{1}{4}\mathrm{Tr}_g\nabla(\|F\|_g^2 \cdot g) \\
&= \sum_{j,k}\left(-(\nabla_{e_j}^\otimes F)(e_j^\vee, e_k)F(\cdot, e_k^\vee) - F(e_j^\vee, e_k^\vee)(\nabla_{e_j}^\otimes F)(\cdot, e_k)\right)
\end{aligned}
$$

$$+\frac{1}{4} \cdot 2g(\nabla^{\otimes}F, F)$$

$$= \sum_{j,k}\Big(-(\nabla^{\otimes}_{e_j}F)(e_j^{\vee}, e_k)F(\cdot, e_k^{\vee}) - F(e_j^{\vee}, e_k^{\vee})(\nabla^{\otimes}_{e_j}F)(\cdot, e_k)$$

$$+\frac{1}{2}(\nabla^{\otimes}F)(e_j, e_k)F(e_j^{\vee}, e_k^{\vee})\Big)$$

$$= \sum_{j,k}\Big((\nabla^{\otimes}_{e_j}F)(e_j^{\vee}, e_k)F(e_k^{\vee}, \cdot) + \frac{1}{2}F(e_j^{\vee}, e_k^{\vee})(\nabla^{\otimes}_{e_j}F)(e_k, \cdot)$$

$$+\frac{1}{2}F(e_j^{\vee}, e_k^{\vee})(\nabla^{\otimes}_{e_k}F)(\cdot, e_j) + \frac{1}{2}(\nabla^{\otimes}F)(e_j^{\vee}, e_k^{\vee})F(e_j, e_k)\Big).$$

Because

$$dF(e_j, e_k, X) = (\nabla^{\otimes}_{e_j}F)(e_k, X) + (\nabla^{\otimes}_{e_k}F)(X, e_j) + (\nabla^{\otimes}_X F)(e_j, e_k),$$

the assertion follows. As

$$\sum_{j<k} F(e_j^{\vee}, e_k^{\vee})dF(e_j, e_k, X) = \langle \iota_X dF, F \rangle_\wedge = -\langle *\iota_X dF, *F \rangle_\wedge$$

$$= -\langle X^{\flat} \wedge *dF, *F \rangle_\wedge = -\langle *dF, \iota_X * F \rangle_\wedge$$

$$= \langle *d * *F, \iota_X * F \rangle_\wedge \stackrel{8.3.5}{=} \langle (\mathrm{div}\,(*F)^{\flat})^{\flat}, \iota_X * F \rangle$$

$$= -(*F)(\mathrm{div}\,(*F)^{\flat}, X),$$

the second formula follows. $\qquad\square$

Bibliography

[A]   Alexandrov, A.D.: Mappings of Spaces with Families of Cones and Space-Time Transformations. *Ann. Mat. Pura Appl.* **103** (1975), 229–257.

[AhBo]   Aharonov, Y.; Bohm, D.: Significance of electromagnetic potentials. *Phys. Rev* **115** (1959), 485–491.

[Am]   Ambrose, W.: Parallel translation of Riemannian curvature. *Ann. of Math.* **64** (1956), 337–363.

[AmSi]   Ambrose, W.; Singer, I.M.: On homogeneous Riemannian manifolds. *Duke Math. J.* **25** (1958), 647–669.

[AO]   Alexandrov, A.D.; Ovchinnikova, V.V.: Notes on the foundations of relativity theory. *Vestnik Leningrad. Univ.* **11** (1953), 95–110.

[Bär]   Bär, Christian: *Elementary Differential Geometry*. Cambridge Univ. Press, 2010.

[Besse] Besse, Arthur L.: *Einstein Manifolds*. Reprint of the 1987 edition. Class. in Math. Springer, Berlin-Heidelberg, 2008.

[BereNi] Berestovskii V.N.; Nikonorov Yu.G.: On δ-homogeneous Riemannian manifolds. *Diff. Geom. Appl.* **26** (2008), 514–535.

[Be]   Berger, Marcel: *A Panoramic View of Riemannian Geometry*. Springer, Berlin-Heidelberg, 2003.

[BGV]   Berline, Nicole; Getzler, Ezra; Vergne Michèle: *Heat kernels and Dirac Operators*. Springer Grundlehren **298**, 1992.

[Bour]   Bourbaki, Nicolas: *Groupes et algèbres de Lie*, Chapitre 1. Hermann, 1972.

[BoTu]   Bott, Raoul; Tu, Loring W.: *Differential Forms in Algebraic Topology*. GTM **82**. Springer, New York-Berlin, 1982.

[BQ]   Beckman, F.S.; Quarles, D.A., Jr. On isometries of Euclidean spaces. *Proc. Amer. Math. Soc.* **4** (1953), 810–815.

[BtD]   Bröcker, Theodor; tom Dieck, Tammo: *Representations of Compact Lie Groups*. GTM **98**, Springer, New York 1985.

[BtD2]   Bröcker, Theodor; tom Dieck, Tammo: *Introduction to Differential Topology*. Cambridge Univ. Press, 1982.

© The Editor(s) (if applicable) and The Author(s), under exclusive license to Springer-Verlag GmbH, DE, part of Springer Nature 2024
K. Köhler, *Differential Geometry and Homogeneous Spaces*, Universitext,
https://doi.org/10.1007/978-3-662-69721-4

[Car]   Cartan, Élie: Espaces à connexion affine, projective et conforme. *Acta Math.* **48** (1926), 1–42.

[Chr]   Christoffel, Elwin Bruno: Über die Transformation der homogenen Differentialausdrücke zweiten Grades. *J. für die Reine und Angew. Math.* **70** (1869), 46–70.

[Chern]   Chern, Shiing-Shen: A Simple Intrinsic Proof of the Gauss-Bonnet Formula for Closed Riemannian Manifolds. *Ann. Math.* **45** (1944), 747–752.

[ChEb]   Cheeger, Jeff; Ebin, David G.: *Comparison Theorems in Riemannian Geometry.* Revised Ed., AMS Chelsea Publishing, Providence, 2008.

[Dar]   Darboux, G.: Sur le théorème fondamental de la géométrie projective. *Clebsch Ann.* **XVII** (1880), 55–62.

[deR]   de Rham, G: Sur l'Analysis situs des variétés à n dimensions. *J. Math. Pures Appl.* **X** (1931),115–200.

[doC]   do Carmo, Manfredo P.: *Differential Geometry of Curves and Surfaces.* Prentice-Hall, Inc., Englewood Cliffs, N.J., 1976.

[DoK]   Donaldson, S.K.; Kronheimer, P.B.: *The Geometry of Four-Manifolds.* Oxford University Press, New York, 1990.

[Du]   Dundas, Bjorn Ian: *Differential Topology.* Johns Hopkins Univ., 2002.

[FH]   Fulton, William; Harris, Joe: *Representation Theory. A First Course.* GTM **129**. Springer, New York, 1991.

[Fi]   Fischer, Gerd: *Analytische Geometrie.* 7th Ed., Vieweg, Braunschweig-Wiesbaden 2001.

[FrKr]   Fröhlicher, Alfred; Kriegl, Andreas: *Linear Spaces and Differentiation Theory.* Wiley-Interscience 1988.

[GHL]   Gallot, Sylvestre; Hulin, Dominique; Lafontaine, Jacques: *Riemannian Geometry.* 3rd Ed., Springer, Berlin, 2004.

[Gauß]   Gauß, C.F.: *Disquisitiones generales circa superficies curvas.* 1828.

[Giu]   Giulini, Domenico: The Rich Structure of Minkowski Space. In Petkov, Vesselin (Ed.): *Minkowski Spacetime: A Hundred Years Later.* Fund. Th. of Physics **165**, Springer 2010.

[Gl]   Gleason, Andrew M.: Groups without small subgroups. *Ann. of Math.* **56** (1952), 193–212.

[Go1]   Gompf, Robert E.: Three exotic $\mathbb{R}^4$'s and other anomalies. *J. Diff. Geom.* **18** (1983), 317–328.

[Go2]   Gompf, Robert E.: An infinite set of exotic $\mathbb{R}^4$'s. *J. Diff. Geom.* **21** (1985), 283–300.

[H]   Hilbert, D.: Die Grundlagen der Physik. (Erste Mitteilung.). *Gött. Nachr.* (1915), 395–407. Die Grundlagen der Physik. (Zweite Mitteilung.) *Gött. Nachr.* (1917), 53–76.

[Ha]   Hatcher, Allen: *Algebraic topology.* Cambridge Univ. Press (2002).

[HCV]   Hilbert, David; Cohn-Vossen, Stefan: *Geometry and the Imagination.* Chelsea 1952.

[HE]   Hawking, S.; Ellis, G.: *Large Scale Structure of Spacetime.* Cambridge Univ. Press 1973.

[Hel]   Helgason: *Differential Geometry, Lie Groups, And Symmetric Spaces*. Acad. Press 1978.

[Her]   Hermann, Robert: A sufficient condition that a mapping of Riemannian manifolds be a fibre bundle. *Proc. AMS* **11** (1960), 236–242.

[Hi]   Hirsch, Morris W.: *Differential Topology*. Springer GTM **33**, 1976.

[Hopf]   Hopf, Heinz.: Vektorfelder in n-dimensionalen Mannigfaltigkeiten. *Math. Ann.* **96** (1926), 225–249.

[HoRi]   Hopf, Heinz; Rinow, W.: Über den Begriff der vollständigen differential-geometrischen Fläche. *Comm. Math. Helv.* **3** (1931), 209–225.

[Kal]   Kaluza, T.: Zum Unitätsproblem der Physik. *Sitzungsberichte Preußische Akad. Wiss.* (1921), 966–972.

[Kl1]   Klingenberg, Wilhelm: *A Course in Differential Geometry*. GTM **51**. Springer, Berlin-New York, 1978.

[Kl2]   Klingenberg, Wilhelm: *Riemannian Geometry*. 2nd Ed., de Gruyter Studies in Mathematics, 1. Walter de Gruyter & Co., Berlin, 1995.

[Ko]   Kobayashi, Shoshichi: *Transformation Groups in Differential Geometry*. Reprint of the 1972 edition. Classics in Mathematics. Springer, Berlin, 1995.

[KoN]   Kobayashi, Shoshichi; Nomizu, Katsumi: *Foundations of Differential Geometry. I& II*. Reprint of the 1969, 1963 originals. Wiley Classics Library. John Wiley & Sons, Inc., New York, 1996.

[KMS]   Kolář, Ivan; Michor, Peter W.; Slovák, Jan: *Natural Operations in Differential Geometry*. Springer, Berlin, 1993.

[Kü]   Kühnel, Wolfgang: *Differential Geometry: Curves – Surfaces – Manifolds*. 3rd ed., AMS Student Math. Lib. **77**, 2015.

[L]   Lee, John M.: *Introduction to Smooth Manifolds*. 2nd edition. GTM **218**. Springer, New York, 2013.

[Levi]   Levi-Civita, Tullio: Nozione di parallelismo in una varietà qualunque e consequente specificazione geometrica della curvatura Riemanniana. *Rend. Circ. Mat. Palermo* **42** (1917), 73–205.

[Loh]   Lohkamp, Joachim: Metrics of negative Ricci curvature. *Ann. of Math.* **140** (1994), 655–683.

[Loos]   Loos, Ottmar: *Symmetric Spaces 1: General Theory, 2: Compact Spaces and Classification*. W. A. Benjamin, Amsterdam, 1969.

[M]   Milnor, John: On manifolds homeomorphic to the 7-sphere. *Ann. of Math.* **64** (1956), 399–405.

[MaQ]   Mathai Varghese; Quillen, Daniel: Superconnections, Thom classes and equivariant differential forms. *Topology* **25** (1986), 85–110.

[MeMi]   Gil-Medrano, Olga; Michor, Peter W.: The Riemannian manifold of all Riemannian metrics. *Quart. J. of Math. (Oxford)* **42** (1991), 183–202.

[Miln]   Milnor, John: Curvatures of left invariant metrics on lie groups. *Adv. Math.* **21** (1976), 293–329.

[Mink]   Minkowski, H.: Die Grundgleichungen für die elektromagnetischen Vorgänge in bewegten Körpern. *Gött. Nachr.* (1908), 53–111.

[MoZi]   Montgomery, Deane; Zippin, Leo: Small subgroups of finite-dimensional groups. *Ann. of Math.* **56** (1952). 213–241.

[MTW]   Misner, C.; Thorne, K.; Wheeler, J.: *Gravitation*. Freeman 1973.

[MySt]   Myers, S.B.; Steenrod, N.E.: The group of isometries of a Riemannian manifold. *Ann. of Math. (2)* **40** (1939), 400–416.

[ON1]   O'Neill, Barrett: The fundamental equations of a submersion. *Michigan Math. J.* **13** (1966), 459–469.

[ON2]   O'Neill, Barrett: *Semi-Riemannian Geometry. With Applications to Relativity*. Pure Appl. Math. **103**. Acad. Press, Inc., New York, 1983.

[ON3]   O'Neill, Barrett: *The Geometry of Kerr Black Holes*. A K Peters, CRC Press, 1992.

[Pal]   Palais, Richard S.: Natural operations on differential forms, *Trans. AMS* **92** (1959), 125–141

[RCLC]   Ricci, Gregorio; Levi-Civita, Tullio: Méthodes de calcul différentiel absolu et leurs applications. *Math. Ann.* **54** (1900), 125–201.

[Rie]   Riemann, Bernhard: *Über die Hypothesen, welche der Geometrie zu Grunde liegen* (Klassische Texte der Wissenschaft). Springer Spektrum, Heidelberg 2013.

[Sam]   Samelson, Hans: Differential Forms, the Early Days; or the Stories of Deahna's Theorem and of Volterra's Theorem. *Am. Math. Monthly* **108** (2001), 522–530.

[Sp]   Spivak, Michael: *A Comprehensive Introduction to Differential Geometry*. 2nd edition. Publish or Perish, Inc., Wilmington, Del., 1979.

[Stee]   Steenrod, Norman: *The Topology of Fibre Bundles*. Princeton Univ. Press, Princeton 1951.

[SWu]   Sachs R.K.; Wu, H.: *General Relativity for Mathematicians*. Springer, New York 1977.

[TaWh]   Taylor, Edwin F.; Wheeler, A. John: *Spacetime Physics: Introduction to Special Relativity*. W.H. Freeman, 1992.

[Varad]   Varadarajan, V.S.: *Lie Groups, Lie Algebras, and Their Representation*. Springer, New York 1984.

[Vi]   Vilms, Jaak: Totally geodesic maps. *J. Differential Geometry* **4** (1970), 73–79.

[Volt]   Volterra, Vito: Delle variabili complesse negli iperspazii. *Rend. Accad. dei Lincei*, ser. IV, **V** (1889), 158–165, 291–299; Sulle funzione conjugate. *Rend. Accad. dei Lincei*, ser. IV, **VI** (1889), 158–169.

[Wald]   Wald, R.M.: *General Relativity*. Univ. of Chicago Press 1984.

[War]   Warner, Frank W.: *Foundations of Differentiable Manifolds and Lie Groups*. Corr. reprint of the 1971 edition. GTM **94**. Springer, New York-Berlin, 1983.

[Whi]   Whitehead, J.H.C.: Convex regions in the geometry of paths. *Q. J. Math., Oxf. Ser.* **3** (1932), 33–42.

[Whit]   Whitney, H.: The Self-Intersections of a Smooth n-Manifold in $2n$-Space, *Ann. Math., 2nd Series*, **45** (1944), 220–246.

[Wi]   Wigner, E.P.: *Group Theory: And Its Application to the Quantum Mechanics of Atomic Spectra*. Acad. Press 2016.

[Wolf1]   Wolf, Joseph: *Spaces of Constant Curvature*. Amer. Math. Soc., 1967.

[Wolf2]   Wolf, Joseph: The geometry and structure of isotropy irreducible homogeneous spaces. *Acta Math.* **120** (1968), 59–148. ; corr., *Acta. Math.* **152** (1984), 141–142.

[Wü]   Wüstner, Michael: A Connected Lie Group Equals the Square of the Exponential Image. *J. Lie Th.* **13** (2003), 307–309.

[WZ]   Wang, McKenzie; Ziller, Wolfgang: On isotropy irreducible Riemannian manifolds. *Acta Math.* **166** (1991), 223–261.

Subject Index

Notation Index